Lecture Notes in Mathematics

Volume 2299

This series reports on new developments in all areas of mathematics and their applications - quickly, informally and at a high level. Mathematical texts analysing new developments in modelling and numerical simulation are welcome. The type of material considered for publication includes:

1. Research monographs
2. Lectures on a new field or presentations of a new angle in a classical field
3. Summer schools and intensive courses on topics of current research.

Texts which are out of print but still in demand may also be considered if they fall within these categories. The timeliness of a manuscript is sometimes more important than its form, which may be preliminary or tentative.

Titles from this series are indexed by Scopus, Web of Science, Mathematical Reviews, and zbMATH.

Yuichiro Hoshi • Shinichi Mochizuki

Topics Surrounding the Combinatorial Anabelian Geometry of Hyperbolic Curves II

Tripods and Combinatorial Cuspidalization

Springer

Yuichiro Hoshi
RIMS
Kyoto University
Kyoto, Japan

Shinichi Mochizuki
RIMS
Kyoto University
Kyoto, Japan

ISSN 0075-8434 ISSN 1617-9692 (electronic)
Lecture Notes in Mathematics
ISBN 978-981-19-1095-1 ISBN 978-981-19-1096-8 (eBook)
https://doi.org/10.1007/978-981-19-1096-8

Mathematics Subject Classification: Primary 14H30; Secondary 14H10

This Springer imprint is published by the registered company Springer Nature Singapore Pte Ltd.
The registered company address is: 152 Beach Road, #21-01/04 Gateway East, Singapore 189721, Singapore

Preface

Let Σ be a subset of the set of prime numbers which is either equal to the entire set of prime numbers or of cardinality one. In the present book, we continue our study of the pro-Σ fundamental groups of hyperbolic curves and their associated configuration spaces over algebraically closed fields in which the primes of Σ are invertible. The starting point of the theory of the present book is a *combinatorial anabelian result* which, unlike results obtained in previous papers, allows one to *eliminate* the hypothesis that *cuspidal inertia subgroups* are *preserved* by the isomorphism in question. This result allows us to [partially] generalize **combinatorial cuspidalization** results obtained in previous papers to the case of outer automorphisms of pro-Σ fundamental groups of configuration spaces that *do not necessarily preserve the cuspidal inertia subgroups* of the various one-dimensional subquotients of such a fundamental group. Such partial combinatorial cuspidalization results allow one in effect to reduce issues concerning the **anabelian geometry** of **configuration spaces** to issues concerning the anabelian geometry of **hyperbolic curves**. These results also allow us, in the case of configuration spaces of sufficiently large dimension, to give **purely group-theoretic** characterizations of the **cuspidal inertia subgroups** of the various one-dimensional subquotients of the pro-Σ fundamental group of a configuration space. We then turn to the study of **tripod synchronization**, i.e., roughly speaking, the phenomenon that an outer automorphism of the pro-Σ fundamental group of a log configuration space associated to a stable log curve typically induces the **same** outer automorphism on the various subquotients of such a fundamental group determined by **tripods** [i.e., copies of the projective line minus three points]. Our study of tripod synchronization allows us to show that outer automorphisms of *pro-*Σ fundamental groups of configuration spaces exhibit somewhat **different behavior** from the behavior that may be observed—as a consequence of the classical **Dehn-Nielsen-Baer theorem**—in the case of *discrete* fundamental groups. Other applications of the theory of tripod synchronization include a result concerning **commuting profinite Dehn multi-twists** that, a priori, arise from distinct *semi-graphs of anabelioids of pro-*Σ *PSC-type* structures [i.e., the profinite analogue of the notion of a *decomposition of a hyperbolic topological surface into hyperbolic subsurfaces*, such as "pants"], as well as the computation, in terms of a certain **scheme-theoretic fundamental group**, of the *purely combinatorial/group-theoretic commensurator* of the group of **profinite Dehn multi-twists**. Finally, we show that the condition that an outer automorphism of the pro-Σ fundamental group of a stable log curve *lift* to an outer automorphism of the pro-Σ fundamental group of the corresponding n-th log configuration space, where $n \geq 2$ is an integer, is compatible, in a suitable sense, with **localization** on the dual graph of the stable log curve. This localizability property, together with the theory of tripod synchronization, is applied to construct a **purely combinatorial analogue** of the natural outer **surjection** from the étale fundamental group of the moduli stack of hyperbolic curves over $\mathbb{Q}$ to the **absolute Galois group** of $\mathbb{Q}$.

The first author was supported by Grant-in-Aid for Scientific Research (C), No. 24540016, Japan Society for the Promotion of Science.

Kyoto, Japan
December 2021

Yuichiro Hoshi
Shinichi Mochizuki

Introduction

Let $\Sigma \subseteq \mathfrak{Primes}$ be a subset of the set of prime numbers $\mathfrak{Primes}$ which is either equal to $\mathfrak{Primes}$ or of cardinality one. In the present monograph, we continue our study of the *pro-*Σ *fundamental groups* of hyperbolic curves and their associated configuration spaces over algebraically closed fields in which the primes of Σ are invertible [cf. [MzTa, CmbCsp, NodNon, CbTpI]].

Before proceeding, we review some fundamental notions that play a central role in the present monograph. We shall say that a scheme X over an algebraically closed field k is a *semi-stable curve* if X is connected and proper over k, and, moreover, for each closed point x of X, the completion of the local ring $\mathcal{O}_{X,x}$ is isomorphic over k either to $k[[t]]$ or to $k[[t_1, t_2]]/(t_1 t_2)$, where t, t_1, and t_2 are indeterminates. We shall say that a scheme X over a scheme S is a *semi-stable curve* if the structure morphism $X \to S$ is flat, and, moreover, every geometric fiber of $X \to S$ is a semi-stable curve. We shall say that a pair (X, D) consisting of a scheme X over a scheme S and a [possibly empty] closed subscheme $D \subseteq X$ is a *pointed stable curve* over S if the following conditions are satisfied: X is a semi-stable curve over S; D is contained in the smooth locus of the structure morphism $X \to S$ and étale over S; the invertible sheaf $\omega_{X/S}(D)$—where we write $\omega_{X/S}$ for the dualizing sheaf of X/S—is *relatively ample* [relative to the morphism $X \to S$]. We shall say that a scheme X over a scheme S is a *hyperbolic curve* over S if there exists a pointed stable curve (Y, E) over S such that Y is smooth over S, and, moreover, X is isomorphic to $Y \setminus E$ over S.

It is well-known [cf. [SGA1], Exposé V, §7] that if X is a connected locally noetherian scheme, and $\overline{x} \to X$ is a geometric point of X, then the category Fét(X) consisting of X-schemes Z whose structure morphism is finite and étale and [necessarily finite étale] X-morphisms forms a *Galois category*, for which the functor from Fét(X) to the category of finite sets given by $Z \mapsto Z \times_X \overline{x}$ is a *fundamental functor* [cf. [SGA1], Exposé V, Définition 5.1]. Thus, it follows from the general theory of Galois categories [cf. the discussion following [SGA1], Exposé V, Remarque 5.10] that one may associate, to the Galois category Fét(X) equipped with the above fundamental functor, the "fundamental pro-group" of the Galois category Fét(X) equipped with the above fundamental functor, which we shall refer to as the *étale fundamental group* $\pi_1(X, \overline{x})$ of $(X, \overline{x})$. If X is a *normal scheme*, $\overline{K}$ is an *algebraic closure* of the *function field* K of X, and $\overline{x}$ is the tautological geometric point of X determined by $\overline{K}$, then $\pi_1(X, \overline{x})$ may be naturally identified with the *quotient* of $\text{Gal}(\overline{K}/K)$ determined by the union of finite subextensions $K \subseteq L \subseteq \overline{K}$ such that the normalization of X in L is *finite étale* over X [cf. [SGA1], Exposé I, Corollaire 10.3]. Since [one verifies easily that] the étale fundamental group is, in a natural sense, independent, up to inner automorphism, of the choice of the basepoint, i.e., the geometric point "$\overline{x}$", we shall omit mention of the basepoint throughout the present monograph.

Let G be a topological group. Then we shall write $\text{Aut}(G)$ for the group of [continuous] automorphisms of G, $\text{Inn}(G) \subseteq \text{Aut}(G)$ for the group of inner automorphisms of G, and $\text{Out}(G) \stackrel{\text{def}}{=} \text{Aut}(G)/\text{Inn}(G)$ for the group of [continuous] *outomorphisms* [i.e., *outer automorphisms*] of G. Thus, an *outer automorphism* of G is an automorphism of G considered up to composition with an inner automorphism.

Let k be a field, k^{sep} a separable closure of k, and X a geometrically connected scheme of finite type over k. Write $G_k \stackrel{\text{def}}{=} \text{Gal}(k^{\text{sep}}/k)$ for the absolute Galois group of k. Then it is well-known [cf. [SGA1], Exposé IX, Théorème 6.1] that the natural morphisms of schemes $X \times_k k^{\text{sep}} \to X \to \text{Spec}\, k$ determine an exact sequence of profinite groups

$$1 \longrightarrow \pi_1(X \times_k k^{\text{sep}}) \longrightarrow \pi_1(X) \longrightarrow G_k \longrightarrow 1.$$

Write Δ_X for the *maximal pro-*Σ *quotient* of the étale fundamental group $\pi_1(X \times_k k^{\text{sep}})$ of $X \times_k k^{\text{sep}}$ and Π_X for the quotient of the étale fundamental group $\pi_1(X)$ of X by the normal closed subgroup of $\pi_1(X)$ determined by the kernel of the natural surjection $(\pi_1(X) \hookleftarrow)\ \pi_1(X \times_k k^{\text{sep}}) \twoheadrightarrow \Delta_X$. Then the above displayed exact sequence determines an exact sequence of profinite groups

$$1 \longrightarrow \Delta_X \longrightarrow \Pi_X \longrightarrow G_k \longrightarrow 1.$$

Next, observe that the above displayed exact sequence induces a natural *action of* Π_X *on* Δ_X by conjugation, i.e., a homomorphism $\Pi_X \to \text{Aut}(\Delta_X)$, which restricts to the tautological homomorphism $\Delta_X \to \text{Inn}(\Delta_X)$. Thus, by considering the respective quotients by Δ_X, we obtain an *outer action of* G_k *on* Δ_X, i.e., a homomorphism

$$G_k \longrightarrow \text{Out}(\Delta_X).$$

This *outer action* is one of the main objects of study in *anabelian geometry*.

In the situation of the preceding paragraph, if X is a hyperbolic curve over k, then each *cusp* of X [i.e., each geometric point of the smooth compactification of X whose image is not contained in X] determines a conjugacy class of closed subgroups of Δ_X [i.e., the *inertia* subgroup(s) associated to the cusp], each member of which we shall refer to as a *cuspidal inertia subgroup* of Δ_X. Now suppose further that k is the field of fractions of a complete regular local ring R, and that every element of Σ is invertible in R. Suppose, moreover, that X has a *stable model over* R, i.e., that there exists a *pointed stable curve* (Y, E) over $S \stackrel{\text{def}}{=} \operatorname{Spec} R$ such that X is isomorphic to $(Y \setminus E) \times_R k$ over k. Then *combinatorial anabelian geometry* may be described as the study of the combinatorial geometric properties of the *irreducible components* and *nodes* [i.e., singular points] of the geometric fiber of (Y, E) over the unique closed point of S by means of the *purely group-theoretic properties* of the outer action of G_k—or, alternatively, various natural subquotients of G_k—on Δ_X. Here, we observe that this geometric fiber of (Y, E) over the unique closed point of S may be regarded as a sort of *degeneration* of the hyperbolic curve X.

Let k be an algebraically closed field of characteristic $\notin \Sigma$ and X a hyperbolic curve over k. For each positive integer m, write

- X_m for the *m-th configuration space* of X, i.e., the open subscheme of the fiber product of m copies of X over k obtained by removing the various diagonals;
- Π_m for the maximal pro-Σ quotient of the étale fundamental group $\pi_1(X_m)$ of X_m;
- $X_0 \stackrel{\text{def}}{=} \operatorname{Spec} k$ and $\Pi_0 \stackrel{\text{def}}{=} \{1\}$.

Let n be a positive integer. We shall think of the factors of X_n as *labeled by the indices* $1, \ldots, n$. Thus, for $E \subseteq \{1, \ldots, n\}$ a subset of cardinality $n - m$, where m is a nonnegative integer, we have a projection morphism $X_n \to X_m$ obtained by forgetting the factors that belong to E, hence also an induced *outer surjection* $\Pi_n \twoheadrightarrow \Pi_m$, i.e., a surjection considered up to composition with an inner automorphism. Normal closed subgroups $\operatorname{Ker}(\Pi_n \twoheadrightarrow \Pi_m) \subseteq \Pi_n$ obtained in this way will be referred to as *fiber subgroups of* Π_n *of length* $n - m$ [cf. [MzTa], Definition 2.3, (iii)]. Write

$$X_n \longrightarrow X_{n-1} \longrightarrow \ldots \longrightarrow X_m \longrightarrow \ldots \longrightarrow X_1 \longrightarrow X_0$$

for the projections obtained by forgetting, successively, the factors labeled by indices $> m$ [as m ranges over the nonnegative integers $\leq n$]. Thus, we obtain a sequence of outer surjections

$$\Pi_n \twoheadrightarrow \Pi_{n-1} \twoheadrightarrow \cdots \twoheadrightarrow \Pi_m \twoheadrightarrow \cdots \twoheadrightarrow \Pi_1 \twoheadrightarrow \Pi_0.$$

For each nonnegative integer $m \leq n$, write $K_m \stackrel{\text{def}}{=} \operatorname{Ker}(\Pi_n \twoheadrightarrow \Pi_m)$. Thus, we have a filtration of subgroups

$$\{1\} = K_n \subseteq K_{n-1} \subseteq \cdots \subseteq K_m \subseteq \cdots \subseteq K_1 \subseteq K_0 = \Pi_n.$$

In the situation of the previous paragraph, let Y be a hyperbolic curve over k and ${}^Y n$ a positive integer. Write ${}^Y \Pi_{{}^Y n}$ for the "Π_n" that occurs in the case where we take "(X, n)" to be $(Y, {}^Y n)$. Let $\alpha : \Pi_n \xrightarrow{\sim} {}^Y\Pi_{Y_n}$ be a(n) [continuous] outer isomorphism. Then we shall say that

- α is *PF-admissible* [cf. [CbTpI], Definition 1.4, (i)] if α induces a bijection between the set of fiber subgroups of Π_n and the set of fiber subgroups of ${}^Y\Pi_{Y_n}$;
- α is *PC-admissible* [cf. [CbTpI], Definition 1.4, (ii), as well as Lemma 3.2, (i), of the present monograph] if, for each positive integer $a \le n$, $\alpha(K_a) \subseteq {}^Y\Pi_{Y_n}$ is a fiber subgroup of ${}^Y\Pi_{Y_n}$ of length ${}^Y n - a$, and, moreover, the ${}^Y\Pi_{Y_n}$-conjugacy-orbit of isomorphisms $K_{a-1}/K_a \xrightarrow{\sim} \alpha(K_{a-1})/\alpha(K_a)$ determined by α induces a bijection between the set of conjugacy classes of cuspidal inertia subgroups of K_{a-1}/K_a and the set of conjugacy classes of cuspidal inertia subgroups of $\alpha(K_{a-1})/\alpha(K_a)$ [where we note that it follows immediately from the various definitions involved that the profinite group K_{a-1}/K_a (respectively,. $\alpha(K_{a-1})/\alpha(K_a)$) is equipped with a natural structure of *pro-Σ surface group*—cf. [MzTa], Definition 1.2];
- α is *PFC-admissible* [cf. [CbTpI], Definition 1.4, (iii)] if α is PF-admissible and PC-admissible.

Suppose, moreover, that $(X, n) = (Y, {}^Y n)$, which thus implies that α is a(n) [continuous] *outomorphism* of $\Pi_n = {}^Y\Pi_{Y_n}$. Then we shall say that

- α is *F-admissible* [cf. [CmbCsp], Definition 1.1, (ii)] if $\alpha(K) = K$ for every fiber subgroup K of Π_n;
- α is *C-admissible* [cf. [CmbCsp], Definition 1.1, (ii)] if α is PC-admissible, and $\alpha(K_a) = K_a$ for each nonnegative integer $a \le n$;
- α is *FC-admissible* [cf. [CmbCsp], Definition 1.1, (ii)] if α is F-admissible and C-admissible.

One central theme of the present monograph is the issue of ***n*-cuspidalizability** [cf. Definition 3.20], i.e., the issue of the extent to which a given isomorphism between the pro-Σ fundamental groups of a pair of hyperbolic curves *lifts* [necessarily *uniquely*, up to a permutation of factors—cf. [NodNon], Theorem B] to a PFC-admissible [cf. [CbTpI], Definition 1.4, (iii)] isomorphism between the pro-Σ fundamental groups of the corresponding n-th configuration spaces, for $n \ge 1$ a positive integer. In this context, we recall that both the *algebraic* and the *anabelian* geometry of such configuration spaces revolves around the behavior of the various *diagonals* that are removed from direct products of copies of the given curve in order to construct these configuration spaces. From this point of view, it is perhaps natural to think of the issue of n-cuspidalizability as a sort of *abstract profinite analogue* of the notion of ***n*-differentiability** in the theory of differential manifolds. In particular, it is perhaps natural to think of the theory of the present monograph [as well as of [MzTa, CmbCsp, NodNon, CbTpI]] as a sort of **abstract profinite analogue** of the classical theory constituted by the **differential topology of surfaces**.

Next, we recall that, to a substantial extent, the theory of **combinatorial cuspidalization** [i.e., the issue of n-cuspidalizability] developed in [CmbCsp] may be thought of as an *essentially formal consequence* of the **combinatorial anabelian result** obtained in [CmbGC], Corollary 2.7, (iii). In a similar vein, the generalization of this theory of [CmbCsp] that is summarized in [NodNon], Theorem B, may be regarded as an essentially formal consequence of the combinatorial anabelian result given in [NodNon], Theorem A. The development of the theory of the present monograph follows this pattern to a substantial extent. That is to say, in Chap. 1, we begin the development of the theory of the present monograph by proving a *fundamental combinatorial anabelian result* [cf. Theorem 1.9], which generalizes the combinatorial anabelian results given in [CmbGC], Corollary 2.7, (iii); [NodNon], Theorem A. A substantial portion of the main results obtained in the remainder of the present monograph may be understood as consisting of various *applications* of Theorem 1.9.

By comparison to the combinatorial anabelian results of [CmbGC], Corollary 2.7, (iii); [NodNon], Theorem A, the *main technical feature* of the combinatorial anabelian result given in Theorem 1.9 of the present monograph is that it allows one, to a substantial extent, to

eliminate the **group-theoretic cuspidality** *hypothesis*

—i.e., the assumption to the effect that the isomorphism between pro-Σ fundamental groups of stable log curves under consideration [that is to say, in effect, an isomorphism between the pro-Σ fundamental groups of certain degenerations of hyperbolic curves] necessarily *preserves cuspidal inertia subgroups*—that plays a *central role* in the proofs of earlier combinatorial anabelian results. In Chap. 2, we apply Theorem 1.9 to obtain the following [partial] **combinatorial cuspidalization** result [cf. Theorem 2.3, (i), (ii); Corollary 3.22], which [partially] generalizes [NodNon], Theorem B.

Theorem A (Partial Combinatorial Cuspidalization for F-admissible Outomorphisms) *Let (g, r) be a pair of nonnegative integers such that $2g - 2 + r > 0$; n a positive integer; Σ a set of prime numbers which is either equal to the set of all prime numbers or of cardinality one; X a* **hyperbolic curve** *of type (g, r) over an algebraically closed field of characteristic $\notin \Sigma$; X_n the n-th* **configuration space** *of X; Π_n the maximal pro-Σ quotient of the fundamental group of X_n;*

$$\mathrm{Out}^{\mathrm{F}}(\Pi_n) \subseteq \mathrm{Out}(\Pi_n)$$

the subgroup of **F-admissible** *outomorphisms [i.e., roughly speaking, outer automorphisms that preserve the fiber subgroups—cf. the discussion preceding Theorem A; [CmbCsp], Definition 1.1, (ii), for more details] of Π_n;*

$$\mathrm{Out}^{\mathrm{FC}}(\Pi_n) \subseteq \mathrm{Out}^{\mathrm{F}}(\Pi_n)$$

the subgroup of **FC-admissible** *outomorphisms [i.e., roughly speaking, outer automorphisms that preserve the fiber subgroups and the cuspidal inertia subgroups*

—*cf. the discussion preceding Theorem A; [CmbCsp], Definition 1.1, (ii), for more details] of* Π_n. *Then the following hold:*

(i) Write

$$n_{\mathrm{inj}} \stackrel{\mathrm{def}}{=} \begin{cases} 1 & \text{if } r \neq 0, \\ 2 & \text{if } r = 0, \end{cases} \qquad n_{\mathrm{bij}} \stackrel{\mathrm{def}}{=} \begin{cases} 3 & \text{if } r \neq 0, \\ 4 & \text{if } r = 0. \end{cases}$$

If $n \geq n_{\mathrm{inj}}$ *(respectively,* $n \geq n_{\mathrm{bij}}$*), then the natural homomorphism*

$$\mathrm{Out}^{\mathrm{F}}(\Pi_{n+1}) \longrightarrow \mathrm{Out}^{\mathrm{F}}(\Pi_n)$$

induced by the projections $X_{n+1} \to X_n$ *obtained by forgetting any one of the* $n+1$ *factors of* X_{n+1} *[cf. [CbTpI], Theorem A, (i)] is* **injective** *(respectively,* **bijective***).*

(ii) Write

$$n_{\mathrm{FC}} \stackrel{\mathrm{def}}{=} \begin{cases} 2 & \text{if } (g, r) = (0, 3), \\ 3 & \text{if } (g, r) \neq (0, 3) \text{and} r \neq 0, \\ 4 & \text{if } r = 0. \end{cases}$$

If $n \geq n_{\mathrm{FC}}$*, then it holds that*

$$\mathrm{Out}^{\mathrm{FC}}(\Pi_n) = \mathrm{Out}^{\mathrm{F}}(\Pi_n).$$

(iii) Suppose that $(g, r) \notin \{(0, 3); (1, 1)\}$*. Then the natural* **injection** *[cf. [NodNon], Theorem B]*

$$\mathrm{Out}^{\mathrm{FC}}(\Pi_2) \hookrightarrow \mathrm{Out}^{\mathrm{FC}}(\Pi_1)$$

induced by the projections $X_2 \to X_1$ *obtained by forgetting either of the two factors of* X_2 *is* **not surjective***.*

Here, we remark that the **non-surjectivity** discussed in Theorem A, (iii), is, in fact, obtained as a consequence of the theory of *tripod synchronization* developed in Chap. 3 [cf. the discussion preceding Theorem C below]. This non-surjectivity is *remarkable* in that it yields an important example of *substantially different behavior* in the theory of profinite fundamental groups of hyperbolic curves from the corresponding theory in the *discrete case*. That is to say, in the case of the classical discrete fundamental group of a hyperbolic topological surface, the **surjectivity** of the corresponding homomorphism may be derived as an essentially formal consequence of the well-known **Dehn-Nielsen-Baer theorem** in the theory of topological surfaces [cf. the discussion of Remark 3.22.1, (i)]. In particular, it constitutes an important *"counterexample"* to the *"line of reasoning"* [i.e., for instance, of the sort which appears in the final paragraph of [Lch], §1; the discussion between [Lch], Theorem 5.1, and [Lch], Conjecture 5.2] that one should expect essentially analogous behavior in the theory of profinite fundamental groups of hyperbolic curves to the relatively well understood behavior observed classically in the theory of discrete fundamental groups of topological surfaces [cf. the discussion of Remark 3.22.1, (iii)].

Theorem A leads naturally to the following strengthening of the result obtained in [CbTpI], Theorem A, (ii), concerning the **group-theoreticity** of the **cuspidal inertia subgroups** of the various one-dimensional subquotients of a configuration space group [cf. Corollary 2.4].

Theorem B (PFC-admissibility of Outomorphisms) *In the notation of Theorem A, write*

$$\mathrm{Out}^{\mathrm{PF}}(\Pi_n) \subseteq \mathrm{Out}(\Pi_n)$$

for the subgroup of **PF-admissible** *outomorphisms [i.e., roughly speaking, outer automorphisms that preserve the fiber subgroups up to a possible permutation of the factors —cf. the discussion preceding Theorem A; [CbTpI], Definition 1.4, (i), for more details] and*

$$\mathrm{Out}^{\mathrm{PFC}}(\Pi_n) \subseteq \mathrm{Out}^{\mathrm{PF}}(\Pi_n)$$

for the subgroup of **PFC-admissible** *outomorphisms [i.e., roughly speaking, outer automorphisms that preserve the fiber subgroups and the cuspidal inertia subgroups up to a possible permutation of the factors —cf. the discussion preceding Theorem A; [CbTpI], Definition 1.4, (iii), for more details]. Let us regard the symmetric group on n letters* $\mathfrak{S}_n$ *as a subgroup of* $\mathrm{Out}(\Pi_n)$ *via the natural inclusion* $\mathfrak{S}_n \hookrightarrow \mathrm{Out}(\Pi_n)$ *obtained by permuting the various factors of* X_n. *Finally, suppose that* $(g, r) \notin \{(0, 3); (1, 1)\}$. *Then the following hold:*

(i) We have an equality

$$\mathrm{Out}(\Pi_n) = \mathrm{Out}^{\mathrm{PF}}(\Pi_n).$$

If, moreover, $(r, n) \neq (0, 2)$, *then we have equalities*

$$\mathrm{Out}(\Pi_n) = \mathrm{Out}^{\mathrm{PF}}(\Pi_n) = \mathrm{Out}^{\mathrm{F}}(\Pi_n) \times \mathfrak{S}_n.$$

(ii) If either

$$r > 0, \quad n \geq 3$$

or

$$n \geq 4,$$

then we have equalities

$$\mathrm{Out}(\Pi_n) = \mathrm{Out}^{\mathrm{PFC}}(\Pi_n) = \mathrm{Out}^{\mathrm{FC}}(\Pi_n) \times \mathfrak{S}_n.$$

The partial combinatorial cuspidalization of Theorem A has natural applications to the **relative** and **[semi-]absolute anabelian geometry of configuration spaces** [cf. Corollaries 2.5, 2.6], which generalize the theory of [AbsTpI], §1. Roughly speaking, these results allow one, in a wide variety of cases, to reduce issues concerning the relative and [semi-]absolute anabelian geometry of *configuration spaces* to the corresponding issues concerning the relative and [semi-]absolute anabelian geometry of *hyperbolic curves*. Also, we remark that in this context, we obtain a purely *scheme-theoretic* result [cf. Lemma 2.7] that states, roughly speaking, that the theory of isomorphisms [of schemes!] between configuration spaces associated to hyperbolic curves may be reduced to the theory of isomorphisms [of schemes!] between hyperbolic curves.

In Chap. 3, we take up the study of [the group-theoretic versions of] the various **tripods** [i.e., copies of the projective line minus three points] that occur in the various one-dimensional fibers of the log configuration spaces associated to a *stable log curve* [cf. the discussion entitled "*Curves*" in [CbTpI], §0]. Roughly speaking, these tripods either occur in the original stable log curve or arise as the result of *blowing up various* ***cusps*** *or* ***nodes*** that occur in the one-dimensional fibers of log configuration spaces of *lower dimension* [cf. Figure 1]. In fact, a substantial portion of Chap. 3 is devoted precisely to the theory of *classification* of the various tripods that occur in the one-dimensional fibers of the log configuration spaces associated to a stable log curve [cf. Lemmas 3.6, 3.8]. This leads naturally to the study of the phenomenon of **tripod synchronization**, i.e., roughly speaking, the phenomenon that an outomorphism [that is to say, an outer automorphism] of the pro-Σ fundamental group of a log configuration space associated to a stable log curve typically induces the **same** outer automorphism on the various [group-theoretic] tripods that occur in subquotients of such a fundamental group [cf. Theorems 3.16, 3.17, 3.18]. The phenomenon of tripod synchronization, in turn, leads naturally to the definition of the **tripod homomorphism** [cf. Definition 3.19], which may be thought of as the homomorphism obtained by associating to an [FC-admissible] outer automorphism of the pro-Σ fundamental group of the n-th log configuration space associated to a stable log curve, where $n \geq 3$ is a positive integer, the outer automorphism induced on a [group-theoretic] **central tripod**, i.e., roughly speaking, a tripod that arises, in the case where $n = 3$ and the given stable log curve has no nodes, by *blowing up the intersection of the three diagonal divisors* of the direct product of three copies of the curve.

Theorem C (Synchronization of Tripods in Three or More Dimensions) *Let* (g, r) *be a pair of nonnegative integers such that* $2g-2+r > 0$; n *a positive integer;* Σ *a set of prime numbers which is either equal to the set of all prime numbers or of cardinality one;* k *an algebraically closed field of characteristic* $\notin \Sigma$; $(\mathrm{Spec}\, k)^{\log}$ *the log scheme obtained by equipping* $\mathrm{Spec}\, k$ *with the log structure determined by*

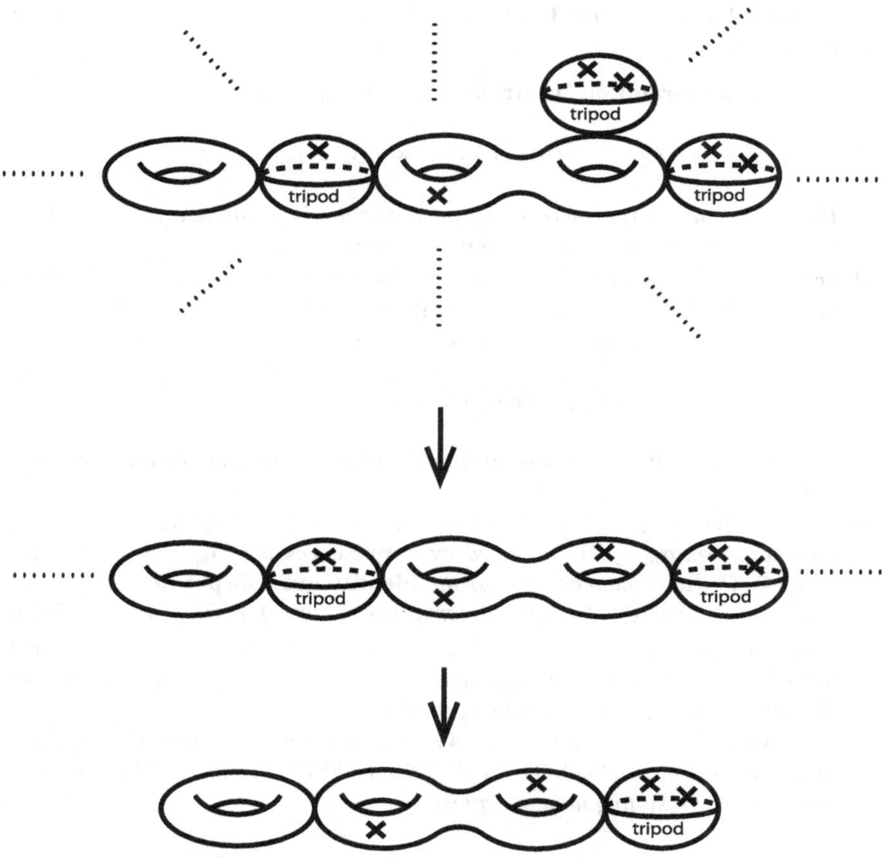

Fig. 1 Tripods in the various fibers of a configuration space

the fs chart $\mathbb{N} \to k$ *that maps* $1 \mapsto 0$*;* $X^{\log} = X_1^{\log}$ *a* **stable log curve** *of type* (g, r) *over* $(\mathrm{Spec}\, k)^{\log}$*. Write* $\mathcal{G}$ *for the semi-graph of anabelioids of pro-*Σ *PSC-type determined by the stable log curve* $X^{\log}$*. For each positive integer* i*, write* $X_i^{\log}$ *for the* i*-th* **log configuration space** *of the stable log curve* $X^{\log}$ *[cf. the discussion entitled "Curves" in "Notations and Conventions"];* Π_i *for the maximal pro-*Σ *quotient of the kernel of the natural surjection* $\pi_1(X_i^{\log}) \twoheadrightarrow \pi_1((\mathrm{Spec}\, k)^{\log})$*. Let* $T \subseteq \Pi_m$ *be a* $\{\mathbf{1}, \cdots, \mathbf{m}\}$**-tripod** *of* Π_n *[cf. Definition 3.3, (i)] for* m *a positive integer* $\leq n$*. Suppose that* $n \geq 3$*. Let*

$$\Pi^{\mathrm{tpd}} \subseteq \Pi_3$$

be a **1−central{1, 2, 3}−tripod** *of* Π_n *[cf. Definitions 3.3, (i); 3.7, (ii)]. Then the following hold:*

(i) The **commensurator** *and* **centralizer** *of* T *in* Π_m *satisfy the equality*

$$C_{\Pi_m}(T) = T \times Z_{\Pi_m}(T).$$

Thus, if an outomorphism α *of* Π_m *preserves the* Π_m*-conjugacy class of* $T \subseteq \Pi_m$*, then one obtains a "***restriction***"* $\alpha|_T \in \mathrm{Out}(T)$.

(ii) Let $\alpha \in \mathrm{Out}^{\mathrm{FC}}(\Pi_n)$ *be an FC-admissible outomorphism of* Π_n*. Then the outomorphism of* Π_3 *induced by* α **preserves** *the* Π_3*-conjugacy class of* $\Pi^{\mathrm{tpd}} \subseteq \Pi_3$*. In particular, by (i), we obtain a natural homomorphism*

$$\mathfrak{T}_{\Pi^{\mathrm{tpd}}} : \mathrm{Out}^{\mathrm{FC}}(\Pi_n) \longrightarrow \mathrm{Out}(\Pi^{\mathrm{tpd}}).$$

We shall refer to this homomorphism as the **tripod homomorphism** *associated to* Π_n.

(iii) Let $\alpha \in \mathrm{Out}^{\mathrm{FC}}(\Pi_n)$ *be an FC-admissible outomorphism of* Π_n *such that the outomorphism* α_m *of* Π_m *induced by* α **preserves** *the* Π_m*-conjugacy class of* $T \subseteq \Pi_m$ *and induces [cf. (i)] the* **identity automorphism** *of the set of* T*-conjugacy classes of cuspidal inertia subgroups of* T*. Then there exists a* **geometric** *[cf. Definition 3.4, (ii)] outer isomorphism* $\Pi^{\mathrm{tpd}} \xrightarrow{\sim} T$ *with respect to which the outomorphism* $\mathfrak{T}_{\Pi^{\mathrm{tpd}}}(\alpha) \in \mathrm{Out}(\Pi^{\mathrm{tpd}})$ *[cf. (ii)] is* **compatible** *with the outomorphism* $\alpha_m|_T \in \mathrm{Out}(T)$ *[cf. (i)].*

(iv) Suppose, moreover, that either $n \geq 4$ *or* $r \neq 0$*. Then the homomorphism* $\mathfrak{T}_{\Pi^{\mathrm{tpd}}}$ *of (ii) factors through* $\mathrm{Out}^{\mathrm{C}}(\Pi^{\mathrm{tpd}})^{\Delta+} \subseteq \mathrm{Out}(\Pi^{\mathrm{tpd}})$ *[cf. Definition 3.4, (i)], and, moreover, the resulting homomorphism*

$$\mathfrak{T}_{\Pi^{\mathrm{tpd}}} : \mathrm{Out}^{\mathrm{F}}(\Pi_n) = \mathrm{Out}^{\mathrm{FC}}(\Pi_n) \longrightarrow \mathrm{Out}^{\mathrm{C}}(\Pi^{\mathrm{tpd}})^{\Delta+}$$

[cf. Theorem A, (ii)] is **surjective**.

Here, we remark that the **surjectivity** of the tripod homomorphism [cf. Theorem C, (iv)] is obtained [cf. Corollary 4.15] as a consequence of the theory of *glueability of combinatorial cuspidalizations* developed in Chap. 4 [cf. the discussion preceding Theorem F below]. Also, we recall that the *codomain* of this surjective tripod homomorphism

$$\mathrm{Out}^{\mathrm{C}}(\Pi^{\mathrm{tpd}})^{\Delta+}$$

may be identified with the [pro-Σ] **Grothendieck-Teichmüller group** GT^{Σ} [cf. the discussion of [CmbCsp], Remark 1.11.1]. Since GT^{Σ} may be thought of as a sort of **abstract combinatorial approximation** *of the absolute Galois group* $G_{\mathbb{Q}}$ of the rational number field $\mathbb{Q}$, it is thus natural to think of the surjective tripod homomorphism

$$\mathrm{Out}^{\mathrm{F}}(\Pi_n) \twoheadrightarrow \mathrm{Out}^{\mathrm{C}}(\Pi^{\mathrm{tpd}})^{\Delta+}$$

of Theorem C, (iv), as a sort of **abstract combinatorial version** of the natural surjective outer homomorphism

$$\pi_1((\mathcal{M}_{g,[r]})_{\mathbb{Q}}) \twoheadrightarrow G_{\mathbb{Q}}$$

induced on étale fundamental groups by the structure morphism $(\mathcal{M}_{g,[r]})_{\mathbb{Q}} \to \mathrm{Spec}\,(\mathbb{Q})$ of the moduli stack $(\mathcal{M}_{g,[r]})_{\mathbb{Q}}$ of hyperbolic curves of type (g, r) [cf. the discussion of Remark 3.19.1]. In particular, the *kernel* of the tripod homomorphism—which we denote by

$$\mathrm{Out}^{\mathrm{F}}(\Pi_n)^{\mathrm{geo}}$$

—may be thought of as a sort of abstract combinatorial analogue of the ***geometric*** *étale fundamental group* of $(\mathcal{M}_{g,[r]})_{\mathbb{Q}}$ [i.e., the kernel of the natural outer homomorphism $\pi_1((\mathcal{M}_{g,[r]})_{\mathbb{Q}}) \twoheadrightarrow G_{\mathbb{Q}}$].

One interesting application of the theory of tripod synchronization is the following. Fix a pro-Σ fundamental group of a hyperbolic curve. Recall the notion of a **nondegenerate profinite Dehn multi-twist** [cf. [CbTpI], Definition 4.4; [CbTpI], Definition 5.8, (ii)] associated to a structure of *semi-graph of anabelioids of pro-*Σ *PSC-type* on such a fundamental group. Here, we recall that such a structure may be thought of as a sort of profinite analogue of the notion of a *decomposition of a hyperbolic topological surface into hyperbolic subsurfaces* [i.e., such as "pants"]. Then the following result asserts that, under certain technical conditions, any such nondegenerate profinite Dehn multi-twist that **commutes** with another nondegenerate profinite Dehn multi-twist associated to some given **totally degenerate** semi-graph of anabelioids of pro-Σ PSC-type [cf. [CbTpI], Definition 2.3, (iv)] necessarily arises from a structure of semi-graph of anabelioids of pro-Σ PSC-type that is **"co-Dehn"** to, i.e., arises by applying a *deformation* to, the given totally degenerate semi-graph of anabelioids of pro-Σ PSC-type [cf. Corollary 3.25]. This sort of result is reminiscent of topological results concerning subgroups of the *mapping class group* generated by pairs of *positive Dehn multi-twists* [cf. [Ishi, HT]].

Theorem D (Co-Dehn-ness of Degeneration Structures in the Totally Degenerate Case) *In the notation of Theorem C, for* $i = 1, 2$, *let* $Y_i^{\log}$ *be a stable log curve over* $(\mathrm{Spec}\, k)^{\log}$; $\mathcal{H}_i$ *the "*$\mathcal{G}$*" that occurs in the case where we take "*$X^{\log}$*" to be* $Y_i^{\log}$; $(\mathcal{H}_i, S_i, \phi_i)$ *a* **3−cuspidalizable degeneration structure** *on* $\mathcal{G}$ *[cf. Definition 3.23, (i), (v)];* $\alpha_i \in \mathrm{Out}(\Pi_{\mathcal{G}})$ *a* **nondegenerate** $(\mathcal{H}_i, S_i, \phi_i)$*-Dehn multi-twist of* $\mathcal{G}$ *[cf. Definition 3.23, (iv)]. Suppose that* α_1 **commutes** *with* α_2*, and that* $\mathcal{H}_2$ *is* **totally degenerate** *[cf. [CbTpI], Definition 2.3, (iv)]. Suppose, moreover, that one of the following conditions is satisfied:*

(i) $r \neq 0$.
(ii) α_1 *and* α_2 *are* **positive definite** *[cf. Definition 3.23, (iv)].*

Then $(\mathcal{H}_1, S_1, \phi_1)$ *is* **co-Dehn** *to* $(\mathcal{H}_2, S_2, \phi_2)$ *[cf. Definition 3.23, (iii)], or, equivalently [since* $\mathcal{H}_2$ *is* **totally degenerate***],* $(\mathcal{H}_2, S_2, \phi_2) \preceq (\mathcal{H}_1, S_1, \phi_1)$ *[cf. Definition 3.23, (ii)].*

Another interesting application of the theory of tripod synchronization is to the computation, in terms of a certain **scheme-theoretic fundamental group**, of the *purely combinatorial* commensurator of the subgroup of profinite Dehn multi-twists in the group of 3-cuspidali- zable, FC-admissible, "geometric" outer automorphisms of the pro-Σ fundamental group of a **totally degenerate** stable log curve [cf. Corollary 3.27]. Here, we remark that the scheme-theoretic [or, perhaps more precisely, "log algebraic stack-theoretic"] fundamental group that appears is, roughly speaking, the pro-Σ geometric fundamental group of a formal neighborhood, in the corresponding logarithmic moduli stack, of the point determined by the given totally degenerate stable log curve. In particular, this computation may also be regarded as a sort of **purely combinatorial algorithm** for constructing this scheme-theoretic fundamental group [cf. Remark 3.27.1].

Theorem E (Commensurator of Profinite Dehn Multi-twists in the Totally Degenerate Case) *In the notation of Theorem C [so* $n \geq 3$*], suppose further that if* $r = 0$*, then* $n \geq 4$*. Also, we assume that* $\mathcal{G}$ *is* **totally degenerate** *[cf. [CbTpI], Definition 2.3, (iv)]. Write* $s\colon \mathrm{Spec}\, k \to (\overline{\mathcal{M}}_{g,[r]})_k \stackrel{\mathrm{def}}{=} (\overline{\mathcal{M}}_{g,[r]})_{\mathrm{Spec}\, k}$ *[cf. the discussion entitled "Curves" in "Notations and Conventions"] for the underlying (1-)morphism of algebraic stacks of the classifying (1-)morphism* $(\mathrm{Spec}\, k)^{\log} \to (\overline{\mathcal{M}}^{\log}_{g,[r]})_k \stackrel{\mathrm{def}}{=} (\overline{\mathcal{M}}^{\log}_{g,[r]})_{\mathrm{Spec}\, k}$ *[cf. the discussion entitled "Curves" in "Notations and Conventions"] of the stable log curve* $X^{\log}$ *over* $(\mathrm{Spec}\, k)^{\log}$*;* $\widetilde{\mathcal{N}}^{\log}_s$ *for the log scheme obtained by equipping* $\widetilde{\mathcal{N}}_s \stackrel{\mathrm{def}}{=} \mathrm{Spec}\, k$ *with the log structure induced, via s, by the log structure of* $(\overline{\mathcal{M}}^{\log}_{g,[r]})_k$*;* $\mathcal{N}^{\log}_s$ *for the log stack obtained by forming the [stack-theoretic] quotient of the log scheme* $\widetilde{\mathcal{N}}^{\log}_s$ *by the natural action of the finite k-group "*$s \times_{(\overline{\mathcal{M}}_{g,[r]})_k} s$*", i.e., the fiber product over* $(\overline{\mathcal{M}}_{g,[r]})_k$ *of two copies of s;* $\mathcal{N}_s$ *for the underlying stack of the log stack* $\mathcal{N}^{\log}_s$*;* $I_{\mathcal{N}_s} \subseteq \pi_1(\mathcal{N}^{\log}_s)$ *for the closed subgroup of the log fundamental group* $\pi_1(\mathcal{N}^{\log}_s)$ *of* $\mathcal{N}^{\log}_s$ *given by the kernel of the natural surjection* $\pi_1(\mathcal{N}^{\log}_s) \twoheadrightarrow \pi_1(\mathcal{N}_s)$ *[induced by the (1-)morphism* $\mathcal{N}^{\log}_s \to \mathcal{N}_s$ *obtained by forgetting the log structure];* $\pi_1^{(\Sigma)}(\mathcal{N}^{\log}_s)$ *for the quotient of* $\pi_1(\mathcal{N}^{\log}_s)$ *by the kernel of the natural surjection from* $I_{\mathcal{N}_s}$ *to its maximal pro-*Σ *quotient* $I^{\Sigma}_{\mathcal{N}_s}$*. Then we have an equality*

$$N_{\mathrm{Out}^{\mathrm{F}}(\Pi_n)^{\mathrm{geo}}}(\mathrm{Dehn}(\mathcal{G})) = C_{\mathrm{Out}^{\mathrm{F}}(\Pi_n)^{\mathrm{geo}}}(\mathrm{Dehn}(\mathcal{G}))$$

and a **natural commutative diagram** *of profinite groups*

$$\begin{array}{ccccccccc} 1 & \longrightarrow & I_{\mathcal{N}_s}^{\Sigma} & \longrightarrow & \pi_1^{(\Sigma)}(\mathcal{N}_s^{\log}) & \longrightarrow & \pi_1(\mathcal{N}_s) & \longrightarrow & 1 \\ & & \downarrow & & \downarrow & & \downarrow & & \\ 1 & \longrightarrow & \mathrm{Dehn}(\mathcal{G}) & \longrightarrow & C_{\mathrm{Out}^{\mathrm{F}}(\Pi_n)^{\mathrm{geo}}}(\mathrm{Dehn}(\mathcal{G})) & \longrightarrow & \mathrm{Aut}(\mathbb{G}) & \longrightarrow & 1 \end{array}$$

[cf. Definition 3.1, (ii), concerning the notation "$\mathbb{G}$"]—where the horizontal sequences are **exact***, and the vertical arrows are* **isomorphisms***. Moreover,* $\mathrm{Dehn}(\mathcal{G})$ *is* **open** *in* $C_{\mathrm{Out}^{\mathrm{F}}(\Pi_n)^{\mathrm{geo}}}(\mathrm{Dehn}(\mathcal{G}))$.

In Chap. 4, we show, under suitable technical conditions, that an automorphism of the pro-Σ fundamental group of the log configuration space associated to a stable log curve necessarily *preserves the* ***graph-theoretic structure*** of the various one-dimensional fibers of such a log configuration space [cf. Theorem 4.7]. This allows us to verify the **glueability of combinatorial cuspidalizations**, i.e., roughly speaking, that, for $n \geq 2$ a positive integer, the datum of an *n-cuspidalizable* outer automorphism of the pro-Σ fundamental group of a stable log curve is *equivalent*, up to possible composition with a profinite Dehn multi-twist, to the datum of a collection of n-cuspidalizable automorphisms of the pro-Σ fundamental groups of the various *irreducible components* of the given stable log curve that satisfy a certain *gluing condition* involving the induced outer actions on *tripods* [cf. Theorem 4.14].

Theorem F (Glueability of Combinatorial Cuspidalizations) *In the notation of Theorem C, write*

$$\mathrm{Out}^{\mathrm{FC}}(\Pi_n)^{\mathrm{brch}} \subseteq \mathrm{Out}^{\mathrm{FC}}(\Pi_n)$$

for the closed subgroup of $\mathrm{Out}^{\mathrm{FC}}(\Pi_n)$ *consisting of FC-admissible outomorphisms* α *of* Π_n *such that the outomorphism of* Π_1 *determined by* α *induces the identity automorphism of* $\mathrm{Vert}(\mathcal{G})$, $\mathrm{Node}(\mathcal{G})$*, and, moreover, fixes each of the branches of every node of* $\mathcal{G}$ *[cf. Definition 4.6, (i)];*

$$\mathrm{Glu}(\Pi_n) \subseteq \prod_{v \in \mathrm{Vert}(\mathcal{G})} \mathrm{Out}^{\mathrm{FC}}((\Pi_v)_n)$$

for the closed subgroup of $\prod_{v \in \mathrm{Vert}(\mathcal{G})} \mathrm{Out}^{\mathrm{FC}}((\Pi_v)_n)$ *consisting of "***glueable***" collections of outomorphisms of the groups "$(\Pi_v)_n$" [cf. Definition 4.9, (iii)]. Then we have a* **natural exact sequence** *of profinite groups*

$$1 \longrightarrow \mathrm{Dehn}(\mathcal{G}) \longrightarrow \mathrm{Out}^{\mathrm{FC}}(\Pi_n)^{\mathrm{brch}} \longrightarrow \mathrm{Glu}(\Pi_n) \longrightarrow 1.$$

This glueability result may, alternatively, be thought of as a result that asserts the **localizability** [i.e., relative to localization on the dual semi-graph of the given stable log curve] of the notion of ***n*-cuspidalizability**. In this context, it is of interest to observe that this glueability result may be regarded as a natural generalization, to the case of *n-cuspidalizability* for $n \geq 2$, of the glueability result obtained in [CbTpI], Theorem B, (iii), in the *"1-cuspidalizable"* case, which is derived as a consequence of the theory of *localizability* [i.e., relative to localization on the dual semi-graph of the given stable log curve] and *synchronization* of **cyclotomes** developed in [CbTpI], §3, §4. From this point of view, it is also of interest to observe that the *sufficiency* portion of [the equivalence that constitutes] this glueability result [i.e., Theorem F] may be thought of as a sort of *"converse"* to the theory of *tripod synchronizations* developed in Chap. 3 [i.e., of which the *necessity* portion of this glueability result is, in essence, a *formal consequence*—cf. the proof of Lemma 4.10, (ii)]. Indeed, the bulk of the proof given in Chap. 4 of Theorem 4.14 is devoted to the *sufficiency* portion of this result, which is verified by means of a detailed combinatorial analysis [cf. the proof of [CbTpI], Proposition 4.10, (ii)] of the **noncyclically primitive** and **cyclically primitive** cases [cf. Lemmas 4.12, 4.13; Figures 4.1, 4.2, 4.3].

Finally, we apply this glueability result to derive a **cuspidalization theorem**—i.e., in the spirit of and generalizing the corresponding results of [AbsCsp], Theorem 3.1; [Hsh], Theorem 0.1; [Wkb], Theorem C [cf. Remark 4.16.1]—for *geometrically pro-l fundamental groups of stable log curves over finite fields* [cf. Corollary 4.16]. That is to say, in the case of stable log curves over finite fields,

> the condition of *compatibility with the* ***Galois action*** is sufficient to imply the ***n*-cuspidalizability** of arbitrary isomorphisms between the geometric pro-l fundamental groups, for $n \geq 1$.

In this context, it is of interest to recall that **strong anabelian results** [i.e., in the style of the *"Grothendieck Conjecture"*] for such geometrically pro-l fundamental groups of stable log curves over finite fields are **not known** in general, at the time of writing. On the other hand, we observe that in the case of **totally degenerate** stable log curves over finite fields, such "strong anabelian results" may be obtained under *certain technical conditions* [cf. Corollary 4.17; Remarks 4.17.1, 4.17.2].

Notations and Conventions

Sets: If S is a set, then we shall denote by $\#S$ the *cardinality* of S.

Groups: We shall refer to an element of a group as *trivial* (respectively, *nontrivial*) if it is (respectively, is not) equal to the identity element of the group. We shall refer to a nonempty subset of a group as *trivial* (respectively, *nontrivial*) if it is (respectively, is not) equal to the set whose unique element is the identity element of the group.

Topological groups: Let G be a topological group and $J, H \subseteq G$ closed subgroups. Then we shall write

$$Z_J(H) \overset{\text{def}}{=} \{ j \in J \mid jh = hj \text{foranyh} \in H \} = Z_G(H) \cap J$$

for the *centralizer* of H in J,

$$Z(G) \stackrel{\text{def}}{=} Z_G(G)$$

for the *center* of G, and

$$Z_J^{\text{loc}}(H) \stackrel{\text{def}}{=} \varinjlim Z_J(U) \subseteq J$$

—where the inductive limit is over all open subgroups $U \subseteq H$ of H—for the *"local centralizer"* of H in J. We shall write $Z^{\text{loc}}(G) \stackrel{\text{def}}{=} Z_G^{\text{loc}}(G)$ for the *"local center"* of G. Thus, a profinite group G is *slim* [cf. the discussion entitled "*Topological groups*" in [CbTpI], §0] if and only if $Z^{\text{loc}}(G) = \{1\}$.

Rings: If R is a commutative ring with unity, then we shall write R^* for the multiplicative group of invertible elements of R.

Curves: Let g, r_1, r_2 be nonnegative integers such that $2g - 2 + r_1 + r_2 > 0$. Then we shall write $\overline{\mathcal{M}}_{g,[r_1]+r_2}$ for the *moduli stack of pointed stable curves of type* $(g, r_1 + r_2)$, where the first r_1 marked points are regarded as *unordered*, but the last r_2 marked points are regarded as *ordered*, over $\mathbb{Z}$; $\mathcal{M}_{g,[r_1]+r_2} \subseteq \overline{\mathcal{M}}_{g,[r_1]+r_2}$ for the open substack of $\overline{\mathcal{M}}_{g,[r_1]+r_2}$ that parametrizes *smooth curves*; $\overline{\mathcal{M}}^{\text{log}}_{g,[r_1]+r_2}$ for the log stack obtained by equipping $\overline{\mathcal{M}}_{g,[r_1]+r_2}$ with the log structure associated to the divisor with normal crossings $\overline{\mathcal{M}}_{g,[r_1]+r_2} \setminus \mathcal{M}_{g,[r_1]+r_2} \subseteq \overline{\mathcal{M}}_{g,[r_1]+r_2}$; $\overline{\mathcal{C}}_{g,[r_1]+r_2} \to \overline{\mathcal{M}}_{g,[r_1]+r_2}$ for the *tautological stable curve* over $\overline{\mathcal{M}}_{g,[r_1]+r_2}$; $\overline{\mathcal{D}}_{g,[r_1]+r_2} \subseteq \overline{\mathcal{C}}_{g,[r_1]+r_2}$ for the corresponding *tautological divisor of cusps* of $\overline{\mathcal{C}}_{g,[r_1]+r_2} \to \overline{\mathcal{M}}_{g,[r_1]+r_2}$. Then the divisor given by the union of $\overline{\mathcal{D}}_{g,[r_1]+r_2}$ with the inverse image in $\overline{\mathcal{C}}_{g,[r_1]+r_2}$ of the divisor $\overline{\mathcal{M}}_{g,[r_1]+r_2} \setminus \mathcal{M}_{g,[r_1]+r_2} \subseteq \overline{\mathcal{M}}_{g,[r_1]+r_2}$ determines a log structure on $\overline{\mathcal{C}}_{g,[r_1]+r_2}$; write $\overline{\mathcal{C}}^{\text{log}}_{g,[r_1]+r_2}$ for the resulting log stack. Thus, we obtain a (1-)morphism of log stacks $\overline{\mathcal{C}}^{\text{log}}_{g,[r_1]+r_2} \to \overline{\mathcal{M}}^{\text{log}}_{g,[r_1]+r_2}$. We shall write $\mathcal{C}_{g,[r_1]+r_2} \subseteq \overline{\mathcal{C}}_{g,[r_1]+r_2}$ for the interior of $\overline{\mathcal{C}}^{\text{log}}_{g,[r_1]+r_2}$ [cf. the discussion entitled "*Log schemes*" in [CbTpI], §0]. In particular, we obtain a (1-)morphism of stacks $\mathcal{C}_{g,[r_1]+r_2} \to \mathcal{M}_{g,[r_1]+r_2}$. Moreover, for a nonnegative integer r such that $2g - 2 + r > 0$, we shall write $\overline{\mathcal{M}}_{g,[r]} \stackrel{\text{def}}{=} \overline{\mathcal{M}}_{g,[r]+0}$; $\mathcal{M}_{g,[r]} \stackrel{\text{def}}{=} \mathcal{M}_{g,[r]+0}$; $\overline{\mathcal{M}}^{\text{log}}_{g,[r]} \stackrel{\text{def}}{=} \overline{\mathcal{M}}^{\text{log}}_{g,[r]+0}$; $\overline{\mathcal{C}}_{g,[r]} \stackrel{\text{def}}{=} \overline{\mathcal{C}}_{g,[r]+0}$; $\overline{\mathcal{D}}_{g,[r]} \stackrel{\text{def}}{=} \overline{\mathcal{D}}_{g,[r]+0}$; $\overline{\mathcal{C}}^{\text{log}}_{g,[r]} \stackrel{\text{def}}{=} \overline{\mathcal{C}}^{\text{log}}_{g,[r]+0}$; $\mathcal{C}_{g,[r]} \stackrel{\text{def}}{=} \mathcal{C}_{g,[r]+0}$. In particular, the stack $\mathcal{M}_{g,[r]}$ may be regarded as a *moduli stack of hyperbolic curves of type* (g, r) over $\mathbb{Z}$. If S is a scheme, then we shall denote by means of a *subscript* S the result of base-changing via the structure morphism $S \to \operatorname{Spec} \mathbb{Z}$ the various log stacks of the above discussion.

Let (g, r) be a pair of nonnegative integers such that $2g-2+r>0$; n a positive integer; $X^{\log}$ a *stable log curve* [cf. the discussion entitled "*Curves*" in [CbTpI], §0] of type (g, r) over a log scheme $S^{\log}$. Then we shall refer to the log scheme obtained by pulling back the (1-)morphism $\overline{\mathcal{M}}^{\log}_{g,[r]+n} \to \overline{\mathcal{M}}^{\log}_{g,[r]}$ given by forgetting the last n [ordered] points via the classifying (1-)morphism $S^{\log} \to \overline{\mathcal{M}}_{g,[r]}$ of $X^{\log}$ as the *n-th log conguration space* of $X^{\log}$.

Contents

Chapter 1
Combinatorial Anabelian Geometry in the Absence of Group-Theoretic Cuspidality

In this chapter, we discuss various combinatorial versions of the Grothendieck Conjecture for outer representations of *NN*- and *IPSC*-type [cf. Theorem 1.9 below]. These Grothendieck Conjecture-type results may be regarded as *generalizations* of [NodNon], Corollary 4.2; [NodNon], Remark 4.2.1, that may be applied to isomorphisms that are *not necessarily group-theoretically cuspidal*. For instance, we prove [cf. Theorem 1.9, (ii), below] that any isomorphism between outer representations of *IPSC-type* [cf. [NodNon], Definition 2.4, (i)] is necessarily *group-theoretically verticial*, i.e., roughly speaking, preserves the verticial subgroups.

A basic reference for the theory of *semi-graphs of anabelioids of PSC-type* is [CmbGC]. We shall use the terms "*semi-graph of anabelioids of PSC-type*", "*PSC-fundamental group of a semi-graph of anabelioids of PSC-type*", "*finite étale covering of semi-graphs of anabelioids of PSC-type*", "*vertex*", "*edge*", "*node*", "*cusp*", "*verticial subgroup*", "*edge-like subgroup*", "*nodal subgroup*", "*cuspidal subgroup*", and "*sturdy*" as they are defined in [CmbGC], Definition 1.1 [cf. also Remark 1.1.2 below]. Also, we shall apply the various notational conventions established in [NodNon], Definition 1.1, and refer to the "PSC-fundamental group of a semi-graph of anabelioids of PSC-type" simply as the "fundamental group" [of the semi-graph of anabelioids of PSC-type]. That is to say, we shall refer to the maximal pro-Σ quotient of the fundamental group of a semi-graph of anabelioids of pro-Σ PSC-type [as a semi-graph of anabelioids!] as the "fundamental group of the semi-graph of anabelioids of PSC-type".

In this chapter, let Σ be a nonempty set of prime numbers and $\mathcal{G}$ a semi-graph of anabelioids of pro-Σ PSC-type. Write $\mathbb{G}$ for the underlying semi-graph of $\mathcal{G}$, $\Pi_{\mathcal{G}}$ for the [pro-Σ] fundamental group of $\mathcal{G}$, and $\widetilde{\mathcal{G}} \to \mathcal{G}$ for the universal covering of $\mathcal{G}$ corresponding to $\Pi_{\mathcal{G}}$. Then since the fundamental group $\Pi_{\mathcal{G}}$ of $\mathcal{G}$ is *topologically finitely generated*, the profinite topology of $\Pi_{\mathcal{G}}$ induces [profinite] topologies on $\mathrm{Aut}(\Pi_{\mathcal{G}})$ and $\mathrm{Out}(\Pi_{\mathcal{G}})$ [cf. the discussion entitled "*Topological groups*" in [CbTpI], §0]. If, moreover, we write $\mathrm{Aut}(\mathcal{G})$ for the automorphism

Y. Hoshi, S. Mochizuki, *Topics Surrounding the Combinatorial Anabelian Geometry of Hyperbolic Curves II*, Lecture Notes in Mathematics 2299,
https://doi.org/10.1007/978-981-19-1096-8_1

group of $\mathcal{G}$, then, by the discussion preceding [CmbGC], Lemma 2.1, the natural homomorphism

$$\mathrm{Aut}(\mathcal{G}) \longrightarrow \mathrm{Out}(\Pi_{\mathcal{G}})$$

is an *injection with closed image*. [Here, we recall that an automorphism of a semi-graph of anabelioids consists of an automorphism of the underlying semi-graph, together with a compatible system of isomorphisms between the various anabelioids at each of the vertices and edges of the underlying semi-graph which are compatible with the various morphisms of anabelioids associated to the branches of the underlying semi-graph—cf. [SemiAn], Definition 2.1; [SemiAn], Remark 2.4.2.] Thus, by equipping $\mathrm{Aut}(\mathcal{G})$ with the topology induced via this homomorphism by the topology of $\mathrm{Out}(\Pi_{\mathcal{G}})$, we may regard $\mathrm{Aut}(\mathcal{G})$ as being equipped with the structure of a *profinite group*.

Definition 1.1 We shall say that an element $\gamma \in \Pi_{\mathcal{G}}$ of $\Pi_{\mathcal{G}}$ is *verticial* (respectively, *edge-like*; *nodal*; *cuspidal*) if γ is contained in a verticial (respectively, an edge-like; a nodal; a cuspidal) subgroup of $\Pi_{\mathcal{G}}$.

Remark 1.1.1 Let $\gamma \in \Pi_{\mathcal{G}}$ be a *nontrivial* [cf. the discussion entitled "*Groups*" in "Notations and Conventions"] element of $\Pi_{\mathcal{G}}$. If $\gamma \in \Pi_{\mathcal{G}}$ is *edge-like* [cf. Definition 1.1], then it follows from [NodNon], Lemma 1.5, that there exists a *unique edge* $\widetilde{e} \in \mathrm{Edge}(\widetilde{\mathcal{G}})$ such that $\gamma \in \Pi_{\widetilde{e}}$. If $\gamma \in \Pi_{\mathcal{G}}$ is *verticial*, but *not nodal* [cf. Definition 1.1], then it follows from [NodNon], Lemma 1.9, (i), that there exists a *unique vertex* $\widetilde{v} \in \mathrm{Vert}(\widetilde{\mathcal{G}})$ such that $\gamma \in \Pi_{\widetilde{v}}$.

Remark 1.1.2 Here, we take the opportunity to correct an *unfortunate misprint* in [CmbGC]. In the final sentence of [CmbGC], Definition 1.1, (ii), the phrase "rank ≥ 2" should read "rank > 2". In particular, we shall say that $\mathcal{G}$ is *sturdy* if the abelianization of the image, in the quotient $\Pi_{\mathcal{G}} \twoheadrightarrow \Pi_{\mathcal{G}}^{\mathrm{unr}}$ of $\Pi_{\mathcal{G}}$ by the normal closed subgroup normally topologically generated by the edge-like subgroups, of every verticial subgroup of $\Pi_{\mathcal{G}}$ is free of rank > 2 over $\widehat{\mathbb{Z}}^{\Sigma}$. Here, we note in passing that $\mathcal{G}$ is *sturdy* if and only if every vertex of $\mathcal{G}$ is of genus ≥ 2 [cf. [CbTpI], Definition 2.3, (iii)].

Lemma 1.2 (Existence of a Certain Connected Finite Étale Covering) *Let n be a positive integer which is a product [possibly with multiplicities!] of primes $\in \Sigma$; $\widetilde{e}_1, \widetilde{e}_2 \in \mathrm{Edge}(\widetilde{\mathcal{G}})$; $\widetilde{v} \in \mathrm{Vert}(\widetilde{\mathcal{G}})$. Write $e_1 \stackrel{\mathrm{def}}{=} \widetilde{e}_1(\mathcal{G})$, $e_2 \stackrel{\mathrm{def}}{=} \widetilde{e}_2(\mathcal{G})$, and $v \stackrel{\mathrm{def}}{=} \widetilde{v}(\mathcal{G})$. Suppose that the following conditions are satisfied:*

- *(i)* $\mathcal{G}$ *is* **untangled** *[cf. [NodNon], Definition 1.2].*
- *(ii) If e_1 is a* **node**, *then the following condition holds: Let $w, w' \in \mathcal{V}(e_1)$ be the two* **distinct** *elements of $\mathcal{V}(e_1)$ [cf. (i)]. Then $\#(\mathcal{N}(w) \cap \mathcal{N}(w')) \geq 3$.*
- *(iii) If e_1 is a* **cusp**, *then the following condition holds: Let $w \in \mathcal{V}(e_1)$ be the unique element of $\mathcal{V}(e_1)$. Then $\#\mathcal{C}(w) \geq 3$.*
- *(iv)* $e_1 \neq e_2$.
- *(v)* $v \notin \mathcal{V}(e_1)$.

Then there exists a finite étale Galois subcovering $\mathcal{G}' \to \mathcal{G}$ of $\widetilde{\mathcal{G}} \to \mathcal{G}$ such that n **divides** $[\Pi_{\widetilde{e}_1} : \Pi_{\widetilde{e}_1} \cap \Pi_{\mathcal{G}'}]$*, and, moreover,* $\Pi_{\widetilde{e}_2}, \Pi_{\widetilde{v}} \subseteq \Pi_{\mathcal{G}'}$.

Proof Suppose that e_1 is a *node* (respectively, *cusp*). Write $\mathbb{H}$ for the [uniquely determined] sub-semi-graph of *PSC-type* [cf. [CbTpI], Definition 2.2, (i)] of $\mathbb{G}$ whose set of vertices is $= \mathcal{V}(e_1) = \{w, w'\}$ [cf. condition (ii)] (respectively, $= \{w\}$ [cf. condition (iii)]). Now it follows from condition (ii) (respectively, (iii)) that there exists an $e_3 \in \mathrm{Node}(\mathcal{G}|_{\mathbb{H}}) = \mathcal{N}(w) \cap \mathcal{N}(w')$ (respectively, $\in \mathrm{Cusp}(\mathcal{G}|_{\mathbb{H}}) \cap \mathrm{Cusp}(\mathcal{G}) = \mathcal{C}(w)$) [cf. [CbTpI], Definition 2.2, (ii)] such that $e_3 \neq e_2$. Moreover, again by applying condition (ii) (respectively, (iii)), together with the well-known structure of the abelianization of the fundamental group of a smooth curve over an algebraically closed field of characteristic $\notin \Sigma$, we conclude that there exists a finite étale Galois covering $\mathcal{G}'_{\mathbb{H}} \to \mathcal{G}|_{\mathbb{H}}$ that arises from a normal open subgroup of $\Pi_{\mathcal{G}|_{\mathbb{H}}}$ and which is *unramified* at every element of $\mathrm{Edge}(\mathcal{G}|_{\mathbb{H}}) \setminus \{e_1, e_3\}$ and *totally ramified* at e_1, e_3 with ramification indices *divisible* by n. Now since $\mathcal{G}'_{\mathbb{H}} \to \mathcal{G}|_{\mathbb{H}}$ is *unramified* at every element of $\mathrm{Cusp}(\mathcal{G}|_{\mathbb{H}}) \cap \mathrm{Node}(\mathcal{G})$, one may extend this covering to a finite étale Galois subcovering $\mathcal{G}' \to \mathcal{G}$ of $\widetilde{\mathcal{G}} \to \mathcal{G}$ which restricts to the *trivial* covering over every vertex u of $\mathcal{G}$ such that $u \neq w, w'$ (respectively, $u \neq w$). Moreover, it follows immediately from the construction of $\mathcal{G}' \to \mathcal{G}$ that n *divides* $[\Pi_{\widetilde{e}_1} : \Pi_{\widetilde{e}_1} \cap \Pi_{\mathcal{G}'}]$, and $\Pi_{\widetilde{e}_2}, \Pi_{\widetilde{v}} \subseteq \Pi_{\mathcal{G}'}$. This completes the proof of Lemma 1.2. □

Lemma 1.3 (Product of Edge-Like Elements) *Let $\gamma_1, \gamma_2 \in \Pi_{\mathcal{G}}$ be two* **nontrivial edge-like** *elements of $\Pi_{\mathcal{G}}$ [cf. Definition 1.1]. Write $\widetilde{e}_1, \widetilde{e}_2 \in \mathrm{Edge}(\widetilde{\mathcal{G}})$ for the unique elements of $\mathrm{Edge}(\widetilde{\mathcal{G}})$ such that $\gamma_1 \in \Pi_{\widetilde{e}_1}$, $\gamma_2 \in \Pi_{\widetilde{e}_2}$ [cf. Remark 1.1.1]. Suppose that the following conditions are satisfied:*

(i) For every positive integer n, it holds that $\gamma_1^n \gamma_2^n$ is **verticial**.
(ii) $\widetilde{e}_1 \neq \widetilde{e}_2$.

Then there exists a [necessarily unique—cf. [NodNon], Remark 1.8.1, (iii)] $\widetilde{v} \in \mathrm{Vert}(\widetilde{\mathcal{G}})$ such that $\{\widetilde{e}_1, \widetilde{e}_2\} \subseteq \mathcal{E}(\widetilde{v})$; in particular, it holds that $\gamma_1\gamma_2 \in \Pi_{\widetilde{v}}$.

Proof Since $\widetilde{e}_1 \neq \widetilde{e}_2$ [cf. condition (ii)], one verifies easily that there exists a finite étale Galois subcovering $\mathcal{H} \to \mathcal{G}$ of $\widetilde{\mathcal{G}} \to \mathcal{G}$ that satisfies the following conditions:

(1) $\widetilde{e}_1(\mathcal{H}) \neq \widetilde{e}_2(\mathcal{H})$.
(2) $\mathcal{H}$ is *untangled* [cf. [NodNon], Definition 1.2; [NodNon], Remark 1.2.1, (i), (ii)].
(3) For $i \in \{1, 2\}$, if $\widetilde{e}_i \in \mathrm{Node}(\widetilde{\mathcal{G}})$, then the following holds: Let $w, w' \in \mathcal{V}(\widetilde{e}_i(\mathcal{H}))$ be the two *distinct* elements of $\mathcal{V}(\widetilde{e}_i(\mathcal{H}))$ [cf. (ii)]. Then $\#(\mathcal{N}(w) \cap \mathcal{N}(w')) \geq 3$.
(4) For $i \in \{1, 2\}$, if $\widetilde{e}_i \in \mathrm{Cusp}(\widetilde{\mathcal{G}})$, then the following holds: Let $w \in \mathcal{V}(\widetilde{e}_i(\mathcal{H}))$ be the *unique* element of $\mathcal{V}(\widetilde{e}_i(\mathcal{H}))$. Then $\#\mathcal{C}(w) \geq 3$.

Now it is immediate that there exists a positive integer m such that $\gamma_1^m \in \Pi_{\widetilde{e}_1} \cap \Pi_{\mathcal{H}}$, $\gamma_2^m \in \Pi_{\widetilde{e}_2} \cap \Pi_{\mathcal{H}}$. Let $\widetilde{v} \in \mathrm{Vert}(\widetilde{\mathcal{G}})$ be such that $\gamma_1^m \gamma_2^m \in \Pi_{\widetilde{v}}$ [cf. condition (i)].

Suppose that $\widetilde{v}(\mathcal{H}) \notin \mathcal{V}(\widetilde{e}_1(\mathcal{H}))$. Then it follows from Lemma 1.2 that there exists a finite étale Galois subcovering $\mathcal{H}' \to \mathcal{H}$ of $\widetilde{\mathcal{G}} \to \mathcal{H}$ such that $\gamma_1^m \notin \Pi_{\mathcal{H}'}$, and, moreover, $\Pi_{\widetilde{e}_2} \cap \Pi_{\mathcal{H}}, \Pi_{\widetilde{v}} \cap \Pi_{\mathcal{H}} \subseteq \Pi_{\mathcal{H}'}$. But this implies that γ_2^m, $\gamma_1^m \gamma_2^m \in \Pi_{\mathcal{H}'}$, hence that $\gamma_1^m \in \Pi_{\mathcal{H}'}$, a *contradiction*. In particular, it holds that $\widetilde{v}(\mathcal{H}) \in \mathcal{V}(\widetilde{e}_1(\mathcal{H}))$; a similar argument implies that $\widetilde{v}(\mathcal{H}) \in \mathcal{V}(\widetilde{e}_2(\mathcal{H}))$, hence that $\mathcal{V}(\widetilde{e}_1(\mathcal{H})) \cap \mathcal{V}(\widetilde{e}_2(\mathcal{H})) \neq \emptyset$. Thus, by applying this argument to a suitable system of connected finite étale coverings of $\mathcal{H}$, we conclude that $\mathcal{V}(\widetilde{e}_1) \cap \mathcal{V}(\widetilde{e}_2) \neq \emptyset$, i.e., that there exists a $\widetilde{v} \in \mathrm{Vert}(\widetilde{\mathcal{G}})$ such that $\{\widetilde{e}_1, \widetilde{e}_2\} \subseteq \mathcal{E}(\widetilde{v})$. Then since $\Pi_{\widetilde{e}_1}, \Pi_{\widetilde{e}_2} \subseteq \Pi_{\widetilde{v}}$, it follows immediately that $\gamma_1 \gamma_2 \in \Pi_{\widetilde{v}}$. This completes the proof of Lemma 1.3. □

Proposition 1.4 (Group-Theoretic Characterization of Closed Subgroups of Edge-Like Subgroups) *Let $H \subseteq \Pi_{\mathcal{G}}$ be a closed subgroup of $\Pi_{\mathcal{G}}$. Then the following conditions are equivalent:*

(i) H is contained in an **edge-like subgroup**.
(ii) An open subgroup of H is contained in an **edge-like subgroup**.
(iii) Every element of H is **edge-like** *[cf. Definition 1.1].*
(iv) There exists a connected finite étale subcovering $\mathcal{G}^\dagger \to \mathcal{G}$ of $\widetilde{\mathcal{G}} \to \mathcal{G}$ such that for any connected finite étale subcovering $\mathcal{G}' \to \mathcal{G}$ of $\widetilde{\mathcal{G}} \to \mathcal{G}$ that factors through $\mathcal{G}^\dagger \to \mathcal{G}$, the image of the composite

$$H \cap \Pi_{\mathcal{G}'} \hookrightarrow \Pi_{\mathcal{G}'} \twoheadrightarrow \Pi_{\mathcal{G}'}^{\mathrm{ab/edge}}$$

—where we write $\Pi_{\mathcal{G}'}^{\mathrm{ab/edge}}$ for the **torsion-free** *[cf. [CmbGC], Remark 1.1.4] quotient of the abelianization $\Pi_{\mathcal{G}'}^{\mathrm{ab}}$ by the closed subgroup topologically generated by the images in $\Pi_{\mathcal{G}'}^{\mathrm{ab}}$ of the edge-like subgroups of $\Pi_{\mathcal{G}'}$—is* **trivial**.

Proof The implications (i) ⇒ (ii) ⇒ (iv) are immediate. The equivalence (iii) ⇔ (iv) follows immediately from [NodNon], Lemma 1.6. Thus, to complete the verification of Proposition 1.4, it suffices to verify the implication (iii) ⇒ (i). To this end, suppose that condition (iii) holds. First, we observe that, to verify the implication (iii) ⇒ (i), it suffices to verify the following assertion:

Claim 1.4.A: Let $\gamma_1, \gamma_2 \in H$ be *nontrivial* elements. Write $\widetilde{e}_1, \widetilde{e}_2 \in \mathrm{Edge}(\widetilde{\mathcal{G}})$ for the *unique* elements of $\mathrm{Edge}(\widetilde{\mathcal{G}})$ such that $\gamma_1 \in \Pi_{\widetilde{e}_1}, \gamma_2 \in \Pi_{\widetilde{e}_2}$ [cf. Remark 1.1.1]. Then $\widetilde{e}_1 = \widetilde{e}_2$.

To verify Claim 1.4.A, let us observe that it follows from condition (iii) that, for every positive integer n, it holds that $\gamma_1^n \gamma_2^n$ is *edge-like*, hence *verticial*. Thus, it follows immediately from Lemma 1.3 that there exists an element $\widetilde{v} \in \mathrm{Vert}(\widetilde{\mathcal{G}})$ such that $\{\widetilde{e}_1, \widetilde{e}_2\} \subseteq \mathcal{E}(\widetilde{v})$; in particular, it holds that $\gamma_1, \gamma_2 \in \Pi_{\widetilde{v}}$. Thus, to complete the verification of Claim 1.4.A, we may assume without loss of generality—by replacing $\Pi_{\mathcal{G}}$, H by $\Pi_{\widetilde{v}}$, $\Pi_{\widetilde{v}} \cap H$, respectively—that $\mathrm{Node}(\mathcal{G}) = \emptyset$ [so $\widetilde{e}_1$, $\widetilde{e}_2 \in \mathrm{Cusp}(\widetilde{\mathcal{G}})$]. Moreover, we may assume without loss of generality—by replacing

$\Pi_{\mathcal{G}}$ (respectively, γ_1, γ_2) by a suitable open subgroup of $\Pi_{\mathcal{G}}$ (respectively, suitable powers of γ_1, γ_2)—that $\#\mathrm{Cusp}(\mathcal{G}) \geq 4$. Thus, it follows immediately from the well-known structure of the abelianization of the fundamental group of a smooth curve over an algebraically closed field of characteristic $\notin \Sigma$ that the direct product of *any* 3 *cuspidal inertia subgroups* of $\Pi_{\mathcal{G}}$ associated to *distinct* cusps of $\mathcal{G}$ maps *injectively* to the abelianization $\Pi_{\mathcal{G}}^{\mathrm{ab}}$ of $\Pi_{\mathcal{G}}$. In particular, since $\gamma_1\gamma_2$ is *edge-like*, hence *cuspidal*, we conclude, by considering the cuspidal inertia subgroups that contain γ_1, γ_2, and $\gamma_1\gamma_2$, that $\widetilde{e}_1 = \widetilde{e}_2$. This completes the proof of Claim 1.4.A, hence also of the implication (iii) $\Rightarrow$ (i). This completes the proof of Proposition 1.4. □

Proposition 1.5 (Group-Theoretic Characterization of Closed Subgroups of Verticial Subgroups) *Let $H \subseteq \Pi_{\mathcal{G}}$ be a closed subgroup of $\Pi_{\mathcal{G}}$. Then the following conditions are equivalent:*

(i) H is contained in a **verticial subgroup**.
(ii) An open subgroup of H is contained in a **verticial subgroup**.
(iii) Every element of H is **verticial** *[cf. Definition 1.1].*
(iv) There exists a connected finite étale subcovering $\mathcal{G}^\dagger \to \mathcal{G}$ of $\widetilde{\mathcal{G}} \to \mathcal{G}$ such that for any connected finite étale subcovering $\mathcal{G}' \to \mathcal{G}$ of $\widetilde{\mathcal{G}} \to \mathcal{G}$ that factors through $\mathcal{G}^\dagger \to \mathcal{G}$, the image of the composite

$$H \cap \Pi_{\mathcal{G}'} \hookrightarrow \Pi_{\mathcal{G}'} \twoheadrightarrow \Pi_{\mathcal{G}'}^{\text{ab-comb}}$$

—where we write $\Pi_{\mathcal{G}'}^{\text{ab-comb}}$ for the **torsion-free** *[cf. [CmbGC], Remark 1.1.4] quotient of the abelianization $\Pi_{\mathcal{G}'}^{\mathrm{ab}}$ by the closed subgroup topologically generated by the images in $\Pi_{\mathcal{G}'}^{\mathrm{ab}}$ of the verticial subgroups of $\Pi_{\mathcal{G}'}$—is* **trivial**.

Proof The implications (i) $\Rightarrow$ (ii) $\Rightarrow$ (iv) are immediate. Next, we verify the implication (iv) $\Rightarrow$ (iii). Suppose that condition (iv) holds. Let $\gamma \in H$. Then to verify that γ is *verticial*, we may assume without loss of generality—by replacing H by the procyclic subgroup of H topologically generated by γ—that H is *procyclic*. Now the implication (iv) $\Rightarrow$ (iii) follows immediately from a similar argument to the argument applied in the proof of the implication (ii) $\Rightarrow$ (i) of [NodNon], Lemma 1.6, in the *edge-like* case. Here, we note that unlike the *edge-like* case, there is a slight complication arising from the fact [cf. [NodNon], Lemma 1.9, (i)] that an element $\widetilde{v} \in \mathrm{Vert}(\widetilde{\mathcal{G}})$ is not necessarily *uniquely determined* by the condition that $H \subseteq \Pi_{\widetilde{v}}$, i.e., there may exist distinct $\widetilde{v}_1, \widetilde{v}_2 \in \mathcal{V}(\widetilde{e})$ for some $\widetilde{e} \in \mathrm{Node}(\widetilde{\mathcal{G}})$ such that $H \subseteq \Pi_{\widetilde{e}} = \Pi_{\widetilde{v}_1} \cap \Pi_{\widetilde{v}_2}$. On the other hand, this phenomenon is, in fact, *irrelevant* to the argument in question, since $\Pi_{\mathcal{G}}$ does not contain any elements that fix, but permute the branches of, $\widetilde{e}$. This completes the proof of the implication (iv) $\Rightarrow$ (iii).

Finally, we verify the implication (iii) $\Rightarrow$ (i). Suppose that condition (iii) holds. Now if every element of H is *edge-like*, then the implication (iii) $\Rightarrow$ (i) follows from the implication (iii) $\Rightarrow$ (i) of Proposition 1.4, together with the fact that every edge-like subgroup is contained in a verticial subgroup. Thus, to verify the implication (iii) $\Rightarrow$ (i), we may assume without loss of generality that there exists an element $\gamma_1 \in H$ of H that is *not edge-like*. Write $\widetilde{v}_1 \in \mathrm{Vert}(\widetilde{\mathcal{G}})$ for the *unique* element of $\mathrm{Vert}(\widetilde{\mathcal{G}})$ such that $\gamma_1 \in \Pi_{\widetilde{v}_1}$ [cf. Remark 1.1.1].

Now we claim the following assertion:

Claim 1.5.A: $H \subseteq \Pi_{\widetilde{v}_1}$.

Indeed, let $\gamma_2 \in H$ be a *nontrivial* element of H. If $\gamma_2 = \gamma_1$, then $\gamma_2 \in \Pi_{\widetilde{v}_1}$. Thus, we may assume without loss of generality that $\gamma_1 \neq \gamma_2$. Write $\gamma \stackrel{\mathrm{def}}{=} \gamma_1\gamma_2^{-1}$.

Next, suppose that γ_2 is *not edge-like*. Write $\widetilde{v}_2 \in \mathrm{Vert}(\widetilde{\mathcal{G}})$ for the *unique* element of $\mathrm{Vert}(\widetilde{\mathcal{G}})$ such that $\gamma_2 \in \Pi_{\widetilde{v}_2}$ [cf. Remark 1.1.1]. Let $\mathcal{H} \to \mathcal{G}$ be a connected finite étale subcovering of $\widetilde{\mathcal{G}} \to \mathcal{G}$. Then since *neither* γ_1 *nor* γ_2 is *edge-like*, one verifies easily—by applying the implication (iv) $\Rightarrow$ (i) of Proposition 1.4 to the closed subgroups of $\Pi_{\mathcal{G}}$ topologically generated by γ_1, γ_2, respectively—that there exist a connected finite étale subcovering $\mathcal{H}' \to \mathcal{H}$ of $\widetilde{\mathcal{G}} \to \mathcal{H}$ and a positive integer n such that $\gamma_1^n, \gamma_2^n \in \Pi_{\mathcal{H}'} \subseteq \Pi_{\mathcal{H}}$, and, moreover, the images of $\gamma_1^n, \gamma_2^n \in \Pi_{\mathcal{H}'}$ via the natural surjection $\Pi_{\mathcal{H}'} \twoheadrightarrow \Pi_{\mathcal{H}'}^{\mathrm{ab/edge}}$ [cf. the notation of Proposition 1.4, (iv)] are *nontrivial*. Thus, it follows from the existence of the natural *split injection*

$$\bigoplus_{v \in \mathrm{Vert}(\mathcal{H}')} \Pi_v^{\mathrm{ab/edge}} \longrightarrow \Pi_{\mathcal{H}'}^{\mathrm{ab/edge}}$$

of [NodNon], Lemma 1.4, together with the fact that $\gamma_1^n\gamma_2^n \in \Pi_{\mathcal{H}'}$ is *verticial* [cf. condition (iii)], that $\widetilde{v}_1(\mathcal{H}') = \widetilde{v}_2(\mathcal{H}')$, hence that $\widetilde{v}_1(\mathcal{H}) = \widetilde{v}_2(\mathcal{H})$. Therefore, by allowing the subcovering $\mathcal{H} \to \mathcal{G}$ of $\widetilde{\mathcal{G}} \to \mathcal{G}$ to *vary*, we conclude that $\widetilde{v}_1 = \widetilde{v}_2$; in particular, it holds that $\gamma_2 \in \Pi_{\widetilde{v}_1}$.

Next, suppose that γ_2 is *edge-like*, but that γ is *not edge-like*. Then, by applying the argument of the preceding paragraph concerning γ_2 to γ, we conclude that γ, hence also γ_2, is contained in $\Pi_{\widetilde{v}_1}$.

Next, suppose that both γ_2 and γ are *edge-like*. Write $\widetilde{e}_2, \widetilde{e} \in \mathrm{Edge}(\widetilde{\mathcal{G}})$ for the *unique* elements of $\mathrm{Edge}(\widetilde{\mathcal{G}})$ such that $\gamma_2 \in \Pi_{\widetilde{e}_2}$, $\gamma \in \Pi_{\widetilde{e}}$ [cf. Remark 1.1.1]. Then since γ_1 is *not edge-like*, it follows immediately that $\widetilde{e}_2 \neq \widetilde{e}$. Moreover, it follows from condition (iii) that for any positive integer n, the element $\gamma_2^n\gamma^n$ is *verticial*. Thus, it follows immediately from Lemma 1.3 that there exists a *unique* $\widetilde{v} \in \mathrm{Vert}(\widetilde{\mathcal{G}})$ such that $\{\widetilde{e}_2, \widetilde{e}\} \subseteq \mathcal{E}(\widetilde{v})$, $\gamma_1 = \gamma\gamma_2 \in \Pi_{\widetilde{v}}$. On the other hand, since $\widetilde{v}_1 \in \mathrm{Vert}(\widetilde{\mathcal{G}})$ is *uniquely determined* by the condition that $\gamma_1 \in \Pi_{\widetilde{v}_1}$, we thus conclude that $\widetilde{v}_1 = \widetilde{v}$, hence that $\gamma_2 \in \Pi_{\widetilde{e}_2} \subseteq \Pi_{\widetilde{v}_1}$, as desired. This completes the proof of Claim 1.5.A and hence also of the implication (iii) $\Rightarrow$ (i). □

Theorem 1.6 (Section Conjecture-Type Result for Outer Representations of SNN-, IPSC-Type) *Let Σ be a nonempty set of prime numbers, $\mathcal{G}$ a semi-graph of anabelioids of pro-Σ PSC-type, and $I \to$* Aut($\mathcal{G}$) *an outer representation of* **SNN-type** *[cf. [NodNon], Definition 2.4, (iii)]. Write $\Pi_{\mathcal{G}}$ for the [pro-Σ] fundamental* *group of* $\mathcal{G}$ *and* $\Pi_I \stackrel{\mathrm{def}}{=} \Pi_{\mathcal{G}} \stackrel{\mathrm{out}}{\rtimes} I$ *[cf. the discussion entitled "Topological groups" in [CbTpI], §0]; thus, we have a natural exact sequence of profinite groups*

$$1 \longrightarrow \Pi_{\mathcal{G}} \longrightarrow \Pi_I \longrightarrow I \longrightarrow 1.$$

Write $\mathrm{Sect}(\Pi_I/I)$ *for the set of sections of the natural surjection* $\Pi_I \twoheadrightarrow I$*. Then the following hold:*

(i) For any $\widetilde{v} \in \mathrm{Vert}(\widetilde{\mathcal{G}})$*, the composite* $I_{\widetilde{v}} \hookrightarrow \Pi_I \twoheadrightarrow I$ *[cf. [NodNon], Definition 2.2, (i)] is an* **isomorphism**. *In particular,* $I_{\widetilde{v}} \subseteq \Pi_I$ *determines an element* $s_{\widetilde{v}} \in \mathrm{Sect}(\Pi_I/I)$*; thus, we have a map*

$$\begin{array}{ccc} \mathrm{Vert}(\widetilde{\mathcal{G}}) & \longrightarrow & \mathrm{Sect}(\Pi_I/I) \\ \widetilde{v} & \mapsto & s_{\widetilde{v}}. \end{array}$$

Finally, the following equalities concerning centralizers of subgroups of Π_I *in* $\Pi_{\mathcal{G}}$ *[cf. the discussion entitled "Topological groups" in "Notations and Conventions"] hold:* $Z_{\Pi_{\mathcal{G}}}(s_{\widetilde{v}}(I)) = Z_{\Pi_{\mathcal{G}}}(I_{\widetilde{v}}) = \Pi_{\widetilde{v}}$.

(ii) The map of (i) is **injective**.

(iii) If, moreover, $I \to \mathrm{Aut}(\mathcal{G})$ *is of* **IPSC-type** *[cf. [NodNon], Definition 2.4, (i)], then, for any* $s \in \mathrm{Sect}(\Pi_I/I)$*, the centralizer* $Z_{\Pi_{\mathcal{G}}}(s(I))$ *is contained in a* **verticial subgroup**.

(iv) Let $s \in \mathrm{Sect}(\Pi_I/I)$*. Consider the following two conditions:*

(1) The section s *is contained in the image of the map of (i), i.e.,* $s = s_{\widetilde{v}}$ *for some* $\widetilde{v} \in \mathrm{Vert}(\widetilde{\mathcal{G}})$.

(2) $Z_{\Pi_{\mathcal{G}}}(Z_{\Pi_{\mathcal{G}}}(s(I))) = \{1\}$.

Then we have an implication

$$(1) \Longrightarrow (2).$$

If, moreover, $I \to \mathrm{Aut}(\mathcal{G})$ *is of* **IPSC-type**, *then we have an equivalence*

$$(1) \Longleftrightarrow (2).$$

Proof First, we verify assertion (i). The fact that the composite $I_{\widetilde{v}} \hookrightarrow \Pi_I \twoheadrightarrow I$ is an *isomorphism* follows from condition (2′) of [NodNon], Definition 2.4, (ii). On the other hand, the equalities $Z_{\Pi_{\mathcal{G}}}(s_{\widetilde{v}}(I)) = Z_{\Pi_{\mathcal{G}}}(I_{\widetilde{v}}) = \Pi_{\widetilde{v}}$ follow from [NodNon], Lemma 3.6, (i). This completes the proof of assertion (i). Assertion (ii) follows immediately from the final equalities of assertion (i), together with [NodNon], Lemma 1.9, (ii). Next, we verify assertion (iii). Write $H \stackrel{\mathrm{def}}{=} Z_{\Pi_{\mathcal{G}}}(s(I))$. Then it follows immediately from [CmbGC], Proposition 2.6, together with the definition of $H = Z_{\Pi_{\mathcal{G}}}(s(I))$, that for any connected finite étale subcovering $\mathcal{G}' \to \mathcal{G}$ of $\widetilde{\mathcal{G}} \to \mathcal{G}$, the image of the composite

$$H \cap \Pi_{\mathcal{G}'} \hookrightarrow \Pi_{\mathcal{G}'} \twoheadrightarrow \Pi_{\mathcal{G}'}^{\text{ab-comb}}$$

[cf. the notation of Proposition 1.5, (iv)] is *trivial*. Thus, it follows from the implication (iv) $\Rightarrow$ (i) of Proposition 1.5 that H is contained in a *verticial subgroup*. This completes the proof of assertion (iii).

Finally, we verify assertion (iv). To verify the implication (1) $\Rightarrow$ (2), suppose that condition (1) holds. Then since $Z_{\Pi_{\mathcal{G}}}(s_{\widetilde{v}}(I)) = Z_{\Pi_{\mathcal{G}}}(I_{\widetilde{v}}) = \Pi_{\widetilde{v}}$ [cf. assertion (i)] is *commensurably terminal* in $\Pi_{\mathcal{G}}$ [cf. [CmbGC], Proposition 1.2, (ii)] and *center-free* [cf. [CmbGC], Remark 1.1.3], we conclude that $Z_{\Pi_{\mathcal{G}}}(Z_{\Pi_{\mathcal{G}}}(s_{\widetilde{v}}(I))) = Z_{\Pi_{\mathcal{G}}}(\Pi_{\widetilde{v}}) = \{1\}$. This completes the proof of the implication (1) $\Rightarrow$ (2). Next, suppose that $I \to \mathrm{Aut}(\mathcal{G})$ is of *IPSC-type*, and that condition (2) holds. Then it follows from assertion (iii) that there exists a $\widetilde{v} \in \mathrm{Vert}(\widetilde{\mathcal{G}})$ such that $H \stackrel{\text{def}}{=} Z_{\Pi_{\mathcal{G}}}(s(I)) \subseteq \Pi_{\widetilde{v}}$, so $I_{\widetilde{v}} \subseteq Z_{\Pi_I}(H)$. On the other hand, since $s(I) \subseteq Z_{\Pi_I}(H)$, and $Z_{\Pi_{\mathcal{G}}}(H) = Z_{\Pi_{\mathcal{G}}}(Z_{\Pi_{\mathcal{G}}}(s(I))) = \{1\}$ [cf. condition (2)], i.e., the composite of natural homomorphisms $Z_{\Pi_I}(H) \hookrightarrow \Pi_I \twoheadrightarrow I$ is *injective*, it follows that $s(I) = Z_{\Pi_I}(H) \supseteq I_{\widetilde{v}}$. Since $I_{\widetilde{v}}$ and $s(I)$ may be obtained as the images of sections, we thus conclude that $I_{\widetilde{v}} = s(I)$, i.e., $s = s_{\widetilde{v}}$. This completes the proof of the implication (2) $\Rightarrow$ (1), hence also of assertion (iv). □

Remark 1.6.1 Recall that in the case of outer representations of NN-type, the *period matrix is not necessarily nondegenerate* [cf. [CbTpI], Remark 5.9.2]. In particular, the argument applied in the proof of Theorem 1.6, (iii)—which depends, in an essential way, on the fact that, in the case of *outer representations of IPSC-type*, the period matrix is *nondegenerate* [cf. the proof of [CmbGC], Proposition 2.6]—cannot be applied in the case of outer representations of NN-type. Nevertheless, the question of whether or not Theorem 1.6, (iii), as well as the application of Theorem 1.6, (iii), given in Corollary 1.7, (ii), below, may be generalized to the case of outer representations of NN-type remains a topic of interest to the authors.

Corollary 1.7 (Group-Theoretic Characterization of Verticial Subgroups for Outer Representations of IPSC-Type) *In the notation of Theorem 1.6, let us refer to a closed subgroup of* $\Pi_{\mathcal{G}}$ *as a* **section-centralizer** *if it may be written in the form* $Z_{\Pi_{\mathcal{G}}}(s(I))$ *for some* $s \in \mathrm{Sect}(\Pi_I/I)$. *Let* $H \subseteq \Pi_{\mathcal{G}}$ *be a closed subgroup of* $\Pi_{\mathcal{G}}$. *Then the following hold:*

(i) Suppose that H *is a* **section-centralizer** *such that* $Z_{\Pi_{\mathcal{G}}}(H) = \{1\}$. *Then the following conditions on a section* $s \in \mathrm{Sect}(\Pi_I/I)$ *are equivalent:*

(i-1) $H = Z_{\Pi_{\mathcal{G}}}(s(I))$.
(i-2) $s(I) \subseteq Z_{\Pi_I}(H)$.
(i-3) $s(I) = Z_{\Pi_I}(H)$.

(ii) Consider the following three conditions:

(ii-1) H *is a* **verticial subgroup**.
(ii-2) H *is a* **section-centralizer** *such that* $Z_{\Pi_{\mathcal{G}}}(H) = \{1\}$.
(ii-3) H *is a* **maximal section-centralizer**.

Then we have implications

$$\text{(ii-1)} \Longrightarrow \text{(ii-2)} \Longrightarrow \text{(ii-3)}.$$

If, moreover, $I \to \mathrm{Aut}(\mathcal{G})$ *is of* **IPSC-type** *[cf. [NodNon], Definition 2.4, (i)], then we have equivalences*

$$\text{(ii-1)} \Longleftrightarrow \text{(ii-2)} \Longleftrightarrow \text{(ii-3)}.$$

Proof First, we verify assertion (i). The implication (i-1) $\Rightarrow$ (i-2) is immediate. To verify the implication (i-2) $\Rightarrow$ (i-3), suppose that condition (i-2) holds. Then since $Z_{\Pi_I}(H) \cap \Pi_{\mathcal{G}} = Z_{\Pi_{\mathcal{G}}}(H) = \{1\}$, the composite $Z_{\Pi_I}(H) \hookrightarrow \Pi_I \twoheadrightarrow I$ is *injective*. Thus, since the composite $s(I) \hookrightarrow Z_{\Pi_I}(H) \hookrightarrow \Pi_I \twoheadrightarrow I$ is an *isomorphism*, it follows immediately that condition (i-3) holds. This completes the proof of the implication (i-2) $\Rightarrow$ (i-3). Finally, to verify the implication (i-3) $\Rightarrow$ (i-1), suppose that condition (i-3) holds. Then since H is a *section-centralizer*, there exists a $t \in \mathrm{Sect}(\Pi_I/I)$ such that $H = Z_{\Pi_{\mathcal{G}}}(t(I))$. In particular, $t(I) \subseteq Z_{\Pi_I}(H) = s(I)$ [cf. condition (i-3)]. We thus conclude that $t = s$, i.e., that condition (i-1) holds. This completes the proof of assertion (i).

Next, we verify assertion (ii). The implication (ii-1) $\Rightarrow$ (ii-2) follows immediately from Theorem 1.6, (i), (iv). To verify the implication (ii-2) $\Rightarrow$ (ii-3), suppose that H satisfies condition (ii-2); let $s \in \mathrm{Sect}(\Pi_I/I)$ be such that $H \subseteq Z_{\Pi_{\mathcal{G}}}(s(I))$. Then it follows immediately that $s(I) \subseteq Z_{\Pi_I}(H)$. Thus, it follows immediately from the equivalence (i-1) $\Leftrightarrow$ (i-2) of assertion (i) that $H = Z_{\Pi_{\mathcal{G}}}(s(I))$. This completes the proof of the implication (ii-2) $\Rightarrow$ (ii-3). Finally, observe that the implication (ii-3) $\Rightarrow$ (ii-1) in the case where $I \to \mathrm{Aut}(\mathcal{G})$ is of *IPSC-type* follows immediately from Theorem 1.6, (iii), together with the fact that every verticial subgroup is a *section-centralizer* [cf. the implication (ii-1) $\Rightarrow$ (ii-2) verified above]. This completes the proof of Corollary 1.7. □

Lemma 1.8 (Group-Theoretic Characterization of Verticial Subgroups for Outer Representations of SNN-Type) *Let* $H \subseteq \Pi_{\mathcal{G}}$ *be a closed subgroup of* $\Pi_{\mathcal{G}}$ *and* $I \to \mathrm{Aut}(\mathcal{G})$ *an outer representation of* **SNN-type** *[cf. [NodNon], Definition 2.4, (iii)]. Write* $\Pi_I \stackrel{\text{def}}{=} \Pi_{\mathcal{G}} \stackrel{\text{out}}{\rtimes} I$ *[cf. the discussion entitled "Topological groups" in [CbTpI], §0]; thus, we have a natural exact sequence of profinite groups*

$$1 \longrightarrow \Pi_{\mathcal{G}} \longrightarrow \Pi_I \longrightarrow I \longrightarrow 1.$$

Suppose that $\mathcal{G}$ is **untangled** *[cf. [NodNon], Definition 1.2]. Then H is a* **verticial subgroup** *if and only if H satisfies the following four conditions:*

(i) The composite $I_H \stackrel{\text{def}}{=} Z_{\Pi_I}(H) \hookrightarrow \Pi_I \twoheadrightarrow I$ is an **isomorphism**.
(ii) It holds that $H = Z_{\Pi_{\mathcal{G}}}(I_H)$.
(iii) For any $\gamma \in \Pi_{\mathcal{G}}$, it holds that $\gamma \in H$ if and only if $H \cap (\gamma \cdot H \cdot \gamma^{-1}) \neq \{1\}$.
(iv) H contains a **nontrivial verticial** *element of $\Pi_{\mathcal{G}}$ [cf. Definition 1.1].*

Proof If H is a *verticial subgroup*, then it is immediate that condition (iv) is satisfied; moreover, it follows from condition (2′) of [NodNon], Definition 2.4, (ii) (respectively, [NodNon], Lemma 3.6, (i); [NodNon], Remark 1.10.1), that H satisfies condition (i) (respectively, (ii); (iii)). This completes the proof of *necessity*.

To verify *sufficiency*, suppose that H satisfies conditions (i), (ii), (iii), and (iv). It follows from condition (iv) that there exists a $\widetilde{v} \in \mathrm{Vert}(\widetilde{\mathcal{G}})$ such that $J \stackrel{\text{def}}{=} H \cap \Pi_{\widetilde{v}} \neq \{1\}$. If either $J = \Pi_{\widetilde{v}}$ or $J = H$, i.e., either $\Pi_{\widetilde{v}} \subseteq H$ or $H \subseteq \Pi_{\widetilde{v}}$, then it is immediate that either $I_H \subseteq I_{\widetilde{v}}$ or $I_{\widetilde{v}} \subseteq I_H$ [cf. [NodNon], Definition 2.2, (i)]. Thus, it follows from condition (i) [for H and $\Pi_{\widetilde{v}}$] that $I_H = I_{\widetilde{v}}$. But then it follows from condition (ii) [for H and $\Pi_{\widetilde{v}}$] that $H = Z_{\Pi_{\mathcal{G}}}(I_H) = Z_{\Pi_{\mathcal{G}}}(I_{\widetilde{v}}) = \Pi_{\widetilde{v}}$; in particular, H is a *verticial subgroup*.

Thus, we may assume without loss of generality that $J \neq H, \Pi_{\widetilde{v}}$. Let $\gamma \in H \setminus J$. Write $J^\gamma \stackrel{\text{def}}{=} \gamma \cdot J \cdot \gamma^{-1}$. Then we have inclusions

$$\Pi_{\widetilde{v}} \supseteq J \subseteq H \supseteq J^\gamma \subseteq \Pi_{\widetilde{v}^\gamma} \; (= \gamma \cdot \Pi_{\widetilde{v}} \cdot \gamma^{-1}).$$

Now we claim the following assertion:

Claim 1.8.A: $N_{\Pi_{\mathcal{G}}}(J) = J$, $N_{\Pi_{\mathcal{G}}}(J^\gamma) = J^\gamma$.

Indeed, let $\sigma \in N_{\Pi_{\mathcal{G}}}(J)$. Then since $\{1\} \neq J = J \cap (\sigma \cdot J \cdot \sigma^{-1}) \subseteq \Pi_{\widetilde{v}} \cap \Pi_{\widetilde{v}^\sigma}$, it follows from condition (iii) [for $\Pi_{\widetilde{v}}$] that $\sigma \in \Pi_{\widetilde{v}}$. Similarly, since $\{1\} \neq J = J \cap (\sigma \cdot J \cdot \sigma^{-1}) \subseteq H \cap (\sigma \cdot H \cdot \sigma^{-1})$, it follows from condition (iii) [for H] that $\sigma \in H$. Thus, $\sigma \in \Pi_{\widetilde{v}} \cap H = J$. In particular, we obtain that $N_{\Pi_{\mathcal{G}}}(J) = J$. A similar argument implies that $N_{\Pi_{\mathcal{G}}}(J^\gamma) = J^\gamma$. This completes the proof of Claim 1.8.A.

Now the composites $N_{\Pi_I}(J), N_{\Pi_I}(J^\gamma) \hookrightarrow \Pi_I \twoheadrightarrow I$ fit into exact sequences of profinite groups

$$1 \longrightarrow N_{\Pi_{\mathcal{G}}}(J) \longrightarrow N_{\Pi_I}(J) \longrightarrow I,$$

$$1 \longrightarrow N_{\Pi_{\mathcal{G}}}(J^\gamma) \longrightarrow N_{\Pi_I}(J^\gamma) \longrightarrow I.$$

Thus, since we have inclusions

$$I_H = Z_{\Pi_I}(H) \subseteq Z_{\Pi_I}(J) \subseteq N_{\Pi_I}(J),$$

$$I_H = Z_{\Pi_I}(H) \subseteq Z_{\Pi_I}(J^\gamma) \subseteq N_{\Pi_I}(J^\gamma),$$

$$I_{\widetilde{v}} = Z_{\Pi_I}(\Pi_{\widetilde{v}}) \subseteq Z_{\Pi_I}(J) \subseteq N_{\Pi_I}(J),$$

$$I_{\widetilde{v}^\gamma} = Z_{\Pi_I}(\Pi_{\widetilde{v}^\gamma}) \subseteq Z_{\Pi_I}(J^\gamma) \subseteq N_{\Pi_I}(J^\gamma),$$

it follows immediately from Claim 1.8.A, together with condition (i) [for H and $\Pi_{\widetilde{v}}$], that

$$N_{\Pi_I}(J) = J \cdot I_H = J \cdot I_{\widetilde{v}}, \quad N_{\Pi_I}(J^\gamma) = J^\gamma \cdot I_H = J^\gamma \cdot I_{\widetilde{v}^\gamma}.$$

In particular, we obtain that

$$I_H \subseteq N_{\Pi_I}(J) = J \cdot I_{\widetilde{v}} \subseteq \Pi_{\widetilde{v}} \cdot D_{\widetilde{v}} = D_{\widetilde{v}},$$

$$I_H \subseteq N_{\Pi_I}(J^\gamma) = J^\gamma \cdot I_{\widetilde{v}^\gamma} \subseteq \Pi_{\widetilde{v}^\gamma} \cdot D_{\widetilde{v}^\gamma} = D_{\widetilde{v}^\gamma}$$

[cf. [NodNon], Definition 2.2, (i)], i.e., $I_H \subseteq D_{\widetilde{v}} \cap D_{\widetilde{v}^\gamma}$. On the other hand, since $H \ni \gamma \notin J = H \cap \Pi_{\widetilde{v}}$, it follows from condition (iii) [for $\Pi_{\widetilde{v}}$] that $\Pi_{\widetilde{v}^\gamma} \cap \Pi_{\widetilde{v}} = \{1\}$; thus, it follows immediately from the fact that $D_{\widetilde{v}} \cap D_{\widetilde{v}^\gamma} \cap \Pi_{\mathcal{G}} = \Pi_{\widetilde{v}} \cap \Pi_{\widetilde{v}^\gamma} = \{1\}$ [cf. [CmbGC], Proposition 1.2, (ii)], together with condition (i), that $I_H = D_{\widetilde{v}} \cap D_{\widetilde{v}^\gamma}$, which implies, by [NodNon], Proposition 3.9, (iii), that there exists a $\widetilde{w} \in \mathrm{Vert}(\widetilde{\mathcal{G}})$ such that $I_H = I_{\widetilde{w}}$. In particular, it follows from condition (ii) [for H and $\Pi_{\widetilde{w}}$] that $H = Z_{\Pi_{\mathcal{G}}}(I_H) = Z_{\Pi_{\mathcal{G}}}(I_{\widetilde{w}}) = \Pi_{\widetilde{w}}$. Thus, H is a *verticial subgroup*. This completes the proof of Lemma 1.8. □

Theorem 1.9 (Group-Theoretic Verticiality/Nodality of Isomorphisms of Outer Representations of NN-, IPSC-Type) *Let Σ be a nonempty set of prime numbers, $\mathcal{G}$ (respectively, $\mathcal{H}$) a semi-graph of anabelioids of pro-Σ PSC-type, $\Pi_{\mathcal{G}}$ (respectively, $\Pi_{\mathcal{H}}$) the [pro-Σ] fundamental group of $\mathcal{G}$ (respectively, $\mathcal{H}$), $\alpha \colon \Pi_{\mathcal{G}} \xrightarrow{\sim} \Pi_{\mathcal{H}}$ an isomorphism of profinite groups, I (respectively, J) a profinite group, $\rho_I \colon I \to \mathrm{Aut}(\mathcal{G})$ (respectively, $\rho_J \colon J \to \mathrm{Aut}(\mathcal{H})$) a continuous homomorphism, and $\beta \colon I \xrightarrow{\sim} J$ an isomorphism of profinite groups. Suppose that the diagram*

$$\begin{array}{ccc} I & \longrightarrow & \mathrm{Out}(\Pi_{\mathcal{G}}) \\ \beta\downarrow & & \downarrow\mathrm{Out}(\alpha) \\ J & \longrightarrow & \mathrm{Out}(\Pi_{\mathcal{H}}) \end{array}$$

—where the right-hand vertical arrow is the isomorphism induced by α; the upper and lower horizontal arrows are the homomorphisms determined by ρ_I and ρ_J, respectively—commutes. Then the following hold:

(i) Suppose, moreover, that ρ_I, ρ_J are of **NN-type** *[cf. [NodNon], Definition 2.4, (iii)]. Then the following three conditions are equivalent:*

(1) The isomorphism α is **group-theoretically verticial** *[i.e., roughly speaking, preserves verticial subgroups—cf. [CmbGC], Definition 1.4, (iv)].*

(2) The isomorphism α is **group-theoretically nodal** *[i.e., roughly speaking, preserves nodal subgroups—cf. [NodNon], Definition 1.12].*

(3) There exists a **nontrivial** verticial *element $\gamma \in \Pi_{\mathcal{G}}$ such that $\alpha(\gamma) \in \Pi_{\mathcal{H}}$ is* **verticial** *[cf. Definition 1.1].*

(ii) Suppose, moreover, that ρ_I is of **NN-type***, and that ρ_J is of* **IPSC-type** *[cf. [NodNon], Definition 2.4, (i)]. [For example, this will be the case if both ρ_I and ρ_J are of* **IPSC-type***—cf. [NodNon], Remark 2.4.2.] Then α is* **group-theoretically verticial***, hence also [cf. (i)]* **group-theoretically nodal**.

Proof First, we verify assertion (i). The implication (1) $\Rightarrow$ (2) follows from [NodNon], Proposition 1.13. The implication (2) $\Rightarrow$ (3) follows from the fact that any nodal subgroup is contained in a verticial subgroup. [Note that if Node($\mathcal{H}$) $= \emptyset$, then every element of $\Pi_{\mathcal{H}}$ is *verticial*.] Finally, we verify the implication (3) $\Rightarrow$ (1). Suppose that condition (3) holds. Since verticial subgroups are *commensurably terminal* [cf. [CmbGC], Proposition 1.2, (ii)], to verify the implication (3) $\Rightarrow$ (1), by replacing Π_I, Π_J by open subgroups of Π_I, Π_J, we may assume without loss of generality that ρ_I, ρ_J are of *SNN-type* [cf. [NodNon], Definition 2.4, (iii)], and, moreover, that $\mathcal{G}$ and $\mathcal{H}$ are *untangled* [cf. [NodNon], Definition 1.2; [NodNon], Remark 1.2.1, (i), (ii)]. Let $\widetilde{v} \in \mathrm{Vert}(\widetilde{\mathcal{G}})$ be such that $\gamma \in \Pi_{\widetilde{v}}$. Then it is immediate that $\alpha(\Pi_{\widetilde{v}})$ satisfies conditions (i), (ii), and (iii) in the statement of Lemma 1.8. On the other hand, it follows from condition (3) that $\alpha(\Pi_{\widetilde{v}})$ satisfies condition (iv) in the statement of Lemma 1.8. Thus, it follows from Lemma 1.8 that $\alpha(\Pi_{\widetilde{v}}) \subseteq \Pi_{\mathcal{H}}$ is a *verticial subgroup*. Now it follows from [NodNon], Theorem 4.1, that α is *group-theoretically verticial*. This completes the proof of the implication (3) $\Rightarrow$ (1).

Finally, we verify assertion (ii). It is immediate that, to verify assertion (ii)—by replacing I, J by open subgroups of I, J—we may assume without loss of generality that ρ_I is of *SNN-type*. Let $H \subseteq \Pi_{\mathcal{G}}$ be a verticial subgroup of $\Pi_{\mathcal{G}}$. Then it follows from Corollary 1.7, (ii), that H, hence also $\alpha(H)$, is a *maximal section-centralizer* [cf. the statement of Corollary 1.7]. Thus, since ρ_J is of *IPSC-type*, again by Corollary 1.7, (ii), we conclude that $\alpha(H) \subseteq \Pi_{\mathcal{H}}$ is a verticial subgroup of $\Pi_{\mathcal{H}}$. In particular, it follows from [NodNon], Theorem 4.1, together with [NodNon], Remark 2.4.2, that α is *group-theoretically verticial* and *group-theoretically nodal*. This completes the proof of assertion (ii). □

Remark 1.9.1 Thus, Theorem 1.9, (i), may be regarded as a *generalization* of [NodNon], Corollary 4.2. Of course, ideally, one would like to be able to prove that conditions (1) and (2) of Theorem 1.9, (i), hold *automatically* [i.e., as in the case of outer representations of IPSC-type treated in Theorem 1.9, (ii)], without assuming condition (3). Although this topic lies beyond the scope of the present monograph, perhaps progress could be made in this direction if, say, in the case where Σ is either equal to the set of all prime numbers or of cardinality one, one starts with an isomorphism α that arises from a *PF-admissible* [cf. [CbTpI], Definition 1.4, (i)] isomorphism between *configuration space groups* corresponding to m-dimensional configuration spaces [where $m \geq 2$] associated to stable curves that give rise to $\mathcal{G}$ and $\mathcal{H}$, respectively [i.e., one assumes the condition of *"m-cuspidalizability"* discussed in Definition 3.20, below, where we *replace* the condition of "PFC-admissibility" by the condition of "PF-admissibility"]. For instance, if $\mathrm{Cusp}(\mathcal{G}) \neq \emptyset$, then it follows from [CbTpI], Theorem 1.8, (iv); [NodNon], Corollary 4.2, that this condition on α is sufficient to imply that conditions (1) and (2) of Theorem 1.9, (i), hold.

Chapter 2
Partial Combinatorial Cuspidalization for F-Admissible Outomorphisms

In this chapter, we apply the results obtained in the preceding Chap. 1, together with the theory developed by the authors in earlier papers, to prove *combinatorial cuspidalization-type results for F-admissible outomorphisms* [cf. Theorem 2.3, (i), below]. We also show that any F-admissible outomorphism of a configuration space group [arising from a configuration space] of *sufficiently high dimension* [i.e., ≥ 3 in the affine case; ≥ 4 in the proper case] is necessarily *C-admissible*, i.e., preserves the cuspidal inertia subgroups of the various subquotients corresponding to surface groups [cf. Theorem 2.3, (ii), below]. Finally, we discuss applications of these combinatorial anabelian results to the *anabelian geometry of configuration spaces* associated to hyperbolic curves over arithmetic fields [cf. Corollaries 2.5, 2.6, below].

In this chapter, let Σ be a set of prime numbers which is either equal to the set of all prime numbers or of cardinality one; n a positive integer; k an algebraically closed field of characteristic $\notin \Sigma$; X a *hyperbolic curve* of type (g, r) over k. For each positive integer i, write X_i for the i-th *configuration space* of X; Π_i for the maximal pro-Σ quotient of the fundamental group of X_i.

Definition 2.1 Let $\alpha \in \mathrm{Aut}(\Pi_n)$ be an automorphism of Π_n.

(i) Write

$$\{1\} = K_n \subseteq K_{n-1} \subseteq \cdots \subseteq K_2 \subseteq K_1 \subseteq K_0 = \Pi_n$$

for the *standard fiber filtration* on Π_n [cf. [CmbCsp], Definition 1.1, (i)]. For each $m \in \{1, 2, \cdots, n\}$, write C_m for the [finite] set of K_{m-1}/K_m-conjugacy classes of cuspidal inertia subgroups of K_{m-1}/K_m [where we recall that K_{m-1}/K_m is equipped with a natural structure of pro-Σ surface group—cf. [MzTa], Definition 1.2]. Then we shall say that α is *wC-admissible* [i.e., "*weakly C-admissible*"] if α preserves the standard fiber filtration on Π_n and, moreover, satisfies the following conditions:

Y. Hoshi, S. Mochizuki, *Topics Surrounding the Combinatorial Anabelian Geometry of Hyperbolic Curves II*, Lecture Notes in Mathematics 2299,
https://doi.org/10.1007/978-981-19-1096-8_2

- If $m \in \{1, 2, \cdots n - 1\}$, then the automorphism of K_{m-1}/K_m determined by α induces an automorphism of C_m.
- It follows immediately from the various definitions involved that we have a natural injection $C_{n-1} \hookrightarrow C_n$. That is to say, if one thinks of K_{n-2} as the two-dimensional configuration space group associated to some hyperbolic curve, then the image of $C_{n-1} \hookrightarrow C_n$ corresponds to the set of cusps of a fiber [of the two-dimensional configuration space over the hyperbolic curve] that arise from the cusps of the hyperbolic curve. Then the automorphism of K_{n-1} determined by α induces an automorphism of the image of the natural injection $C_{n-1} \hookrightarrow C_n$.

Write

$$\mathrm{Aut}^{\mathrm{wC}}(\Pi_n) \subseteq \mathrm{Aut}(\Pi_n)$$

for the subgroup of *wC-admissible* automorphisms and

$$\mathrm{Out}^{\mathrm{wC}}(\Pi_n) \stackrel{\mathrm{def}}{=} \mathrm{Aut}^{\mathrm{wC}}(\Pi_n)/\mathrm{Inn}(\Pi_n) \subseteq \mathrm{Out}(\Pi_n).$$

We shall refer to an element of $\mathrm{Out}^{\mathrm{wC}}(\Pi_n)$ as a *wC-admissible* outomorphism.

(ii) We shall say that α is *FwC-admissible* if α is *F-admissible* [cf. [CmbCsp], Definition 1.1, (ii)] and *wC-admissible* [cf. (i)]. Write

$$\mathrm{Aut}^{\mathrm{FwC}}(\Pi_n) \subseteq \mathrm{Aut}^{\mathrm{F}}(\Pi_n)$$

for the subgroup of *FwC-admissible* automorphisms and

$$\mathrm{Out}^{\mathrm{FwC}}(\Pi_n) \stackrel{\mathrm{def}}{=} \mathrm{Aut}^{\mathrm{FwC}}(\Pi_n)/\mathrm{Inn}(\Pi_n) \subseteq \mathrm{Out}^{\mathrm{F}}(\Pi_n).$$

We shall refer to an element of $\mathrm{Out}^{\mathrm{FwC}}(\Pi_n)$ as an *FwC-admissible* outomorphism.

(iii) We shall say that α is *DF-admissible* [i.e., "*diagonal-fiber-admissible*"] if α is *F-admissible*, and, moreover, α induces the *same* automorphism of Π_1 relative to the various quotients $\Pi_n \twoheadrightarrow \Pi_1$ by *fiber subgroups of co-length* 1 [cf. [MzTa], Definition 2.3, (iii)]. Write

$$\mathrm{Aut}^{\mathrm{DF}}(\Pi_n) \subseteq \mathrm{Aut}^{\mathrm{F}}(\Pi_n)$$

for the subgroup of *DF-admissible* automorphisms.

Remark 2.1.1 Thus, it follows immediately from the definitions that

$$\text{C-admissible} \implies \text{wC-admissible.}$$

In particular, we have inclusions

$$\begin{array}{ccccccc} \mathrm{Aut}^{\mathrm{FC}}(\Pi_n) & \subset & \mathrm{Aut}^{\mathrm{FwC}}(\Pi_n) & & \mathrm{Out}^{\mathrm{FC}}(\Pi_n) & \subset & \mathrm{Out}^{\mathrm{FwC}}(\Pi_n) \\ \cap & & \cap & & \cap & & \cap \\ \mathrm{Aut}^{\mathrm{C}}(\Pi_n) & \subset & \mathrm{Aut}^{\mathrm{wC}}(\Pi_n) & & \mathrm{Out}^{\mathrm{C}}(\Pi_n) & \subset & \mathrm{Out}^{\mathrm{wC}}(\Pi_n) \end{array}$$

[cf. Definition 2.1, (i), (ii)].

Lemma 2.2 (F-Admissible Automorphisms and Inertia Subgroups) *Let* $\alpha \in \mathrm{Aut}^{\mathrm{F}}(\Pi_n)$ *be an F-admissible automorphism of* Π_n*. Then the following hold:*

(i) There exist $\beta \in \mathrm{Aut}^{\mathrm{DF}}(\Pi_n)$ *[cf. Definition 2.1, (iii)] and* $\iota \in \mathrm{Inn}(\Pi_n)$ *such that* $\alpha = \beta \circ \iota$.

(ii) For each positive integer i*, write* $Z_i^{\log}$ *for the* i*-th log configuration space of* X *[cf. the discussion entitled "Curves" in "Notations and Conventions"];* $U_{Z_i} \subseteq Z_i$ *for the interior of* $Z_i^{\log}$ *[cf. the discussion entitled "Log schemes" in [CbTpI], §0], which may be identified with* X_i*. Let* ϵ *be an irreducible component of the complement* $Z_{n-1} \setminus U_{Z_{n-1}}$ *[cf. [CmbCsp], Proposition 1.3];* $\mathbb{I}_\epsilon \subseteq \Pi_{n-1}$ *an inertia subgroup of* Π_{n-1} *associated to the divisor* ϵ *of* Z_{n-1}*;* $\mathrm{pr}\colon U_{Z_n} \to U_{Z_{n-1}}$ *the projection obtained by forgetting the factor labeled* n*;* $\mathrm{pr}^{\Pi}\colon \Pi_n \twoheadrightarrow \Pi_{n-1}$ *the surjection induced by* pr*;* $\Pi_{n/n-1} \stackrel{\mathrm{def}}{=} \mathrm{Ker}(\mathrm{pr}^{\Pi})$*;* θ *an irreducible component of the fiber of the [uniquely determined] extension* $Z_n \to Z_{n-1}$ *of* pr *over the generic point of* ϵ *[so* θ *naturally determines an irreducible component of the complement* $Z_n \setminus U_{Z_n}$*];* $\mathbb{D}_\theta^{\mathbb{I}} \subseteq \Pi_n \times_{\Pi_{n-1}} \mathbb{I}_\epsilon$ $(\subseteq \Pi_n)$*—where the homomorphism* $\Pi_n \to \Pi_{n-1}$ *implicit in the fiber product is the surjection* $\mathrm{pr}^{\Pi}\colon \Pi_n \twoheadrightarrow \Pi_{n-1}$*—a decomposition subgroup of* $\Pi_n \times_{\Pi_{n-1}} \mathbb{I}_\epsilon$ $(\subseteq \Pi_n)$ *associated to the divisor [naturally determined by]* θ *of* Z_n*;* $\Pi_\theta \stackrel{\mathrm{def}}{=} \mathbb{D}_\theta^{\mathbb{I}} \cap \Pi_{n/n-1}$ *[cf. [CmbCsp], Proposition 1.3, (iv)]. Suppose that the automorphism of* Π_{n-1} *induced by* $\alpha \in \mathrm{Aut}^{\mathrm{F}}(\Pi_n)$ *relative to* pr^{Π} *stabilizes* $\mathbb{I}_\epsilon \subseteq \Pi_{n-1}$*. Then* α ***preserves*** *the* $\Pi_{n/n-1}$*-conjugacy class of* Π_θ*.*

Proof Assertion (i) follows immediately from [CbTpI], Theorem A, (i). Assertion (ii) follows immediately from Theorem 1.9, (ii) [cf. also the proof of [CmbCsp], Proposition 1.3, (iv)]. □

Theorem 2.3 (Partial Combinatorial Cuspidalization for F-Admissible Outomorphisms) *Let* Σ *be a set of prime numbers which is either equal to the set of all prime numbers or of cardinality one;* n *a positive integer;* X *a* **hyperbolic curve** *of type* (g, r) *over an algebraically closed field of characteristic* $\notin \Sigma$*;* X_n *the* n*-th* **configuration space** *of* X*;* Π_n *the maximal pro-*Σ *quotient of the fundamental group of* X_n*;*

$$\mathrm{Out}^{\mathrm{F}}(\Pi_n) \subseteq \mathrm{Out}(\Pi_n)$$

the subgroup of **F-admissible** *outomorphisms [i.e., roughly speaking, outomorphisms that preserve the fiber subgroups—cf. [CmbCsp], Definition 1.1, (ii)] of* Π_n*;*

$$\mathrm{Out}^{\mathrm{FC}}(\Pi_n) \subseteq \mathrm{Out}^{\mathrm{F}}(\Pi_n)$$

the subgroup of **FC-admissible** *outomorphisms [i.e., roughly speaking, outomorphisms that preserve the fiber subgroups and the cuspidal inertia subgroups—cf. [CmbCsp], Definition 1.1, (ii)] of* Π_n*;*

$$(\mathrm{Out}^{\mathrm{FC}}(\Pi_n) \subseteq)\ \mathrm{Out}^{\mathrm{FwC}}(\Pi_n) \subseteq \mathrm{Out}^{\mathrm{F}}(\Pi_n)$$

the subgroup of **FwC-admissible** *outomorphisms [cf. Definition 2.1, (ii); Remark 2.1.1] of* Π_n*. Then the following hold:*

(i) Write

$$n_{\mathrm{inj}} \overset{\mathrm{def}}{=} \begin{cases} 1 & \textit{if } r \neq 0, \\ 2 & \textit{if } r = 0, \end{cases} \qquad n_{\mathrm{bij}} \overset{\mathrm{def}}{=} \begin{cases} 3 & \textit{if } r \neq 0, \\ 4 & \textit{if } r = 0. \end{cases}$$

If $n \geq n_{\mathrm{inj}}$ *(respectively,* $n \geq n_{\mathrm{bij}}$*), then the natural homomorphism*

$$\mathrm{Out}^{\mathrm{F}}(\Pi_{n+1}) \longrightarrow \mathrm{Out}^{\mathrm{F}}(\Pi_n)$$

induced by the projections $X_{n+1} \to X_n$ *obtained by forgetting any one of the* $n+1$ *factors of* X_{n+1} *[cf. [CbTpI], Theorem A, (i)] is* **injective** *(respectively,* **bijective***).*

(ii) Write

$$n_{\mathrm{FC}} \overset{\mathrm{def}}{=} \begin{cases} 2 & \textit{if } (g, r) = (0, 3), \\ 3 & \textit{if } (g, r) \neq (0, 3) \textit{ and } r \neq 0, \\ 4 & \textit{if } r = 0. \end{cases}$$

If $n \geq n_{\mathrm{FC}}$*, then it holds that*

$$\mathrm{Out}^{\mathrm{FC}}(\Pi_n) = \mathrm{Out}^{\mathrm{F}}(\Pi_n).$$

(iii) Write

$$n_{\mathrm{FwC}} \overset{\mathrm{def}}{=} \begin{cases} 2 & \textit{if } r \geq 2, \\ 3 & \textit{if } r = 1, \\ 4 & \textit{if } r = 0. \end{cases}$$

If $n \geq n_{\mathrm{FwC}}$*, then it holds that*

$$\mathrm{Out}^{\mathrm{FwC}}(\Pi_n) = \mathrm{Out}^{\mathrm{F}}(\Pi_n).$$

(iv) Consider the natural inclusion

$$\mathfrak{S}_n \hookrightarrow \mathrm{Out}(\Pi_n)$$

—where we write $\mathfrak{S}_n$ for the symmetric group on n letters—obtained by permuting the various factors of X_n. If $(r,n) \neq (0,2)$, then the image of this inclusion is contained in the **centralizer** $Z_{\mathrm{Out}(\Pi_n)}(\mathrm{Out}^{\mathrm{F}}(\Pi_n))$.

Proof First, we verify assertion (iii) in the case where $n = 2$, which implies that $r \geq 2$ [cf. the statement of assertion (iii)]. To verify assertion (iii) in the case where $n = 2$, it is immediate that it suffices to verify that

$$\mathrm{Aut}^{\mathrm{FwC}}(\Pi_2) = \mathrm{Aut}^{\mathrm{F}}(\Pi_2).$$

Let $\alpha \in \mathrm{Aut}^{\mathrm{F}}(\Pi_2)$. Let us assign the cusps of X the *labels* $a_1, \cdots, a_r$. Now, for each $i \in \{1, \cdots, r\}$, recall that there is a uniquely determined cusp of the geometric generic fiber $X_{2/1}$ of the projection $X_2 \to X$ to the factor labeled 1 that corresponds naturally to the cusp of X labeled a_i; we assign to this uniquely determined cusp the *label* b_i. Thus, there is precisely one cusp of $X_{2/1}$ that has not been assigned a label $\in \{b_1, \cdots, b_r\}$; we assign to this uniquely determined cusp the *label* b_{r+1}. Then since the automorphism of Π_1 induced by α relative to either p_1 or p_2—where we write p_1, p_2 for the surjections $\Pi_2 \twoheadrightarrow \Pi_1$ induced by the projections $X_2 \to X$ to the factors labeled 1, 2, respectively—is *FC-admissible* [cf. [CbTpI], Theorem A, (ii)], it follows from the various definitions involved that, to verify that $\alpha \in \mathrm{Aut}^{\mathrm{FwC}}(\Pi_2)$, it suffices to verify the following assertion:

Claim 2.3.A: For any $b \in \{b_1, \cdots, b_r\}$, if $I_b \subseteq \Pi_{2/1} \stackrel{\mathrm{def}}{=} \mathrm{Ker}(p_1) \subseteq \Pi_2$ is a cuspidal inertia subgroup associated to the cusp labeled b, then $\alpha(I_b)$ is a cuspidal inertia subgroup.

Now observe that to verify Claim 2.3.A, by replacing α by the composite of α with a suitable element of $\mathrm{Aut}^{\mathrm{FC}}(\Pi_2)$ [cf. [CmbCsp], Lemma 2.4], we may assume without loss of generality that the [necessarily FC-admissible] automorphism of Π_1 induced by α relative to p_1, hence also relative to p_2 [cf. [CbTpI], Theorem A, (i)], induces the *identity automorphism* on the set of conjugacy classes of cuspidal inertia subgroups of Π_1.

To verify Claim 2.3.A, let us *fix* $b \in \{b_1, \cdots, b_r\}$, together with a cuspidal inertia subgroup $I_b \subseteq \Pi_{2/1}$ associated to the cusp labeled b of $\Pi_{2/1}$. Also, let us *fix*

- $a \in \{a_1, \cdots, a_r\}$ such that if $b = b_i$ and $a = a_j$, then $i \neq j$ [cf. the assumption that $r \geq 2$!];
- a cuspidal inertia subgroup $I_a \subseteq \Pi_1$ associated to the cusp labeled a of Π_1.

Now observe that since the [necessarily FC-admissible] automorphism of Π_1 induced by α relative to p_1 induces the *identity automorphism* on the set of conjugacy classes of cuspidal inertia subgroups of Π_1, to verify the fact that $\alpha(I_b)$ is a cuspidal inertia subgroup, we may assume without loss of generality [by replacing α by a suitable Π_2-conjugate of α] that the automorphism of Π_1

induced by α relative to p_1 *fixes* I_a. Let $\Pi_{F_a} \subseteq \Pi_{2/1}$ be a *major verticial subgroup* at a [cf. [CmbCsp], Definition 1.4, (ii)] such that $I_b \subseteq \Pi_{F_a}$. Then it follows from Lemma 2.2, (ii), that α fixes the $\Pi_{2/1}$-conjugacy class of Π_{F_a}, i.e., that $\Pi^{\dagger}_{F_a} \stackrel{\mathrm{def}}{=} \alpha(\Pi_{F_a})$ is a $\Pi_{2/1}$-conjugate of Π_{F_a}. Thus, one verifies easily that, to verify that $\alpha(I_b)$ is a cuspidal inertia subgroup, it suffices to verify that the isomorphism $\Pi_{F_a} \xrightarrow{\sim} \Pi^{\dagger}_{F_a}$ induced by α is *group-theoretically cuspidal*—cf. [CmbGC], Definition 1.4, (iv). [Note that it follows immediately from the various definitions involved that Π_{F_a} and $\Pi^{\dagger}_{F_a}$ may be regarded as *pro-Σ fundamental groups of semi-graphs of anabelioids of pro-Σ PSC-type*.] On the other hand, it follows immediately from the various definitions involved that this isomorphism factors as the composite

$$\Pi_{F_a} \xrightarrow{\sim} \Pi_1 \xrightarrow{\sim} \Pi_1 \xleftarrow{\sim} \Pi^{\dagger}_{F_a}$$

—where the first and third arrows are the isomorphisms induced by $p_2 \colon \Pi_2 \twoheadrightarrow \Pi_1$ [cf. [CmbCsp], Definition 1.4, (ii)], and the second arrow is the automorphism induced by α relative to p_2—and that the three arrows appearing in this composite are *group-theoretically cuspidal*. Thus, we conclude that $\alpha(I_b)$ is a cuspidal inertia subgroup. This completes the proof of Claim 2.3.A, hence also of assertion (iii) in the case where $n = 2$.

Next, we verify assertion (ii) in the case where $(g, r, n) = (0, 3, 2)$. In the following, we shall use the notation "a_i" [for $i = 1, 2, 3$] and "b_j" [for $j = 1, 2, 3, 4$] introduced in the proof of assertion (iii) in the case where $n = 2$. Now, to verify assertion (ii) in the case where $(g, r, n) = (0, 3, 2)$, it is immediate that it suffices to verify that

$$\mathrm{Aut}^{\mathrm{FC}}(\Pi_2) = \mathrm{Aut}^{\mathrm{F}}(\Pi_2).$$

Let $\alpha \in \mathrm{Aut}^{\mathrm{F}}(\Pi_2)$. Then let us observe that to verify that $\alpha \in \mathrm{Aut}^{\mathrm{FC}}(\Pi_2)$, by replacing α by the composite of α with a suitable element of $\mathrm{Aut}^{\mathrm{FC}}(\Pi_2)$ [cf. [CmbCsp], Lemma 2.4], we may assume without loss of generality that the [necessarily FC-admissible—cf. [CbTpI], Theorem A, (ii)] automorphism of Π_1 induced by α relative to p_1, hence also relative to p_2 [cf. [CbTpI], Theorem A, (i)]—where we write p_1, p_2 for the surjections $\Pi_2 \twoheadrightarrow \Pi_1$ induced by the projections $X_2 \to X$ to the factors labeled 1, 2, respectively—induces the *identity automorphism* on the set of conjugacy classes of cuspidal inertia subgroups of Π_1. Now it follows from assertion (iii) in the case where $n = 2$ that α is *FwC-admissible*; thus, to verify the fact that α is *FC-admissible*, it suffices to verify the following assertion:

Claim 2.3.B: If $I_{b_4} \subseteq \Pi_{2/1} \stackrel{\mathrm{def}}{=} \mathrm{Ker}(p_1) \subseteq \Pi_2$ is a cuspidal inertia subgroup associated to the cusp labeled b_4, then $\alpha(I_{b_4})$ is a cuspidal inertia subgroup.

On the other hand, as is well-known [cf. e.g., [CbTpI], Lemma 6.10, (ii)], there exists an automorphism of X_2 over X relative to the projection to the factor labeled

1 which switches the cusps on the geometric generic fiber $X_{2/1}$ labeled b_1 and b_4. In particular, there exists an automorphism ι of Π_2 over Π_1 relative to p_1 which switches the respective $\Pi_{2/1}$-conjugacy classes of cuspidal inertia subgroups associated to b_1 and b_4. Write $\beta = \iota^{-1} \circ \alpha \circ \iota$.

Now let us verify that Claim 2.3.B follows from the following assertion:

Claim 2.3.C: $\beta \in \mathrm{Aut}^{\mathrm{F}}(\Pi_2)$.

Indeed, if Claim 2.3.C holds, then it follows from assertion (iii) in the case where $n = 2$ that, for any cuspidal inertia subgroup $I_{b_1} \subseteq \Pi_{2/1}$ associated to the cusp labeled b_1, $\beta(I_{b_1})$ is a cuspidal inertia subgroup. Thus, it follows immediately from our choice of ι that, for any cuspidal inertia subgroup $I_{b_4} \subseteq \Pi_{2/1}$ associated to the cusp labeled b_4, $\alpha(I_{b_4})$ is a cuspidal inertia subgroup. This completes the proof of the assertion that Claim 2.3.C implies Claim 2.3.B.

Finally, we verify Claim 2.3.C. Since α and ι, hence also β, preserve $\Pi_{2/1} \subseteq \Pi_2$, it follows immediately from [CmbCsp], Proposition 1.2, (i), that, to verify Claim 2.3.C, it suffices to verify that β preserves $\Xi_2 \subseteq \Pi_2$ [cf. [CmbCsp], Definition 1.1, (iii)], i.e., the normal closed subgroup of Π_2 topologically normally generated by a cuspidal inertia subgroup associated to b_4. On the other hand, this follows immediately from the fact that α preserves the $\Pi_{2/1}$-conjugacy class of cuspidal inertia subgroups associated to b_1 [cf. assertion (iii) in the case where $n = 2$], together with our choice of ι. This completes the proof of Claim 2.3.C, hence also of assertion (ii) in the case where $(g, r, n) = (0, 3, 2)$.

Next, we verify assertion (ii) in the case where $(g, r, n) \neq (0, 3, 2)$. Thus, $n \geq 3$. Write $\Pi_3^{\dagger}$ (respectively, $\Pi_2^{\dagger}$; $\Pi_1^{\dagger}$) for the kernel of the surjection $\Pi_n \twoheadrightarrow \Pi_{n-3}$ (respectively, $\Pi_{n-1} \twoheadrightarrow \Pi_{n-3}$; $\Pi_{n-2} \twoheadrightarrow \Pi_{n-3}$) induced by the projection obtained by forgetting the factor(s) labeled $n, n-1, n-2$ (respectively, $n-1, n-2$; $n-2$). Here, if $n = 3$, then we set $\Pi_{n-3} = \Pi_0 \stackrel{\mathrm{def}}{=} \{1\}$. Then recall [cf., e.g., the proof of [CmbCsp], Theorem 4.1, (i)] that we have natural isomorphisms

$$\Pi_n \simeq \Pi_3^{\dagger} \stackrel{\mathrm{out}}{\rtimes} \Pi_{n-3}; \quad \Pi_{n-1} \simeq \Pi_2^{\dagger} \stackrel{\mathrm{out}}{\rtimes} \Pi_{n-3}; \quad \Pi_{n-2} \simeq \Pi_1^{\dagger} \stackrel{\mathrm{out}}{\rtimes} \Pi_{n-3}$$

[cf. the discussion entitled "*Topological groups*" in [CbTpI], §0]. Also, we recall [cf. [MzTa], Proposition 2.4, (i)] that one may *interpret* the surjections $\Pi_3^{\dagger} \twoheadrightarrow \Pi_2^{\dagger} \twoheadrightarrow \Pi_1^{\dagger}$ induced by the surjections $\Pi_n \twoheadrightarrow \Pi_{n-1} \twoheadrightarrow \Pi_{n-2}$ as the surjections "$\Pi_3 \twoheadrightarrow \Pi_2 \twoheadrightarrow \Pi_1$" that arise from the projections $X_3 \to X_2 \to X$ in the case of an "X" of type $(g, r+n-3)$. Moreover, one verifies easily that this *interpretation* is compatible with the definition of the various "Out$(-)$'s" involved. Thus, since $n_{\mathrm{FC}} = 4$ if $r = 0$, the above *natural isomorphisms*, together with [CbTpI], Theorem A, (ii), allow one to reduce the *equality* in question to the case where $n = 3$ and $r \neq 0$.

Now one verifies easily that, to verify the *equality* in question in the case where $n = 3$ and $r \neq 0$, it is immediate that it suffices to verify that

$$\mathrm{Aut}^{\mathrm{FC}}(\Pi_3) = \mathrm{Aut}^{\mathrm{F}}(\Pi_3).$$

Let $\alpha \in \mathrm{Aut}^{\mathrm{F}}(\Pi_3)$. Then let us observe that to verify $\alpha \in \mathrm{Aut}^{\mathrm{FC}}(\Pi_3)$, by replacing α by the composite of α with a suitable element of $\mathrm{Aut}^{\mathrm{FC}}(\Pi_3)$ [cf. [CmbCsp], Lemma 2.4], we may assume without loss of generality that the [necessarily FC-admissible—cf. [CbTpI], Theorem A, (ii)] automorphism of Π_1 induced by α relative to q_1, hence also relative to either q_2 or q_3 [cf. [CbTpI], Theorem A, (i)]—where we write q_1, q_2, q_3 for the surjections $\Pi_3 \twoheadrightarrow \Pi_1$ induced by the projections $X_3 \to X$ to the factors labeled 1, 2, 3, respectively—induces the *identity automorphism* on the set of conjugacy classes of cuspidal inertia subgroups of Π_1; in particular, one verifies easily that the [necessarily FC-admissible—cf. [CbTpI], Theorem A, (ii)] automorphism of $\Pi_{2/1}$—where we write $p_1 : \Pi_2 \twoheadrightarrow \Pi_1$ for the surjection induced by the projection $X_2 \to X$ to the factor labeled 1 and $\Pi_{2/1} \overset{\mathrm{def}}{=} \mathrm{Ker}(p_1) \subseteq \Pi_2$—induced by α induces the *identity automorphism* on the set of conjugacy classes of cuspidal inertia subgroups of $\Pi_{2/1}$. Write $X_{2/1}$ (respectively, $X_{3/2}$; $X_{3/1}$) for the geometric generic fiber of the projection $X_2 \to X$ (respectively, $X_3 \to X_2$; $X_3 \to X$) to the factor(s) labeled 1 (respectively, 1, 2; 1). Let us assign the cusps of X the *labels* $a_1, \cdots, a_r$. For each $i \in \{1, \cdots, r\}$, we assign to the cusp of $X_{2/1}$ that corresponds naturally to the cusp of X labeled a_i the *label* b_i. Thus, there is precisely one cusp of $X_{2/1}$ that has not been assigned a label $\in \{b_1, \cdots, b_r\}$; we assign to this uniquely determined cusp the *label* b_{r+1}. For each $i \in \{1, \cdots, r+1\}$, we assign to the cusp of $X_{3/2}$ that corresponds naturally to the cusp of $X_{2/1}$ labeled b_i the *label* c_i. Thus, there is precisely one cusp of $X_{3/2}$ that has not been assigned a label $\in \{c_1, \cdots, c_{r+1}\}$; we assign to this uniquely determined cusp the *label* c_{r+2}. Now it follows from assertion (iii) in the case where $n = 2$, applied to the restriction of α to $\Pi_{3/1} \overset{\mathrm{def}}{=} \mathrm{Ker}(q_1)$, together with [CbTpI], Theorem A, (ii), that α is *FwC-admissible*. Write $q_{12} : \Pi_3 \twoheadrightarrow \Pi_2$ for the surjection induced by the projection $X_3 \to X_2$ to the factors labeled 1, 2; $\Pi_{3/2} \overset{\mathrm{def}}{=} \mathrm{Ker}(q_{12}) \subseteq \Pi_3$. Thus, to verify the fact that α is *FC-admissible*, it suffices to verify the following assertion:

Claim 2.3.D: If $I_{c_{r+2}} \subseteq \Pi_{3/2}$ is a cuspidal inertia subgroup associated to the cusp labeled c_{r+2}, then $\alpha(I_{c_{r+2}})$ is a cuspidal inertia subgroup.

To verify Claim 2.3.D, let us *fix* a cusp labeled $b \in \{b_1, \cdots, b_r\}$ [where we recall that $r \neq 0$], a cuspidal inertia subgroup $I_b \subseteq \Pi_{2/1}$ associated to the cusp labeled b of $X_{2/1}$, and a cuspidal inertia subgroup $I_{c_{r+2}} \subseteq \Pi_{3/2}$ associated to the cusp labeled c_{r+2} of $\Pi_{3/2}$. Now observe that since the [necessarily FC-admissible] automorphism of $\Pi_{2/1}$ induced by α induces the *identity automorphism* on the set of conjugacy classes of cuspidal inertia subgroups of $\Pi_{2/1}$, to verify the assertion that $\alpha(I_{c_{r+2}})$ is a cuspidal inertia subgroup, we may assume without loss of generality [by replacing α by a suitable Π_3-conjugate of α] that the automorphism of $\Pi_{2/1}$ induced by α fixes I_b. Let $\Pi_{E_b} \subseteq \Pi_{3/2}$ be a *minor verticial subgroup*, relative to the two-dimensional configuration space $X_{3/1}$ associated to the hyperbolic curve $X_{2/1}$, at the cusp labeled b [cf. [CmbCsp], Definition 1.4, (ii)] such that $I_{c_{r+2}} \subseteq \Pi_{E_b}$. Then it follows immediately from Lemma 2.2, (ii), that α fixes the $\Pi_{3/2}$-conjugacy class of Π_{E_b}, i.e., that $\Pi^{\dagger}_{E_b} \stackrel{\mathrm{def}}{=} \alpha(\Pi_{E_b})$ is a $\Pi_{3/2}$-conjugate of Π_{E_b}. Thus, one verifies easily that, to verify that $\alpha(I_{c_{r+2}})$ is a cuspidal inertia subgroup, it suffices to verify that the isomorphism $\Pi_{E_b} \xrightarrow{\sim} \Pi^{\dagger}_{E_b}$ induced by α is *group-theoretically cuspidal*—cf. [CmbGC], Definition 1.4, (iv). [Note that it follows immediately from the various definitions involved that Π_{E_b} and $\Pi^{\dagger}_{E_b}$ may be regarded as *pro-Σ fundamental groups of semi-graphs of anabelioids of pro-Σ PSC-type*.] On the other hand, it follows immediately from a similar argument to the argument applied in the discussion concerning the isomorphism of the second display of [CmbCsp], Definition 1.4, (ii), that the composites

$$\Pi_{E_b},\ \Pi^{\dagger}_{E_b} \hookrightarrow \Pi_{3/2} \twoheadrightarrow \Pi_{2/1}$$

—where the second arrow is the surjection determined by the surjection $q_{13}\colon \Pi_3 \twoheadrightarrow \Pi_2$ induced by the projection $X_3 \to X_2$ to the factors labeled 1, 3—are *injective*, and that the $\Pi_{2/1}$-conjugacy class of the image in $\Pi_{2/1}$ of either of these composite injections coincides with the $\Pi_{2/1}$-conjugacy class of a *minor verticial subgroup* at the cusp labeled a_i [where we write $b = b_i$—cf. [CmbCsp], Definition 1.4, (ii)]. In particular, since the automorphism of Π_2 induced by α relative to q_{13} is *FC-admissible* [cf. [CbTpI], Theorem A, (ii)], it follows immediately that the isomorphism $\Pi_{E_b} \xrightarrow{\sim} \Pi^{\dagger}_{E_b}$ induced by α is *group-theoretically cuspidal*. This completes the proof of Claim 2.3.D, hence also of assertion (ii).

Now assertion (iii) in the case where $n \neq 2$ follows immediately from assertion (ii), together with the natural inclusions $\mathrm{Out}^{\mathrm{FC}}(\Pi_n) \subseteq \mathrm{Out}^{\mathrm{FwC}}(\Pi_n) \subseteq \mathrm{Out}^{\mathrm{F}}(\Pi_n)$ [cf. Remark 2.1.1]. This completes the proof of assertion (iii).

Next, we verify assertion (i). The *bijectivity* portion of assertion (i) follows from assertion (ii), together with the *bijectivity* portion of [NodNon], Theorem B. Thus, it suffices to verify the *injectivity* portion of assertion (i). First, we observe that *injectivity* in the case where $(g, r) = (0, 3)$ follows from assertion (ii), together with the *injectivity* portion of [NodNon], Theorem B. Write $\Pi^{\dagger}_2$ (respectively, $\Pi^{\dagger}_1$) for the kernel of the surjection $\Pi_{n+1} \twoheadrightarrow \Pi_{n-1}$ (respectively, $\Pi_n \twoheadrightarrow \Pi_{n-1}$) induced by the projection obtained by forgetting the factor(s) labeled $n+1, n$ (respectively, n). Here, if $n = 1$, then we set $\Pi_{n-1} = \Pi_0 \stackrel{\mathrm{def}}{=} \{1\}$. Then recall [cf. e.g., the proof of [CmbCsp], Theorem 4.1, (i)] that we have natural isomorphisms

$$\Pi_{n+1} \simeq \Pi^{\dagger}_2 \stackrel{\mathrm{out}}{\rtimes} \Pi_{n-1}; \quad \Pi_n \simeq \Pi^{\dagger}_1 \stackrel{\mathrm{out}}{\rtimes} \Pi_{n-1}$$

[cf. the discussion entitled "*Topological groups*" in [CbTpI], §0]. Also, we recall [cf. [MzTa], Proposition 2.4, (i)] that one may *interpret* the surjection $\Pi^{\dagger}_2 \twoheadrightarrow \Pi^{\dagger}_1$ induced by the surjection $\Pi_{n+1} \twoheadrightarrow \Pi_n$ in question as the surjection "$\Pi_2 \twoheadrightarrow \Pi_1$" that arises from the projection $X_2 \to X$ in the case of an "X" of type $(g, r+n-1)$. Moreover, one verifies easily that this *interpretation* is compatible with the definition of the various "Out$(-)$'s" involved. Thus, since $n_{\mathrm{inj}} = 2$ if $r = 0$, the above *natural isomorphisms* allow one to reduce the *injectivity* in question to the case where $n = 1$

and $r \neq 0$. On the other hand, this *injectivity* follows immediately from a similar argument to the argument used in the proof of [CmbCsp], Corollary 2.3, (ii), by replacing [CmbCsp], Proposition 1.2, (iii) (respectively, the non-resp'd portion of [CmbCsp], Proposition 1.3, (iv); [CmbCsp], Corollary 1.12, (i)), in the proof of [CmbCsp], Corollary 2.3, (ii), by Lemma 2.2, (i) (respectively, Lemma 2.2, (ii); the *injectivity* in question in the case where $(g, r) = (0, 3)$, which was verified above). This completes the proof of the *injectivity* portion of assertion (i), hence also of assertion (i).

Finally, assertion (iv) follows immediately from assertion (i), together with a similar argument to the argument applied in the proof of [CmbCsp], Theorem 4.1, (iv). This completes the proof of Theorem 2.3. □

Corollary 2.4 (PFC-Admissibility of Outomorphisms) *In the notation of Theorem 2.3, write*

$$\mathrm{Out}^{\mathrm{PF}}(\Pi_n) \subseteq \mathrm{Out}(\Pi_n)$$

for the subgroup of **PF-admissible** *outomorphisms [i.e., roughly speaking, outomorphisms that preserve the fiber subgroups up to a possible permutation of the factors—cf. [CbTpI], Definition 1.4, (i)] and*

$$\mathrm{Out}^{\mathrm{PFC}}(\Pi_n) \subseteq \mathrm{Out}^{\mathrm{PF}}(\Pi_n)$$

for the subgroup of **PFC-admissible** *outomorphisms [i.e., roughly speaking, outomorphisms that preserve the fiber subgroups and the cuspidal inertia subgroups up to a possible permutation of the factors—cf. [CbTpI], Definition 1.4, (iii)]. Let us regard the symmetric group on n letters $\mathfrak{S}_n$ as a subgroup of* $\mathrm{Out}(\Pi_n)$ *via the natural inclusion of Theorem 2.3, (iv). Finally, suppose that $(g, r) \notin \{(0, 3); (1, 1)\}$. Then the following hold:*

(i) We have an equality

$$\mathrm{Out}(\Pi_n) = \mathrm{Out}^{\mathrm{PF}}(\Pi_n).$$

If, moreover, $(r, n) \neq (0, 2)$, then we have equalities

$$\mathrm{Out}(\Pi_n) = \mathrm{Out}^{\mathrm{PF}}(\Pi_n) = \mathrm{Out}^{\mathrm{F}}(\Pi_n) \times \mathfrak{S}_n$$

[cf. the notational conventions introduced in Theorem 2.3].

(ii) If either

$$r > 0, \quad n \geq 3$$

or

$$n \geq 4,$$

then we have equalities

$$\mathrm{Out}(\Pi_n) = \mathrm{Out}^{\mathrm{PFC}}(\Pi_n) = \mathrm{Out}^{\mathrm{FC}}(\Pi_n) \times \mathfrak{S}_n$$

[cf. the notational conventions introduced in Theorem 2.3].

Proof First, we verify assertion (i). The equality in the first display of assertion (i) follows from [MzTa], Corollary 6.3, together with the assumption that $(g, r) \notin \{(0, 3); (1, 1)\}$. The second equality in the second display of assertion (i) follows from Theorem 2.3, (iv). This completes the proof of assertion (i). Next, we verify assertion (ii). The first equality of assertion (ii) follows immediately from Theorem 2.3, (ii), together with the first equality of assertion (i). The second equality of assertion (ii) follows from [NodNon], Theorem B. This completes the proof of assertion (ii). □

Corollary 2.5 (Anabelian Properties of Hyperbolic Curves and Associated Configuration Spaces I) *Let Σ be a set of prime numbers which is either equal to the set of all prime numbers or of cardinality one; $m \leq n$ positive integers; (g, r) a pair of nonnegative integers such that $2g - 2 + r > 0$; k a field of characteristic $\notin \Sigma$; $\overline{k}$ a separable closure of k; X a* **hyperbolic curve** *of type (g, r) over k. Write $G_k \stackrel{\mathrm{def}}{=} \mathrm{Gal}(\overline{k}/k)$. For each positive integer i, write X_i for the i-th* **configuration space** *of X; $(X_i)_{\overline{k}} \stackrel{\mathrm{def}}{=} X_i \times_k \overline{k}$; Δ_{X_i} for the maximal pro-Σ quotient of the étale fundamental group of $(X_i)_{\overline{k}}$;*

$$\rho^{\Sigma}_{X_i} : G_k \longrightarrow \mathrm{Out}(\Delta_{X_i})$$

for the pro-Σ outer Galois representation associated to X_i; $\mathfrak{S}_i$ for the symmetric group on i letters;

$$\Phi_i : \mathfrak{S}_i \longrightarrow \mathrm{Out}(\Delta_{X_i})$$

for the outer representation arising from the permutations of the factors of X_i. Suppose that the following conditions are satisfied:

(1) $(g, r) \notin \{(0, 3); (1, 1)\}$.
(2) If $(r, n, m) \in \{(0, 2, 1); (0, 2, 2); (0, 3, 1)\}$, then there exists an $l \in \Sigma$ such that k is ***l*-cyclotomically full**, *i.e., the l-adic cyclotomic character of G_k has open image.*

Then the following hold:

(i) Let $\alpha \in \mathrm{Out}(\Delta_{X_n})$. Then there exists a **unique** *element $\sigma_\alpha \in \mathfrak{S}_n$ such that $\alpha \circ \Phi_n(\sigma_\alpha) \in \mathrm{Out}^{\mathrm{F}}(\Delta_{X_n})$ [cf. the notational conventions introduced in Theorem 2.3]. Write*

$$\alpha_m \in \mathrm{Out}^{\mathrm{F}}(\Delta_{X_m})$$

for the outomorphism of Δ_{X_m} *induced by* $\alpha \circ \Phi_n(\sigma_\alpha)$, *relative to the quotient* $\Delta_{X_n} \twoheadrightarrow \Delta_{X_m}$ *by a fiber subgroup of co-length m of* Δ_{X_n}. *[Note that it follows from [CbTpI], Theorem A, (i), that* α_m *does* **not depend** *on the choice of fiber subgroup of co-length m of* Δ_{X_n}.*]*

(ii) *If* $(r, n, m) \in \{(0, 2, 1); (0, 2, 2); (0, 3, 1)\}$, *then*

$$C_{\mathrm{Out}(\Delta_{X_n})}(\mathrm{Im}(\rho^{\Sigma}_{X_n})) \subseteq \mathrm{Out}^{\mathrm{PFC}}(\Delta_{X_n})$$

[cf. the notational conventions introduced in Corollary 2.4].

(iii) *The map*

$$\begin{array}{ccc} \mathrm{Out}(\Delta_{X_n}) & \longrightarrow & \mathrm{Out}(\Delta_{X_m}) \\ \alpha & \mapsto & \alpha_m \end{array}$$

[cf. (i)] determines an **exact sequence** *of homomorphisms of profinite groups*

$$1 \longrightarrow \mathfrak{S}_n \xrightarrow{\Phi_n} \mathrm{Out}^{\mathrm{PFC}}(\Delta_{X_n}) \longrightarrow \mathrm{Out}(\Delta_{X_m})$$

—where the second arrow is a **split injection** *whose image* **commutes** *with* $\mathrm{Out}^{\mathrm{FC}}(\Delta_{X_n})$ *and has* **trivial intersection** *with* $\mathrm{Im}(\rho^{\Sigma}_{X_n})$. *If* $(r, n) \neq (0, 2)$, *then the map* $\alpha \mapsto \alpha_m$ *determines a sequence of homomorphisms of profinite groups*

$$1 \longrightarrow \mathfrak{S}_n \xrightarrow{\Phi_n} \mathrm{Out}(\Delta_{X_n}) \longrightarrow \mathrm{Out}(\Delta_{X_m})$$

—where the second arrow is a **split injection** *whose image* **commutes** *with* $\mathrm{Out}^{\mathrm{F}}(\Delta_{X_n})$ *and has* **trivial intersection** *with* $\mathrm{Im}(\rho^{\Sigma}_{X_n})$*—which is* **exact** *if, moreover,* $(r, n, m) \neq (0, 3, 1)$.

(iv) *Let* $\alpha \in \mathrm{Out}(\Delta_{X_n})$. *If* $(r, n, m) \in \{(0, 2, 1); (0, 3, 1)\}$, *then we suppose further that* $\alpha \in \mathrm{Out}^{\mathrm{PFC}}(\Delta_{X_n})$, *which is the case if, for instance,* $\alpha \in C_{\mathrm{Out}(\Delta_{X_n})}(\mathrm{Im}(\rho^{\Sigma}_{X_n}))$ *[cf. (ii)]. Then it holds that*

$$\alpha \in Z_{\mathrm{Out}(\Delta_{X_n})}(\mathrm{Im}(\rho^{\Sigma}_{X_n}))$$

$$(\textit{respectively},\ N_{\mathrm{Out}(\Delta_{X_n})}(\mathrm{Im}(\rho^{\Sigma}_{X_n}));\ \ C_{\mathrm{Out}(\Delta_{X_n})}(\mathrm{Im}(\rho^{\Sigma}_{X_n})))$$

if and only if

$$\alpha_m \in Z_{\mathrm{Out}(\Delta_{X_m})}(\mathrm{Im}(\rho^{\Sigma}_{X_m}))$$

$$(\textit{respectively},\ N_{\mathrm{Out}(\Delta_{X_m})}(\mathrm{Im}(\rho^{\Sigma}_{X_m}));\ \ C_{\mathrm{Out}(\Delta_{X_m})}(\mathrm{Im}(\rho^{\Sigma}_{X_m}))).$$

(v) *For each positive integer* i, *write* $\mathrm{Aut}_k(X_i)$ *for the group of automorphisms of* X_i *over* k. *Then if the natural homomorphism*

$$\mathrm{Aut}_k(X_m) \longrightarrow Z_{\mathrm{Out}(\Delta_{X_m})}(\mathrm{Im}(\rho^{\Sigma}_{X_m}))$$

is **bijective**, *then the natural homomorphism*

$$\mathrm{Aut}_k(X_n) \longrightarrow Z_{\mathrm{Out}(\Delta_{X_n})}(\mathrm{Im}(\rho^{\Sigma}_{X_n}))$$

is **bijective.**

(vi) *For each positive integer* i, *write* $\mathrm{Aut}((X_i)_{\overline{k}}/k)$ *for the group of automorphisms of* $(X_i)_{\overline{k}}$ *that are compatible with some automorphism of* k; $\mathrm{Aut}^{\rho}(G_k)$ *for the group of automorphisms of* G_k *that preserve* $\mathrm{Ker}(\rho^{\Sigma}_{X_1}) \subseteq G_k$ *[where we note that, by [NodNon], Corollary 6.2, (i), for any positive integer* i, *it holds that* $\mathrm{Ker}(\rho^{\Sigma}_{X_1}) = \mathrm{Ker}(\rho^{\Sigma}_{X_i})$*]. Then if the natural homomorphism*

$$\mathrm{Aut}((X_m)_{\overline{k}}/k) \longrightarrow \mathrm{Aut}^{\rho}(G_k) \times_{\mathrm{Aut}(\mathrm{Im}(\rho^{\Sigma}_{X_m}))} N_{\mathrm{Out}(\Delta_{X_m})}(\mathrm{Im}(\rho^{\Sigma}_{X_m}))$$

is **bijective**, *then the natural homomorphism*

$$\mathrm{Aut}((X_n)_{\overline{k}}/k) \longrightarrow \mathrm{Aut}^{\rho}(G_k) \times_{\mathrm{Aut}(\mathrm{Im}(\rho^{\Sigma}_{X_n}))} N_{\mathrm{Out}(\Delta_{X_n})}(\mathrm{Im}(\rho^{\Sigma}_{X_n}))$$

is **bijective.**

Proof First, we verify assertion (i). The existence of such a σ_α follows from the fact that $\mathrm{Out}(\Delta_{X_n}) = \mathrm{Out}^{\mathrm{PF}}(\Delta_{X_n})$ [cf. Corollary 2.4, (i), together with assumption (1)]. The uniqueness of such a σ_α follows immediately from the easily verified *faithfulness* of the action of $\mathfrak{S}_n$, via Φ_n, on the set of fiber subgroups of Δ_{X_n}. This completes the proof of assertion (i). Next, we verify assertion (ii). Since $\mathrm{Out}(\Delta_{X_n}) = \mathrm{Out}^{\mathrm{PF}}(\Delta_{X_n})$ [cf. Corollary 2.4, (i), together with assumption (1)], assertion (ii) follows immediately from [CmbGC], Corollary 2.7, (i), together with condition (2). This completes the proof of assertion (ii).

Next, we verify assertion (iii). First, let us observe that it follows immediately from the various definitions involved that $\mathrm{Im}(\Phi_n) \subseteq \mathrm{Out}^{\mathrm{PFC}}(\Delta_{X_n})$. Now since $\mathrm{Out}(\Delta_{X_n}) = \mathrm{Out}^{\mathrm{PF}}(\Delta_{X_n})$ [cf. Corollary 2.4, (i), together with assumption (1)], and $\mathrm{Out}^{\mathrm{F}}(\Delta_{X_n})$ is *normalized* by $\mathrm{Out}^{\mathrm{PF}}(\Delta_{X_n})$, one verifies easily [i.e., by considering the action of elements of $\mathrm{Out}^{\mathrm{PF}}(\Delta_{X_n})$ on the set of fiber subgroups of Δ_{X_n}] that the second arrow in either of the two displayed sequences is a *split injection*. Moreover, since [as is easily verified] the outer action of G_k, via $\rho^{\Sigma}_{X_n}$, on Δ_{X_n} *fixes* every fiber subgroup of Δ_{X_n}, it follows immediately from the *faithfulness* of the action of $\mathfrak{S}_n$, via Φ_n, on the set of fiber subgroups of Δ_{X_n} that the image of the second arrow in either of the two displayed sequences has *trivial intersection* with $\mathrm{Im}(\rho^{\Sigma}_{X_n})$. Now it follows from [NodNon], Theorem B, that the image of the second arrow of the first displayed sequence *commutes* with $\mathrm{Out}^{\mathrm{FC}}(\Delta_{X_n})$; in particular, one verifies easily from the various definitions involved [cf. also Corollary 2.4, (i), together with assumption (1)] that the third arrow of the first displayed sequence is a *homomorphism*. If $(r, n) \neq (0, 2)$, then it follows from Corollary 2.4, (i), together with assumption (1), that the image of the second arrow of the second displayed sequence *commutes* with $\mathrm{Out}^{\mathrm{F}}(\Delta_{X_n})$; in particular, one verifies easily from the various definitions involved [cf. also Corollary 2.4, (i), together with assumption (1)] that the third arrow of the second displayed sequence is a *homomorphism*. Now if $(r, m) \neq (0, 1)$, then it follows immediately from the injectivity portion of Theorem 2.3, (i), together with the equality $\mathrm{Out}(\Delta_{X_n}) = \mathrm{Out}^{\mathrm{PF}}(\Delta_{X_n})$ [cf. Corollary 2.4, (i), together with assumption (1)], that the kernel of the third arrow in either of the two displayed sequences is $\mathrm{Im}(\Phi_n)$. Moreover, if $(r, n, m) \in \{(0, 2, 1); (0, 3, 1)\}$, then it follows immediately from the injectivity portion of [NodNon], Theorem B, that the kernel of the third arrow in the first displayed sequence is $\mathrm{Im}(\Phi_n)$. On the other hand, if $(r, m) = (0, 1)$ and $n \notin \{2, 3\}$, then it follows immediately from the injectivity portion of [NodNon], Theorem B, together with Corollary 2.4, (ii), together with assumption (1), that the kernel of the third arrow in either of the two displayed sequences is $\mathrm{Im}(\Phi_n)$. This completes the proof of assertion (iii).

Next, we verify assertion (iv). Now since the permutations of the factors of X_n give rise to *automorphisms of* X_n *over* k, it follows immediately that $\mathrm{Im}(\Phi_n) \subseteq Z_{\mathrm{Out}(\Delta_{X_n})}(\mathrm{Im}(\rho^{\Sigma}_{X_n}))$. In particular, to verify assertion (iv), we may assume without loss of generality—by replacing α by α_n [cf. assertion (i)]—that $\alpha \in \mathrm{Out}^{\mathrm{F}}(\Delta_{X_n})$, and that $m < n$. Then *necessity* follows immediately. On the other hand, *sufficiency* follows immediately from the exact sequences of assertion (iii). This completes the proof of assertion (iv). Assertion (v) (respectively, (vi)) follows immediately from assertions (i), (ii), (iii), (iv), together with Lemma 2.7, (iii), below (respectively, Lemma 2.7, (iv), below). This completes the proof of Corollary 2.5. □

Corollary 2.6 (Anabelian Properties of Hyperbolic Curves and Associated Configuration Spaces II) *Let* Σ *be a set of prime numbers which is either equal to the set of all prime numbers or of cardinality one;* $m \leq n$ *positive integers;* (g_X, r_X), (g_Y, r_Y) *pairs of nonnegative integers such that* $2g_X - 2 + r_X$, $2g_Y - 2 + r_Y > 0$; k_X, k_Y *fields;* $\overline{k}_X$, $\overline{k}_Y$ *separable closures of* k_X, k_Y, *respectively;* X, Y **hyperbolic curves** *of type* (g_X, r_X), (g_Y, r_Y) *over* k_X, k_Y, *respectively. Write* $G_{k_X} \overset{\mathrm{def}}{=} \mathrm{Gal}(\overline{k}_X/k_X)$; $G_{k_Y} \overset{\mathrm{def}}{=} \mathrm{Gal}(\overline{k}_Y/k_Y)$. *For each positive integer* i, *write* X_i, Y_i *for the* i*-th* **configuration spaces** *of* X, Y, *respectively;* $(X_i)_{\overline{k}_X} \overset{\mathrm{def}}{=} X_i \times_{k_X} \overline{k}_X$; $(Y_i)_{\overline{k}_Y} \overset{\mathrm{def}}{=} Y_i \times_{k_Y} \overline{k}_Y$; $\pi_1^{\Sigma}((X_i)_{\overline{k}_X})$, $\pi_1^{\Sigma}((Y_i)_{\overline{k}_Y})$ *for the maximal pro-*Σ *quotients of the étale fundamental groups* $\pi_1((X_i)_{\overline{k}_X})$, $\pi_1((Y_i)_{\overline{k}_Y})$ *of* $(X_i)_{\overline{k}_X}$, $(Y_i)_{\overline{k}_Y}$, *respectively;* $\pi_1^{(\Sigma)}(X_i)$, $\pi_1^{(\Sigma)}(Y_i)$ *for the geometrically pro-*Σ *étale fundamental groups of* X_i, Y_i, *respectively, i.e., the quotients of the étale fundamental groups* $\pi_1(X_i)$, $\pi_1(Y_i)$ *of* X_i, Y_i *by the respective kernels of the natural surjections* $\pi_1((X_i)_{\overline{k}_X}) \twoheadrightarrow \pi_1^{\Sigma}((X_i)_{\overline{k}_X})$, $\pi_1((Y_i)_{\overline{k}_Y}) \twoheadrightarrow \pi_1^{\Sigma}((Y_i)_{\overline{k}_Y})$. *Suppose that the following conditions are satisfied:*

(1) $\{(g_X, r_X); (g_Y, r_Y)\} \cap \{(0, 3); (1, 1)\} = \emptyset$.

(2) *If* (r_X, n, m) *(respectively,* (r_Y, n, m)*) is contained in the set* $\{(0, 2, 1); (0, 2, 2); (0, 3, 1)\}$*, then there exists an* $l \in \Sigma$ *such that* k_X *(respectively,* k_Y*) is* **l-cyclotomically full**, *i.e., the l-adic cyclotomic character of* G_{k_X} *(respectively,* G_{k_Y}*) has open image.*

Then the following hold:

(i) *Let* $\theta : \overline{k}_X \xrightarrow{\sim} \overline{k}_Y$ *be an isomorphism of fields that determines an isomorphism* $k_X \xrightarrow{\sim} k_Y$*. For each positive integer i, write* $\mathrm{Isom}_\theta(X_i, Y_i)$ *for the set of isomorphisms of* X_i *with* Y_i *that are compatible with the isomorphism* $k_X \xrightarrow{\sim} k_Y$ *determined by* θ*;* $\mathrm{Isom}_\theta(\pi_1^{(\Sigma)}(X_i), \pi_1^{(\Sigma)}(Y_i))$ *for the set of isomorphisms of* $\pi_1^{(\Sigma)}(X_i)$ *with* $\pi_1^{(\Sigma)}(Y_i)$ *that are compatible with the isomorphism* $G_{k_X} \xrightarrow{\sim} G_{k_Y}$ *determined by* θ*. Then if the natural map*

$$\mathrm{Isom}_\theta(X_m, Y_m) \longrightarrow \mathrm{Isom}_\theta(\pi_1^{(\Sigma)}(X_m), \pi_1^{(\Sigma)}(Y_m))/\mathrm{Inn}(\pi_1^{\Sigma}((Y_m)_{\overline{k}_Y}))$$

is **bijective**, *then the natural map*

$$\mathrm{Isom}_\theta(X_n, Y_n) \longrightarrow \mathrm{Isom}_\theta(\pi_1^{(\Sigma)}(X_n), \pi_1^{(\Sigma)}(Y_n))/\mathrm{Inn}(\pi_1^{\Sigma}((Y_n)_{\overline{k}_Y}))$$

is **bijective**.

(ii) *For each positive integer i, write* $\mathrm{Isom}((X_i)_{\overline{k}_X}/k_X, (Y_i)_{\overline{k}_Y}/k_Y)$ *for the set of isomorphisms of* $(X_i)_{\overline{k}_X}$ *with* $(Y_i)_{\overline{k}_Y}$ *that are compatible with some field isomorphism of* k_X *with* k_Y*;*

$$\mathrm{Isom}(\pi_1^{(\Sigma)}(X_i)/G_{k_X}, \pi_1^{(\Sigma)}(Y_i)/G_{k_Y})$$

for the set of isomorphisms of $\pi_1^{(\Sigma)}(X_i)$ *with* $\pi_1^{(\Sigma)}(Y_i)$ *that are compatible with some isomorphism of* G_{k_X} *with* G_{k_Y}*. Then if the natural map*

$$\mathrm{Isom}((X_m)_{\overline{k}_X}/k_X, (Y_m)_{\overline{k}_Y}/k_Y)$$

$$\longrightarrow \mathrm{Isom}(\pi_1^{(\Sigma)}(X_m)/G_{k_X}, \pi_1^{(\Sigma)}(Y_m)/G_{k_Y})/\mathrm{Inn}(\pi_1^{\Sigma}((Y_m)_{\overline{k}_Y}))$$

is **bijective**, *then the natural map*

$$\mathrm{Isom}((X_n)_{\overline{k}_X}/k_X, (Y_n)_{\overline{k}_Y}/k_Y)$$

$$\longrightarrow \mathrm{Isom}(\pi_1^{(\Sigma)}(X_n)/G_{k_X}, \pi_1^{(\Sigma)}(Y_n)/G_{k_Y})/\mathrm{Inn}(\pi_1^{\Sigma}((Y_n)_{\overline{k}_Y}))$$

is **bijective**.

Proof Consider assertion (i) (respectively, (ii)). If the set

$$\mathrm{Isom}_\theta(\pi_1^{(\Sigma)}(X_n), \pi_1^{(\Sigma)}(Y_n))/\mathrm{Inn}(\pi_1^{\Sigma}((Y_n)_{\overline{k}_Y}))$$

(respectively,

$$\mathrm{Isom}(\pi_1^{(\Sigma)}(X_n)/G_{k_X}, \pi_1^{(\Sigma)}(Y_n)/G_{k_Y})/\mathrm{Inn}(\pi_1^{\Sigma}((Y_n)_{\overline{k}_Y})))$$

is *empty*, then assertion (i) (respectively, (ii)) is immediate. Thus, we may suppose without loss of generality that this set is *nonempty*. Then one verifies easily from [MzTa], Corollary 6.3, together with condition (1), that the set

$$\mathrm{Isom}_\theta(\pi_1^{(\Sigma)}(X_m), \pi_1^{(\Sigma)}(Y_m))/\mathrm{Inn}(\pi_1^{\Sigma}((Y_m)_{\overline{k}_Y}))$$

(respectively,

$$\mathrm{Isom}(\pi_1^{(\Sigma)}(X_m)/G_{k_X}, \pi_1^{(\Sigma)}(Y_m)/G_{k_Y})/\mathrm{Inn}(\pi_1^{\Sigma}((Y_m)_{\overline{k}_Y})))$$

is *nonempty*. Thus, it follows immediately from the *bijectivity* assumed in assertion (i) (respectively, (ii)) that there exists an isomorphism $X_m \xrightarrow{\sim} Y_m$ that is compatible with the isomorphism $k_X \xrightarrow{\sim} k_Y$ determined by θ (respectively, an isomorphism $(X_m)_{\overline{k}_X} \xrightarrow{\sim} (Y_m)_{\overline{k}_Y}$ that is compatible with some isomorphism $k_X \xrightarrow{\sim} k_Y$). In particular, it follows immediately from Lemma 2.7, (iii), below (respectively, Lemma 2.7, (iv), below) that there exists an isomorphism $X \xrightarrow{\sim} Y$ that is compatible with the isomorphism $k_X \xrightarrow{\sim} k_Y$ determined by θ (respectively, an isomorphism $X \times_{k_X} \overline{k}_X \xrightarrow{\sim} Y \times_{k_Y} \overline{k}_Y$ that is compatible with some isomorphism $k_X \xrightarrow{\sim} k_Y$). Thus, by pulling back the various objects involved via this isomorphism, to verify assertion (i) (respectively, (ii)), we may assume without loss of generality that $(X, k_X, \overline{k}_X, \theta) = (Y, k_Y, \overline{k}_Y, \mathrm{id}_{\overline{k}_X})$ (respectively, $(X, k_X, \overline{k}_X) = (Y, k_Y, \overline{k}_Y)$). Then assertion (i) (respectively, (ii)) follows from Corollary 2.5, (v) (respectively, Corollary 2.5, (vi)). This completes the proof of Corollary 2.6. □

Lemma 2.7 (Isomorphisms Between Configuration Spaces of Hyperbolic Curves) *Let n be a positive integer; (g_X, r_X), (g_Y, r_Y) pairs of nonnegative integers such that $2g_X - 2 + r_X$, $2g_Y - 2 + r_Y > 0$; k_X, k_Y fields; $\overline{k}_X$, $\overline{k}_Y$ separable closures of k_X, k_Y, respectively; X, Y* **hyperbolic curves** *of type (g_X, r_X), (g_Y, r_Y) over k_X, k_Y, respectively. Write X_n, Y_n for the n-th* **configuration spaces** *of X, Y, respectively; $X_{\overline{k}_X} \stackrel{\mathrm{def}}{=} X \times_{k_X} \overline{k}_X$; $Y_{\overline{k}_Y} \stackrel{\mathrm{def}}{=} Y \times_{k_Y} \overline{k}_Y$; $(X_n)_{\overline{k}_X} \stackrel{\mathrm{def}}{=} X_n \times_{k_X} \overline{k}_X$; $(Y_n)_{\overline{k}_Y} \stackrel{\mathrm{def}}{=} Y_n \times_{k_Y} \overline{k}_Y$; $\mathfrak{S}_n$ for the symmetric group on n letters; $\mathrm{Aut}_{k_X}(X_n)$ for the group of automorphisms of X_n over k_X;*

$$\Psi_n : \mathfrak{S}_n \longrightarrow \mathrm{Aut}_{k_X}(X_n)$$

for the action of $\mathfrak{S}_n$ *on* X_n *over* k_X *obtained by permuting the factors of* X_n*. Suppose that* $(g_X, r_X), (g_Y, r_Y) \notin \{(0, 3); (1, 1)\}$*. Then the following hold:*

(i) Let $\alpha\colon X_n \xrightarrow{\sim} Y_n$ *be an isomorphism. Then there exists a* **unique** *isomorphism* $\alpha_0\colon k_Y \xrightarrow{\sim} k_X$ *that is compatible with* α *relative to the structure morphisms of* X_n, Y_n*.*

(ii) Let $\alpha\colon X_n \xrightarrow{\sim} Y_n$ *be an isomorphism. Then there exist a* **unique** *permutation* $\sigma \in \Psi_n(\mathfrak{S}_n) \subseteq \mathrm{Aut}_{k_X}(X_n)$ *and a* **unique** *isomorphism* $\alpha_1\colon X \xrightarrow{\sim} Y$ *that is compatible with* $\alpha \circ \sigma$ *relative to the projections* $X_n \to X$*,* $Y_n \to Y$ *to each of the n factors.*

(iii) Write $\mathrm{Isom}(X_n, Y_n)$ *for the set of isomorphisms of* X_n *with* Y_n*;* $\mathrm{Isom}(X, Y) \stackrel{\mathrm{def}}{=} \mathrm{Isom}(X_1, Y_1)$*. Then the natural map*

$$\mathrm{Isom}(X, Y) \times \Psi_n(\mathfrak{S}_n) \longrightarrow \mathrm{Isom}(X_n, Y_n)$$

is **bijective**.

(iv) Write $\mathrm{Isom}((X_n)_{\overline{k}_X}/k_X, (Y_n)_{\overline{k}_Y}/k_Y)$ *for the set of isomorphisms* $(X_n)_{\overline{k}_X} \xrightarrow{\sim} (Y_n)_{\overline{k}_Y}$ *that are compatible with some isomorphism* $k_Y \xrightarrow{\sim} k_X$*;* $\mathrm{Isom}(X_{\overline{k}_X}/k_X, Y_{\overline{k}_Y}/k_Y) \stackrel{\mathrm{def}}{=} \mathrm{Isom}((X_1)_{\overline{k}_X}/k_X, (Y_1)_{\overline{k}_Y}/k_Y)$*. Then the natural map*

$$\mathrm{Isom}(X_{\overline{k}_X}/k_X, Y_{\overline{k}_Y}/k_Y) \times \Psi_n(\mathfrak{S}_n) \longrightarrow \mathrm{Isom}((X_n)_{\overline{k}_X}/k_X, (Y_n)_{\overline{k}_Y}/k_Y)$$

is **bijective**.

Proof First, we verify assertion (i). Write $(C_n^X)^{\log}$, $(C_n^Y)^{\log}$ for the n-th *log configuration spaces* [cf. the discussion entitled "*Curves*" in "Notations and Conventions"] of [the smooth log curves over k_X, k_Y determined by] X, Y, respectively. Then recall [cf. the discussion at the beginning of [MzTa], §2] that $(C_n^X)^{\log}$, $(C_n^Y)^{\log}$ are *log regular* log schemes whose interiors are *naturally isomorphic* to X_n, Y_n, respectively, and that the underlying schemes C_n^X, C_n^Y of $(C_n^X)^{\log}$, $(C_n^Y)^{\log}$ are *proper* over k_X, k_Y, respectively. Thus, by applying [ExtFam], Theorem A, (1), to the composite

$$X_n \stackrel{\alpha}{\to} Y_n \hookrightarrow C_n^Y \hookrightarrow \overline{\mathcal{M}}_{g_Y, r_Y+n}$$

—where we refer to the discussion entitled "*Curves*" in [CbTpI], §0, concerning the notation "$\overline{\mathcal{M}}_{g_Y, r_Y+n}$"; the third arrow is the natural (1-)morphism arising from the definition of C_n^Y—we conclude that the composite

$$X_n \stackrel{\alpha}{\to} Y_n \hookrightarrow C_n^Y \hookrightarrow \overline{\mathcal{M}}_{g_Y, r_Y+n} \to (\overline{\mathcal{M}}_{g_Y, r_Y+n})^c$$

—where we write $(\overline{\mathcal{M}}_{g_Y, r_Y+n})^c$ for the coarse moduli space associated to $\overline{\mathcal{M}}_{g_Y, r_Y+n}$—factors through the natural open immersion $X_n \hookrightarrow C_n^X$. On the other hand, one verifies immediately that the composite $C_n^Y \hookrightarrow \overline{\mathcal{M}}_{g_Y, r_Y+n} \to (\overline{\mathcal{M}}_{g_Y, r_Y+n})^c$ is *proper* and *quasi-finite*, hence *finite*. In particular, if we write $C^\Gamma \subseteq C_n^X \times_k C_n^Y$ for the scheme-theoretic closure of the *graph* of the composite $X_n \xrightarrow{\alpha} Y_n \hookrightarrow C_n^Y$, then the composite $C^\Gamma \hookrightarrow C_n^X \times_k C_n^Y \xrightarrow{\mathrm{pr}_1} C_n^X$ is a *finite* morphism from an *integral* scheme to a *normal* scheme which induces an *isomorphism* between the respective *function fields*. Thus, we conclude that this composite is an *isomorphism*, hence that α extends uniquely to a morphism $C_n^X \to C_n^Y$. Now recall that C_n^X is *proper*, *geometrically normal*, and *geometrically connected* over k_X. Thus, one verifies immediately, by considering global sections of the respective structure sheaves, that there exists a *unique* homomorphism $\alpha_0 \colon k_Y \to k_X$ that is compatible with α. Moreover, by applying a similar argument to α^{-1}, it follows that α_0 is an *isomorphism*. This completes the proof of assertion (i).

Next, we verify assertion (ii). First, let us observe that, by replacing Y by the result of base-changing Y via $\alpha_0 \colon k_Y \xrightarrow{\sim} k_X$ [cf. assertion (i)], we may assume without loss of generality that $k_Y = k_X$, $\overline{k}_Y = \overline{k}_X$, and that α is an *isomorphism over* k_X. Next, let us observe that it is immediate that σ and α_1 as in the statement of assertion (ii) are *unique*; thus, it remains to verify the *existence* of such σ and α_1. Next, let us observe that it follows immediately from [MzTa], Corollary 6.3, that there exists a permutation $\sigma \in \Psi_n(\mathfrak{S}_n)$ such that if we identify the respective sets of fiber subgroups of Δ_{X_n}, Δ_{Y_n}—where we write Δ_{X_n}, Δ_{Y_n} for the maximal pro-l quotients of the étale fundamental groups of $(X_n)_{\overline{k}_X}$, $(Y_n)_{\overline{k}_X}$, respectively, for some prime number l that is *invertible* in k_X—with the set $2^{\{1,\cdots,n\}}$ [cf. the discussion entitled "*Sets*" in [CbTpI], §0] in the evident way, then the automorphism of the set $2^{\{1,\cdots,n\}}$ induced by the composite $\beta \stackrel{\mathrm{def}}{=} \alpha \circ \sigma$ is the *identity automorphism*. Write $\mathrm{pr}_X \colon X_n \to X$, $\mathrm{pr}_Y \colon Y_n \to Y$ for the projections to the factor labeled n, respectively. Then we claim that the following assertion holds:

> Claim 2.7.A: There exists an isomorphism $\alpha_1 \colon X \xrightarrow{\sim} Y$ that is compatible with β relative to pr_X, pr_Y.

Indeed, write $\Gamma \subseteq X \times_{k_X} Y$ for the scheme-theoretic image via $X_n \times_{k_X} Y \xrightarrow{(\mathrm{pr}_X, \mathrm{id}_Y)} X \times_{k_X} Y$ of the *graph* of the composite $X_n \xrightarrow{\beta} Y_n \xrightarrow{\mathrm{pr}_Y} Y$. Next, let us observe that if Z is an irreducible scheme of finite type over $\overline{k}_X$, then any *nonconstant* [i.e., *dominant*] $\overline{k}_X$-morphism $Z \to Y_{\overline{k}_X}$ induces an *open* homomorphism between the respective fundamental groups. Thus, since the automorphism of the set $2^{\{1,\cdots,n\}}$ induced by β is the *identity automorphism*, it follows immediately that, for any $\overline{k}_X$-valued geometric point $\overline{x}$ of X, if we write F for the geometric fiber of $\mathrm{pr}_X \colon X_n \to X$ at $\overline{x}$, then the composite $F \to (X_n)_{\overline{k}_X} \xrightarrow{\beta_{\overline{k}_X}} (Y_n)_{\overline{k}_X} \xrightarrow{(\mathrm{pr}_Y)_{\overline{k}_X}} Y_{\overline{k}_X}$ is *constant*. In particular, one verifies immediately that Γ is an *integral*, *separated* scheme of *dimension* 1. Thus, since pr_X is *surjective*, *geometrically connected*, *smooth*, and *factors* through the composite $\Gamma \hookrightarrow X \times_{k_X} Y \xrightarrow{\mathrm{pr}_1} X$, it follows immediately that this composite morphism $\Gamma \to X$ is *surjective* and induces an *isomorphism* between the respective *function fields*. Therefore, one concludes easily, by applying Zariski's main theorem, that the composite $\Gamma \hookrightarrow X \times_{k_X} Y \stackrel{\mathrm{pr}_1}{\to} X$ is an *isomorphism*, hence that there exists a *unique* morphism $\alpha_1 : X \to Y$ such that $\mathrm{pr}_Y \circ \beta = \alpha_1 \circ \mathrm{pr}_X$. Moreover, by applying a similar argument to β^{-1}, it follows that α_1 is an *isomorphism*. This completes the proof of Claim 2.7.A.

Write γ for the composite of β with the isomorphism $Y_n \stackrel{\sim}{\to} X_n$ determined by α_1^{-1}. Then it is immediate that γ is an *automorphism of* X_n *over* X relative to pr_X; in particular, the outomorphism of Δ_{X_n} induced by γ is contained in the kernel of the homomorphism $\mathrm{Out}^{\mathrm{F}}(\Delta_{X_n}) \to \mathrm{Out}^{\mathrm{F}}(\Delta_X)$—where we write Δ_X for the maximal pro-l quotient of the étale fundamental group of $X_{\overline{k}_X}$—induced by pr_X. Now, by applying a similar argument to the argument of the proof of Claim 2.7.A, one verifies easily that, for each $i \in \{1, \cdots, n\}$, there exists an automorphism $\gamma_{1,i}$ of X that is compatible with γ relative to the projection $X_n \to X$ to the factor labeled i. [Thus, $\gamma_{1,n} = \mathrm{id}_X$.] Moreover, since, by applying induction on n, we may assume that assertion (ii) has already been verified for $n-1$, it follows immediately that the outomorphism of Δ_{X_n} induced by γ is contained in $\mathrm{Out}^{\mathrm{FC}}(\Delta_{X_n})$, hence in the kernel of the homomorphism $\mathrm{Out}^{\mathrm{FC}}(\Delta_{X_n}) \to \mathrm{Out}^{\mathrm{FC}}(\Delta_X)$ induced by the projections $X_n \to X$ to each of the n factors [cf. [CmbCsp], Proposition 1.2, (iii)]. Therefore, it follows immediately from the argument of the first paragraph of the proof of [LocAn], Theorem 14.1, that, for each $i \in \{1, \cdots, n\}$, $\gamma_{1,i}$ is the *identity automorphism* of X, hence also that γ is the *identity automorphism* of X_n. This completes the proof of assertion (ii).

Assertions (iii), (iv) follow immediately from assertion (ii), together with the various definitions involved. This completes the proof of Lemma 2.7. □

Chapter 3
Synchronization of Tripods

In this chapter, we introduce and study the notion of a *tripod* of the log fundamental group of the log configuration space of a stable log curve [cf. Definition 3.3, (i), below]. In particular, we discuss the phenomenon of *synchronization* among the *various tripods* of the log fundamental group [cf. Theorems 3.17; 3.18, below]. One interesting consequence of this phenomenon of tripod synchronization is a certain *non-surjectivity* result [cf. Corollary 3.22 below]. Finally, we apply the theory of synchronization of tripods to show that, under certain conditions, *commuting profinite Dehn multi-twists* are *"co-Dehn"* [cf. Corollary 3.25 below] and to compute the *commensurator of certain purely combinatorial/group-theoretic groups of profinite Dehn multi-twists* in terms of *scheme theory* [cf. Corollary 3.27 below].

In this chapter, let (g, r) be a pair of nonnegative integers such that $2g-2+r > 0$; n a positive integer; Σ a set of prime numbers which is either the set of all prime numbers or of cardinality one; k an algebraically closed field of characteristic $\notin \Sigma$; $(\mathrm{Spec}\, k)^{\log}$ the log scheme obtained by equipping $\mathrm{Spec}\, k$ with the log structure determined by the fs chart $\mathbb{N} \to k$ that maps $1 \mapsto 0$; $X^{\log} = X_1^{\log}$ a *stable log curve* of type (g, r) over $(\mathrm{Spec}\, k)^{\log}$. For each [possibly empty] subset $E \subseteq \{1, \cdots, n\}$, write

$$X_E^{\log}$$

for the $\#E$-th *log configuration space* of the stable log curve $X^{\log}$ [cf. the discussion entitled "*Curves*" in "Notations and Conventions"], where we think of the factors as being labeled by the elements of $E \subseteq \{1, \cdots, n\}$;

$$\Pi_E$$

Y. Hoshi, S. Mochizuki, *Topics Surrounding the Combinatorial Anabelian Geometry of Hyperbolic Curves II*, Lecture Notes in Mathematics 2299,
https://doi.org/10.1007/978-981-19-1096-8_3

for the maximal pro-Σ quotient of the kernel of the natural surjection $\pi_1(X_E^{\log}) \twoheadrightarrow \pi_1((\mathrm{Spec}\, k)^{\log})$. Thus, by applying a suitable *specialization isomorphism*—cf. the discussion preceding [CmbCsp], Definition 2.1, as well as [CbTpI], Remark 5.6.1—one verifies easily that Π_E is equipped with a natural structure of *pro-Σ configuration space group*—cf. [MzTa], Definition 2.3, (i). For each $1 \leq m \leq n$, write

$$X_m^{\log} \stackrel{\text{def}}{=} X_{\{1,\cdots,m\}}^{\log}; \quad \Pi_m \stackrel{\text{def}}{=} \Pi_{\{1,\cdots,m\}}.$$

Thus, for subsets $E' \subseteq E \subseteq \{1, \cdots, n\}$, we have a projection

$$p_{E/E'}^{\log} \colon X_E^{\log} \to X_{E'}^{\log}$$

obtained by forgetting the factors that belong to $E \setminus E'$. For $E' \subseteq E \subseteq \{1, \cdots, n\}$ and $1 \leq m' \leq m \leq n$, we shall write

$$p_{E/E'}^{\Pi} \colon \Pi_E \twoheadrightarrow \Pi_{E'}$$

for some *fixed* surjection [that belongs to the collection of surjections that constitutes the outer surjection] induced by $p_{E/E'}^{\log}$;

$$\Pi_{E/E'} \stackrel{\text{def}}{=} \mathrm{Ker}(p_{E/E'}^{\Pi}) \subseteq \Pi_E;$$

$$p_{m/m'}^{\log} \stackrel{\text{def}}{=} p_{\{1,\cdots,m\}/\{1,\cdots,m'\}}^{\log} \colon X_m^{\log} \longrightarrow X_{m'}^{\log};$$

$$p_{m/m'}^{\Pi} \stackrel{\text{def}}{=} p_{\{1,\cdots,m\}/\{1,\cdots,m'\}}^{\Pi} \colon \Pi_m \twoheadrightarrow \Pi_{m'};$$

$$\Pi_{m/m'} \stackrel{\text{def}}{=} \Pi_{\{1,\cdots,m\}/\{1,\cdots,m'\}} \subseteq \Pi_m.$$

Finally, recall [cf. the statement of Theorem 2.3, (iv)] the natural inclusion

$$\mathfrak{S}_n \hookrightarrow \mathrm{Out}(\Pi_n)$$

—where we write $\mathfrak{S}_n$ for the symmetric group on n letters—obtained by permuting the various factors of X_n. We shall regard $\mathfrak{S}_n$ as a subgroup of $\mathrm{Out}(\Pi_n)$ by means of this natural inclusion.

Definition 3.1 Let $i \in E \subseteq \{1, \cdots, n\}$; $x \in X_n(k)$ a k-valued geometric point of the underlying scheme X_n of $X_n^{\log}$.

(i) Let $E' \subseteq \{1, \cdots, n\}$ be a subset. Then we shall write $x_{E'} \in X_{E'}(k)$ for the k-valued geometric point of $X_{E'}$ obtained by forming the image of $x \in X_n(k)$ via $p_{\{1,\cdots,n\}/E'}\colon X_n \to X_{E'}$; $x_{E'}^{\log} \stackrel{\mathrm{def}}{=} x_{E'} \times_{X_{E'}} X_{E'}^{\log}$.

(ii) We shall write

$$\mathcal{G}$$

for the semi-graph of anabelioids of pro-Σ PSC-type determined by the stable log curve $X^{\log}$ over $(\operatorname{Spec} k)^{\log}$ [cf. [CmbGC], Example 2.5];

$$\mathbb{G}$$

for the underlying semi-graph of $\mathcal{G}$;

$$\Pi_{\mathcal{G}}$$

for the [pro-Σ] fundamental group of $\mathcal{G}$;

$$\widetilde{\mathcal{G}} \longrightarrow \mathcal{G}$$

for the universal covering of $\mathcal{G}$ corresponding to $\Pi_{\mathcal{G}}$. Thus, we have a natural outer isomorphism

$$\Pi_1 \xrightarrow{\sim} \Pi_{\mathcal{G}}.$$

Throughout our discussion of the objects introduced at the beginning of this chapter, let us *fix an isomorphism* $\Pi_1 \xrightarrow{\sim} \Pi_{\mathcal{G}}$ that belongs to the collection of isomorphisms that constitutes the above natural outer isomorphism.

(iii) We shall write

$$\mathcal{G}_{i \in E, x}$$

for the semi-graph of anabelioids of pro-Σ PSC-type determined by the geometric fiber of the projection $p_{E/(E \setminus \{i\})}^{\log}\colon X_E^{\log} \to X_{E \setminus \{i\}}^{\log}$ over $x_{E \setminus \{i\}}^{\log} \to X_{E \setminus \{i\}}^{\log}$ [cf. (i)];

$$\Pi_{\mathcal{G}_{i \in E, x}}$$

for the [pro-Σ] fundamental group of $\mathcal{G}_{i \in E, x}$. Thus, we have a *natural identification*

$$\mathcal{G} = \mathcal{G}_{i \in \{i\}, x}$$

and a *natural* Π_E*-orbit* [i.e., relative to composition with automorphisms induced by conjugation by elements of Π_E] of *isomorphisms*

$$(\Pi_E \supseteq) \ \Pi_{E/(E\setminus\{i\})} \xrightarrow{\sim} \Pi_{\mathcal{G}_{i\in E,x}}.$$

Throughout our discussion of the objects introduced at the beginning of this chapter, let us *fix an outer isomorphism*

$$\Pi_{E/(E\setminus\{i\})} \xrightarrow{\sim} \Pi_{\mathcal{G}_{i\in E,x}}$$

whose constituent isomorphisms belong to the Π_E-orbit of isomorphisms just discussed.

(iv) Let $v \in \mathrm{Vert}(\mathcal{G}_{i\in E,x})$ (respectively, $e \in \mathrm{Cusp}(\mathcal{G}_{i\in E,x})$; $e \in \mathrm{Node}(\mathcal{G}_{i\in E,x})$; $e \in \mathrm{Edge}(\mathcal{G}_{i\in E,x})$; $z \in \mathrm{VCN}(\mathcal{G}_{i\in E,x})$). Then we shall refer to the image [in Π_E] of a verticial (respectively, a cuspidal; a nodal; an edge-like; a VCN-) subgroup [cf. [CbTpI], Definition 2.1, (i)] of $\Pi_{\mathcal{G}_{i\in E,x}}$ associated to v (respectively, e; e; e; z) via the inverse $\Pi_{\mathcal{G}_{i\in E,x}} \xrightarrow{\sim} \Pi_{E/(E\setminus\{i\})} \subseteq \Pi_E$ of any isomorphism that lifts the *fixed* outer isomorphism discussed in (iii) as a *verticial* (respectively, a *cuspidal*; a *nodal*; an *edge-like*; a *VCN-*) *subgroup of* Π_E *associated to* v (respectively, e; e; e; z). Thus, the notion of a verticial (respectively, a cuspidal; a nodal; an edge-like; a VCN-) subgroup of Π_E associated to v (respectively, e; e; e; z) depends on the choice of the *fixed* outer isomorphism of (iii) [but cf. Lemma 3.2, (i), below, in the case of *cusps*!].

(v) We shall say that a vertex $v \in \mathrm{Vert}(\mathcal{G}_{i\in E,x})$ of $\mathcal{G}_{i\in E,x}$ is a(n) [E-]*tripod* of $X_n^{\log}$ if v is of *type* $(0, 3)$ [cf. [CbTpI], Definition 2.3, (iii)]. If, in this situation, $C(v) \neq \emptyset$, then we shall say that the tripod v is *cusp-supporting*.

(vi) We shall say that a cusp $c \in \mathrm{Cusp}(\mathcal{G}_{i\in E,x})$ of $\mathcal{G}_{i\in E,x}$ is *diagonal* if c does not arise from a cusp of the copy of $X^{\log}$ given by the factor of $X_E^{\log}$ labeled $i \in E$.

Lemma 3.2 (Cusps of Various Fibers) *Let $i \in E \subseteq \{1, \cdots, n\}$; $x \in X_n(k)$. Then the following hold:*

(i) Let $c \in \mathrm{Cusp}(\mathcal{G}_{i\in E,x})$ and $\Pi_c \subseteq \Pi_{\mathcal{G}_{i\in E,x}} \xleftarrow{\sim} \Pi_{E/(E\setminus\{i\})}$ a cuspidal subgroup of $\Pi_{\mathcal{G}_{i\in E,x}} \xleftarrow{\sim} \Pi_{E/(E\setminus\{i\})}$ associated to $c \in \mathrm{Cusp}(\mathcal{G}_{i\in E,x})$. Then any $\boldsymbol{\Pi_E}$***-conjugate*** *of Π_c is, in fact, a* $\boldsymbol{\Pi_{E/(E\setminus\{i\})}}$***-conjugate*** *of Π_c.*

*(ii) Each **diagonal cusp** of $\mathcal{G}_{i\in E,x}$ [cf. Definition 3.1, (vi)] admits a natural label $\in E \setminus \{i\}$. More precisely, for each $j \in E \setminus \{i\}$, there exists a **unique diagonal cusp** of $\mathcal{G}_{i\in E,x}$ that arises from the divisor of the fiber product over k of $\#E$ copies of X consisting of the points whose i-th and j-th factors coincide.*

*(iii) Let $\alpha \in \mathrm{Aut}^{\mathrm{F}}(\Pi_n)$ [cf. [CmbCsp], Definition 1.1, (ii)]. Suppose that either $E \neq \{1, \cdots, n\}$ or $n \geq n_{\mathrm{FC}}$ [cf. Theorem 2.3, (ii)]. Then the outomorphism of $\Pi_{\mathcal{G}_{i\in E,x}} \xleftarrow{\sim} \Pi_{E/(E\setminus\{i\})}$ determined by α is **group-theoretically cuspidal** [cf. [CmbGC], Definition 1.4, (iv)].*

(iv) Let $\alpha \in \mathrm{Aut}^{\mathrm{F}}(\Pi_n)$ and $c \in \mathrm{Cusp}(\mathcal{G}_{i\in E,x})$ a ***diagonal cusp*** *of $\mathcal{G}_{i\in E,x}$. Suppose that the outomorphism of $\Pi_{\mathcal{G}_{i\in E,x}} \overset{\sim}{\leftarrow} \Pi_{E/(E\setminus\{i\})}$ determined by α is* ***group-theoretically cuspidal****. Then this outomorphism* ***preserves*** *the $\Pi_{\mathcal{G}_{i\in E,x}}$-conjugacy class of cuspidal subgroups of $\Pi_{\mathcal{G}_{i\in E,x}} \overset{\sim}{\leftarrow} \Pi_{E/(E\setminus\{i\})}$ associated to $c \in \mathrm{Cusp}(\mathcal{G}_{i\in E,x})$.*

Proof Assertion (i) follows immediately from the [easily verified] fact that the restriction of $p^{\Pi}_{E/(E\setminus\{i\})}\colon \Pi_E \twoheadrightarrow \Pi_{E\setminus\{i\}}$ to the normalizer of Π_c in Π_E is *surjective*. Assertion (ii) follows immediately from the various definitions involved. Next, we verify assertion (iii). If $E \neq \{1, \cdots, n\}$ (respectively, $n \geq n_{\mathrm{FC}}$), then assertion (iii) follows immediately from [CbTpI], Theorem A, (ii) (respectively, Theorem 2.3, (ii), of the present monograph), together with assertion (i). This completes the proof of assertion (iii). Finally, assertion (iv) follows immediately from the definition of *F-admissibility* [cf. also assertion (ii)]. This completes the proof of Lemma 3.2. □

Definition 3.3 Let $E \subseteq \{1, \cdots, n\}$.

(i) We shall say that a closed subgroup $H \subseteq \Pi_E$ of Π_E is a(n) *[E-]tripod* of Π_n if H is a verticial subgroup of Π_E [cf. Definition 3.1, (iv)] associated to a(n) [E-]tripod v of $X_n^{\log}$ [cf. Definition 3.1, (v)]. If, in this situation, the tripod v is cusp-supporting [cf. Definition 3.1, (v)], then we shall say that the tripod H is *cusp-supporting*.

(ii) We shall say that an E-tripod of Π_n [cf. (i)] is *trigonal* if, for every $j \in E$, the image of the tripod via $p^{\Pi}_{E/\{j\}}\colon \Pi_E \twoheadrightarrow \Pi_{\{j\}}$ is trivial.

(iii) Let $T \subseteq \Pi_E$ be an E-tripod of Π_n [cf. (i)] and $E' \subseteq E$. Then we shall say that T is *E'-strict* if the image $p^{\Pi}_{E/E'}(T) \subseteq \Pi_{E'}$ of T via $p^{\Pi}_{E/E'}\colon \Pi_E \twoheadrightarrow \Pi_{E'}$ is an E'-tripod of Π_n, and, moreover, for every $E'' \subsetneq E'$, the image of the E'-tripod $p^{\Pi}_{E/E'}(T)$ via $p^{\Pi}_{E'/E''}\colon \Pi_{E'} \twoheadrightarrow \Pi_{E''}$ is *not* a tripod of Π_n.

(iv) Let h be a positive integer. Then we shall say that an E-tripod T of Π_n [cf. (i)] is *h-descendable* if there exists a subset $E' \subseteq E$ such that the image of T via $p^{\Pi}_{E/E'}\colon \Pi_E \twoheadrightarrow \Pi_{E'}$ is an E'-tripod of Π_n, and, moreover, $\#E' \leq n-h$. [Thus, one verifies immediately that an E-tripod $T \subseteq \Pi_E$ of Π_n is 1-descendable if and only if either $E \neq \{1, \cdots, n\}$ or T fails to be E-strict—cf. (iii).]

Remark 3.3.1 In the notation of Definition 3.1, let $v \in \mathrm{Vert}(\mathcal{G}_{i\in E,x})$ be an *E-tripod* of $X_n^{\log}$ [cf. Definition 3.1, (v)] and $T \subseteq \Pi_E$ an *E-tripod* of Π_n associated to v [cf. Definition 3.3, (i)]. Write F_v for the *irreducible component* of the geometric fiber of $p_{E/(E\setminus\{i\})}\colon X_E \to X_{E\setminus\{i\}}$ at $x_{E\setminus\{i\}}$ corresponding to v; $F_v^{\log}$ for the *log scheme* obtained by equipping F_v with the log structure induced by the log structure of $X_E^{\log}$; n_v for the *rank* of the group-characteristic of $F_v^{\log}$ [cf. [MzTa], Definition 5.1, (i)] at the generic point of F_v. Then it is immediate that the n_v-interior $U_v \subseteq F_v$ of $F_v^{\log}$ [cf. [MzTa], Definition 5.1, (i)] is a *nonempty open subset* of F_v which is *isomorphic* to $\mathbb{P}^1_k \setminus \{0, 1, \infty\}$ over k. Moreover, one verifies easily that if we write $U_v^{\log}$ for the log scheme obtained by equipping U_v with the log structure induced by the log structure of $X_E^{\log}$, then the natural morphism $U_v^{\log} \to U_v$ [obtained by forgetting the log structure of $U_v^{\log}$] determines a *natural outer isomorphism* $T \xrightarrow{\sim} \pi_1^{\Sigma}(U_v)$—where we write "$\pi_1^{\Sigma}(-)$" for the maximal pro-$\Sigma$ quotient of the étale fundamental group of "$(-)$". In particular, we obtain a natural outer isomorphism

$$T \xrightarrow{\sim} \pi_1^{\Sigma}(\mathbb{P}_k^1 \setminus \{0, 1, \infty\})$$

that is well-defined up to composition with an outomorphism of $\pi_1^{\Sigma}(\mathbb{P}_k^1 \setminus \{0, 1, \infty\})$ that arises from an automorphism of $\mathbb{P}_k^1 \setminus \{0, 1, \infty\}$ over k.

Definition 3.4 Let $E \subseteq \{1, \cdots, n\}$.

(i) Let $T \subseteq \Pi_E$ be an E-tripod of Π_n [cf. Definition 3.3, (i)]. Then T may be regarded as the "Π_1" that occurs in the case where we take "$X^{\log}$" to be the smooth log curve associated to $\mathbb{P}_k^1 \setminus \{0, 1, \infty\}$ [cf. Remark 3.3.1]. We shall write

$$\mathrm{Out}^{\mathrm{C}}(T) \subseteq \mathrm{Out}(T)$$

for the [*closed*] subgroup of $\mathrm{Out}(T)$ consisting of *C-admissible* outomorphisms of T [cf. [CmbCsp], Definition 1.1, (ii)];

$$\mathrm{Out}^{\mathrm{C}}(T)^{\mathrm{cusp}} \subseteq \mathrm{Out}^{\mathrm{C}}(T)$$

for the [*closed*] subgroup of $\mathrm{Out}(T)$ consisting of *C-admissible* outomorphisms of T that induce the *identity automorphism* of the set of T-conjugacy classes of cuspidal inertia subgroups;

$$\mathrm{Out}(T)^{\Delta} \subseteq \mathrm{Out}(T)$$

for the *centralizer* of the subgroup [$\simeq \mathfrak{S}_3$, where we write $\mathfrak{S}_3$ for the symmetric group on 3 letters] of $\mathrm{Out}(T)$ consisting of the *outer modular symmetries* [cf. [CmbCsp], Definition 1.1, (vi)];

$$\mathrm{Out}(T)^{+} \subseteq \mathrm{Out}(T)$$

for the [*closed*] subgroup of $\mathrm{Out}(T)$ given by the image of the natural homomorphism $\mathrm{Out}^{\mathrm{F}}(T_2) = \mathrm{Out}^{\mathrm{FC}}(T_2) \to \mathrm{Out}(T)$ [cf. Theorem 2.3, (ii); [CmbCsp], Proposition 1.2, (iii)]—where we write T_2 for the "Π_2" that occurs in the case where we take "$X^{\log}$" to be the smooth log curve associated to $\mathbb{P}_k^1 \setminus \{0, 1, \infty\}$;

$$\mathrm{Out}^{\mathrm{C}}(T)^{\Delta} \stackrel{\mathrm{def}}{=} \mathrm{Out}^{\mathrm{C}}(T) \cap \mathrm{Out}(T)^{\Delta};$$

$$\mathrm{Out}^{\mathrm{C}}(T)^{\Delta+} \stackrel{\mathrm{def}}{=} \mathrm{Out}^{\mathrm{C}}(T)^{\Delta} \cap \mathrm{Out}(T)^{+}$$

[cf. [CmbCsp], Definition 1.11, (i)].

(ii) Let $E' \subseteq \{1, \cdots, n\}$; let $T \subseteq \Pi_E$, $T' \subseteq \Pi_{E'}$ be E-, E'-tripods of Π_n [cf. Definition 3.3, (i)], respectively. Then we shall say that an outer isomorphism $\alpha\colon T \xrightarrow{\sim} T'$ is *geometric* if the composite

$$\pi_1^{\Sigma}(\mathbb{P}_k^1 \setminus \{0, 1, \infty\}) \xleftarrow{\sim} T \xrightarrow[\sim]{\alpha} T' \xrightarrow{\sim} \pi_1^{\Sigma}(\mathbb{P}_k^1 \setminus \{0, 1, \infty\})$$

—where the first and third arrows are natural outer isomorphisms of the sort discussed in Remark 3.3.1—arises from an automorphism of $\mathbb{P}_k^1 \setminus \{0, 1, \infty\}$ over k.

Remark 3.4.1 In the notation of Definition 3.4, (ii), one verifies easily that every *geometric* outer isomorphism $\alpha\colon T \xrightarrow{\sim} T'$ *preserves* cuspidal inertia subgroups and outer modular symmetries [cf. [CmbCsp], Definition 1.1, (vi)], and, moreover, *lifts* to an outer isomorphism $T_2 \xrightarrow{\sim} T_2'$ [i.e., of the corresponding "Π_2's"] that arises from an isomorphism of two-dimensional configuration spaces. In particular, the isomorphism $\mathrm{Out}(T) \xrightarrow{\sim} \mathrm{Out}(T')$ induced by α determines *isomorphisms*

$$\mathrm{Out}^{\mathrm{C}}(T) \xrightarrow{\sim} \mathrm{Out}^{\mathrm{C}}(T'), \quad \mathrm{Out}^{\mathrm{C}}(T)^{\mathrm{cusp}} \xrightarrow{\sim} \mathrm{Out}^{\mathrm{C}}(T')^{\mathrm{cusp}},$$

$$\mathrm{Out}(T)^{\Delta} \xrightarrow{\sim} \mathrm{Out}(T')^{\Delta}, \quad \mathrm{Out}(T)^{+} \xrightarrow{\sim} \mathrm{Out}(T')^{+}$$

[cf. Definition 3.4, (i)].

Lemma 3.5 (Triviality of the Action on the Set of Cusps) *In the notation of Definition 3.4, it holds that* $\mathrm{Out}^{\mathrm{C}}(T)^{\Delta} \subseteq \mathrm{Out}^{\mathrm{C}}(T)^{\mathrm{cusp}}$.

Proof This follows immediately from the [easily verified] fact that $\mathfrak{S}_3$ is *center-free*, together with the various definitions involved. □

Lemma 3.6 (Vertices, Cusps, and Nodes of Various Fibers) *Let $i, j \in E$ be two **distinct** elements of a subset $E \subseteq \{1, \cdots, n\}$; $x \in X_n(k)$. Write $z_{i,j,x} \in$ $\mathrm{VCN}(\mathcal{G}_{j\in E\setminus\{i\},x})$ for the element of $\mathrm{VCN}(\mathcal{G}_{j\in E\setminus\{i\},x})$ on which $x_{E\setminus\{i\}}$ lies, that is to say: If $x_{E\setminus\{i\}}$ is a cusp or node of the geometric fiber of the projection $p^{\log}_{(E\setminus\{i\})/(E\setminus\{i,j\})}\colon X^{\log}_{E\setminus\{i\}} \to X^{\log}_{E\setminus\{i,j\}}$ over $x^{\log}_{E\setminus\{i,j\}}$ corresponding to an edge $e \in$ $\mathrm{Edge}(\mathcal{G}_{j\in E\setminus\{i\},x})$, then $z_{i,j,x} \overset{\mathrm{def}}{=} e$; if $x_{E\setminus\{i\}}$ is neither a cusp nor a node of the geometric fiber of the projection $p^{\log}_{(E\setminus\{i\})/(E\setminus\{i,j\})}\colon X^{\log}_{E\setminus\{i\}} \to X^{\log}_{E\setminus\{i,j\}}$ over $x^{\log}_{E\setminus\{i,j\}}$, but lies on the irreducible component of the geometric fiber corresponding to a vertex $v \in \mathrm{Vert}(\mathcal{G}_{j\in E\setminus\{i\},x})$, then $z_{i,j,x} \overset{\mathrm{def}}{=} v$. Then the following hold:*

*(i) The automorphism of $X_E^{\log}$ determined by permuting the factors labeled i, j induces **natural bijections***

$$\mathrm{Vert}(\mathcal{G}_{j\in E\setminus\{i\},x}) \xrightarrow{\sim} \mathrm{Vert}(\mathcal{G}_{i\in E\setminus\{j\},x});$$

$$\mathrm{Cusp}(\mathcal{G}_{j\in E\setminus\{i\},x}) \xrightarrow{\sim} \mathrm{Cusp}(\mathcal{G}_{i\in E\setminus\{j\},x});$$

$$\mathrm{Node}(\mathcal{G}_{j\in E\setminus\{i\},x}) \xrightarrow{\sim} \mathrm{Node}(\mathcal{G}_{i\in E\setminus\{j\},x}).$$

(ii) Let us write

$$c^{\mathrm{diag}}_{i,j,x} \in \mathrm{Cusp}(\mathcal{G}_{i\in E,x})$$

for the ***diagonal cusp*** *of $\mathcal{G}_{i\in E,x}$ [cf. Definition 3.1, (vi)] labeled $j \in E \setminus \{i\}$ [cf. Lemma 3.2, (ii)]. Then $p^{\log}_{E/(E\setminus\{j\})}: X^{\log}_E \to X^{\log}_{E\setminus\{j\}}$ induces a* ***bijection***

$$\mathrm{Cusp}(\mathcal{G}_{i\in E,x}) \setminus \{c^{\mathrm{diag}}_{i,j,x}\} \xrightarrow{\sim} \mathrm{Cusp}(\mathcal{G}_{i\in E\setminus\{j\},x}).$$

(iii) Suppose that $z_{i,j,x} \in \mathrm{Vert}(\mathcal{G}_{j\in E\setminus\{i\},x})$. Then $p^{\log}_{E/(E\setminus\{j\})}: X^{\log}_E \to X^{\log}_{E\setminus\{j\}}$ induces a ***bijection***

$$\mathrm{Vert}(\mathcal{G}_{i\in E,x}) \xrightarrow{\sim} \mathrm{Vert}(\mathcal{G}_{i\in E\setminus\{j\},x}).$$

(iv) Suppose that $z_{i,j,x} \in \mathrm{Edge}(\mathcal{G}_{j\in E\setminus\{i\},x})$. Then there exists a ***unique vertex***

$$v^{\mathrm{new}}_{i,j,x} \in \mathrm{Vert}(\mathcal{G}_{i\in E,x})$$

such that $p^{\log}_{E/(E\setminus\{j\})}: X^{\log}_E \to X^{\log}_{E\setminus\{j\}}$ induces a ***bijection***

$$\mathrm{Vert}(\mathcal{G}_{i\in E,x}) \setminus \{v^{\mathrm{new}}_{i,j,x}\} \xrightarrow{\sim} \mathrm{Vert}(\mathcal{G}_{i\in E\setminus\{j\},x}).$$

Moreover, $v^{\mathrm{new}}_{i,j,x}$ is of ***type*** $\mathbf{(0,3)}$ *[i.e., $v^{\mathrm{new}}_{i,j,x}$ is an* ***E-tripod*** *of $X^{\log}_n$—cf. Definition 3.1, (v)], and $c^{\mathrm{diag}}_{i,j,x} \in C(v^{\mathrm{new}}_{i,j,x})$ [cf. (ii)]. Finally, any verticial subgroup of Π_E associated to $v^{\mathrm{new}}_{i,j,x}$ surjects, via $p^{\Pi}_{E/(E\setminus\{j\})}$, onto an edge-like subgroup of $\Pi_{E\setminus\{j\}}$ associated to the edge $\in \mathrm{Edge}(\mathcal{G}_{i\in E\setminus\{j\},x})$ determined by $z_{i,j,x} \in \mathrm{Edge}(\mathcal{G}_{j\in E\setminus\{i\},x})$ via the bijections of (i).*

(v) Suppose that $\#E = 3$. Write $h \in E \setminus \{i, j\}$ for the ***unique*** *element of $E \setminus \{i, j\}$. Suppose, moreover, that $z_{i,j,x} = c^{\mathrm{diag}}_{j,h,x} \in \mathrm{Cusp}(\mathcal{G}_{j\in E\setminus\{i\},x})$ [cf. (ii)]. Then the Π_E-conjugacy class of a verticial subgroup of Π_E associated to the vertex $v^{\mathrm{new}}_{i,j,x} \in \mathrm{Vert}(\mathcal{G}_{i\in E,x})$ [cf. (iv)]* ***depends only*** *on i and* ***not*** *on the choice of the pair (j, x). Moreover, these* ***three*** *Π_E-conjugacy classes [cf. the dependence on the choice of $i \in E$] may also be characterized* ***uniquely*** *as the Π_E-conjugacy classes of subgroups of Π_E associated to some* ***trigonal E-tripod*** *of Π_n [cf. Definition 3.3, (ii)].*

Proof First, we verify assertions (i), (ii), (iii), and (iv). To verify assertions (i), (ii), (iii), and (iv)—by replacing $X_E^{\log}$ by the base-change of $p_{E\setminus\{i,j\}}^{\log} : X_E^{\log} \to X_{E\setminus\{i,j\}}^{\log}$ via a suitable morphism of log schemes $(\mathrm{Spec}\, k)^{\log} \to X_{E\setminus\{i,j\}}^{\log}$ whose image lies on $x_{E\setminus\{i,j\}} \in X_{E\setminus\{i,j\}}(k)$ [cf. Definition 3.1, (i)]—we may assume without loss of generality that $\#E = 2$. Then one verifies easily from the various definitions involved that assertions (i), (ii), (iii), and (iv) hold. This completes the proof of assertions (i), (ii), (iii), and (iv). Finally, we consider assertion (v). First, we observe the easily verified fact [cf. assertions (iii), (iv)] that the irreducible component corresponding to an *E-tripod* of $X_n^{\log}$ [cf. Definition 3.1, (v)] that gives rise to a *trigonal E-tripod* of Π_n necessarily *collapses to a point* upon projection to $X_{E'}$ for any $E' \subseteq E$ of cardinality ≤ 2. In light of this observation, it follows immediately [cf. assertions (i), (ii), (iii), (iv)] that any E-tripod of $X_n^{\log}$ that gives rise to a trigonal E-tripod of Π_n arises as a vertex "$v_{i,j,x}^{\mathrm{new}}$" as described in the statement of assertion (v). Now the remainder of assertion (v) follows immediately from the various definitions involved [cf. also the situation discussed in [CmbCsp], Definition 1.8, Proposition 1.9, Corollary 1.10, as well as the discussion, concerning *specialization isomorphisms*, preceding [CmbCsp], Definition 2.1; [CbTpI], Remark 5.6.1]. This completes the proof of Lemma 3.6. □

Definition 3.7 Let $E \subseteq \{1, \cdots, n\}$.

(i) Let v be an E-tripod of $X_n^{\log}$ [cf. Definition 3.1, (v)]; thus, v belongs to $\mathrm{Vert}(\mathcal{G}_{i\in E,x})$ for some choice of $i \in E$ and $x \in X_n(k)$. Let $j \in E \setminus \{i\}$ and $e \in \mathrm{Edge}(\mathcal{G}_{j\in E\setminus\{i\},x})$. Then we shall say that v, or equivalently, an E-tripod of Π_n associated to v [cf. Definition 3.3, (i)], *arises from e* if $e = z_{i,j,x}$ [cf. the statement of Lemma 3.6], and $v = v_{i,j,x}^{\mathrm{new}}$ [cf. Lemma 3.6, (iv)].

(ii) Let $i \in E$. Then we shall say that an E-tripod of Π_n is *i-central* if $\#E = 3$, and, moreover, the tripod is a verticial subgroup of the sort discussed in Lemma 3.6, (v), i.e., the unique, up to Π_E-conjugacy, trigonal E-tripod of Π_n contained in $\Pi_{E/(E\setminus\{i\})}$ [cf. the final portion of Lemma 3.6, (iv)]. We shall say that an E-tripod of Π_n is *central* if it is j-central for some $j \in E$.

Remark 3.7.1 Let $E \subseteq \{1, \cdots, n\}$; $T \subseteq \Pi_E$ an *E-tripod* of Π_n [cf. Definition 3.3, (i)]; $\sigma \in \mathfrak{S}_n \subseteq \mathrm{Out}(\Pi_n)$ [cf. the discussion at the beginning of this chapter]; $\widetilde{\sigma} \in \mathrm{Aut}(\Pi_n)$ a lifting of $\sigma \in \mathfrak{S}_n \subseteq \mathrm{Out}(\Pi_n)$. Write

$$T^{\widetilde{\sigma}} \subseteq \Pi_{\sigma(E)}$$

for the image of $T \subseteq \Pi_E$ by the isomorphism $\Pi_E \xrightarrow{\sim} \Pi_{\sigma(E)}$ determined by $\widetilde{\sigma} \in \mathrm{Aut}(\Pi_n)$.

(i) One verifies easily that $T^{\widetilde{\sigma}} \subseteq \Pi_{\sigma(E)}$ is a *$\sigma(E)$-tripod* of Π_n.

(ii) If, moreover, the equality $\#E = 3$ holds, and T is *i-central* [cf. Definition 3.7, (ii)] for some $i \in E$, then one verifies easily from Lemma 3.6, (v), that $T^{\widetilde{\sigma}} \subseteq \Pi_{\sigma(E)}$ is *$\sigma(i)$-central.*

(iii) In the situation of (ii), let $T' \subseteq \Pi_E$ be a *central E-tripod* of Π_n. Then it follows from Lemma 3.6, (v), that there exist an element $\tau \in \mathfrak{S}_n \subseteq \mathrm{Out}(\Pi_n)$ and a lifting $\tilde{\tau} \in \mathrm{Aut}(\Pi_n)$ of τ such that τ *preserves* the subset $E \subseteq \{1, \cdots, n\}$, and, moreover, the image of $T \subseteq \Pi_E$ by the automorphism of Π_E determined by $\tilde{\tau} \in \mathrm{Aut}(\Pi_n)$ *coincides* with $T' \subseteq \Pi_E$.

Lemma 3.8 (Strict Tripods) *Let $E \subseteq \{1, \cdots, n\}$ and $T \subseteq \Pi_E$ an* ***E-tripod*** *of Π_n [cf. Definition 3.3, (i)] that arises as a verticial subgroup associated to a vertex $v \in \mathrm{Vert}(\mathcal{G}_{i\in E,x})$ for some $i \in \{1, \cdots, n\}$ [which thus implies that $T \subseteq \Pi_{E/(E\setminus\{i\})} \subseteq \Pi_E$]. Then the following hold:*

(i) There exists a [not necessarily unique!] subset $E' \subseteq E$ such that T is ***E′-strict*** *[cf. Definition 3.3, (iii)]. In this situation, $i \in E'$, and, moreover, $p^{\Pi}_{E/E'}: \Pi_E \twoheadrightarrow \Pi_{E'}$ induces an* ***isomorphism*** *$T \xrightarrow{\sim} T_{E'}$ onto an E'-tripod $T_{E'}$ of Π_n.*

(ii) T is ***E-strict*** *if and only if one of the following conditions is satisfied:*

(1) $\#E = 1$.

(2_{C}) $\#E = 2$; $T \subseteq \Pi_E$ is a verticial subgroup of Π_E associated to the vertex $v^{\mathrm{new}}_{i,j,x} \in \mathrm{Vert}(\mathcal{G}_{i\in E,x})$ of Lemma 3.6, (iv), for some choice of (i, j, x) such that $z_{i,j,x} \in \mathrm{Cusp}(\mathcal{G}_{j\in E\setminus\{i\},x})$. [In particular, T ***arises*** *from $z_{i,j,x} \in \mathrm{Cusp}(\mathcal{G}_{j\in E\setminus\{i\},x})$—cf. Definition 3.7, (i).]*

(2_{N}) $\#E = 2$; $T \subseteq \Pi_E$ is a verticial subgroup of Π_E associated to the vertex $v^{\mathrm{new}}_{i,j,x} \in \mathrm{Vert}(\mathcal{G}_{i\in E,x})$ of Lemma 3.6, (iv), for some choice of (i, j, x) such that $z_{i,j,x} \in \mathrm{Node}(\mathcal{G}_{j\in E\setminus\{i\},x})$. [In particular, T ***arises*** *from $z_{i,j,x} \in \mathrm{Node}(\mathcal{G}_{j\in E\setminus\{i\},x})$—cf. Definition 3.7, (i).]*

(3) $\#E = 3$, and T is ***central*** *[cf. Definition 3.7, (ii)].*

(iii) Suppose that T is ***trigonal*** *[cf. Definition 3.3, (ii)]. Then there exists a [not necessarily unique!] subset $E' \subseteq E$ such that $\#E' = 3$, and, moreover, the image of $T \subseteq \Pi_E$ via $p^{\Pi}_{E/E'}: \Pi_E \twoheadrightarrow \Pi_{E'}$ is a* ***central tripod****.*

Proof Assertion (i) follows immediately from the various definitions involved by *induction on $\#E$*, together with the well-known elementary fact that any surjective endomorphism of a topologically finitely generated profinite group is necessarily *bijective*. Next, we verify assertion (ii). First, let us observe that *sufficiency* is immediate. Thus, it remains to verify *necessity*. Suppose that T is *E-strict*. Now one verifies easily that if there exists an element $j \in E \setminus \{i\}$ such that $c^{\mathrm{diag}}_{i,j,x} \notin C(v)$ [cf. Lemma 3.6, (ii)], then it follows immediately that the image of $T \subseteq \Pi_E$ via $p^{\Pi}_{E/(E\setminus\{j\})}: \Pi_E \twoheadrightarrow \Pi_{E\setminus\{j\}}$ is an $(E \setminus \{j\})$-tripod [cf. also Lemma 3.6, (iii), (iv)]. Thus, since T is *E-strict*, we conclude that every cusp of $\mathcal{G}_{i\in E,x}$ that is $\notin C(v)$ is *non-diagonal*. In particular, since v is of *type* $(0, 3)$, it follows immediately from Lemma 3.2, (ii), that $0 \le \#E - 1 \le \#C(v) \le 3$. If $\#C(v) = 0$, then it follows from the inequality $\#E - 1 \le \#C(v)$ that $\#E = 1$, i.e., condition (1) is satisfied. If $\#C(v) = 3$, then one verifies easily that $\#E = 1$, i.e., condition (1) is satisfied. Thus, it remains to verify assertion (ii) in the case where $\#C(v) \in \{1, 2\}$.

Suppose that $\#C(v) = 1$ and $\#E \neq 1$. Then it follows immediately from the inequality $\#E - 1 \le \#C(v)$ that $\#E = 2$. Now let us recall [cf. Lemma 3.2, (ii)] that the number of *diagonal* cusps of $\mathcal{G}_{i\in E,x}$ is $= \#E - 1 = 1$. Moreover, the unique cusp on v is the unique *diagonal* cusp of $\mathcal{G}_{i\in E,x}$ [cf. the argument of the preceding paragraph]. Thus, one verifies easily that T satisfies condition (2_{N}). Next, suppose that $\#C(v) = 2$ and $\#E \neq 1$. Then it follows immediately from the inequality $\#E - 1 \leq \#C(v)$ that $\#E \in \{2, 3\}$. Now let us recall [cf. Lemma 3.2, (ii)] that if $\#E = 2$ (respectively, $\#E = 3$), then the number of *diagonal* cusps of $\mathcal{G}_{i\in E,x}$ is $= \#E - 1$, i.e., 1 (respectively, 2). Moreover, the set of *diagonal* cusp(s) of $\mathcal{G}_{i\in E,x}$ is contained in (respectively, is equal to) $C(v)$ [cf. the argument of the preceding paragraph]. Thus, one verifies easily that T satisfies condition (2_{C}) (respectively, (3)). This completes the proof of assertion (ii).

Finally, we verify assertion (iii). It follows from assertion (i) that there exists a subset $E' \subseteq E$ such that T is *E'-strict*. Moreover, it follows immediately from the definition of a trigonal tripod that the E'-tripod given by the image $p^{\Pi}_{E/E'}(T) \subseteq \Pi_{E'}$ is *trigonal*. On the other hand, if the E'-tripod $p^{\Pi}_{E/E'}(T)$ satisfies any of conditions (1), (2_{C}), (2_{N}) of assertion (ii), then one verifies easily that $p^{\Pi}_{E/E'}(T)$ is *not trigonal* [cf. the final portion of Lemma 3.6, (iv)]. Thus, $p^{\Pi}_{E/E'}(T)$ satisfies condition (3) of assertion (ii); in particular, $p^{\Pi}_{E/E'}(T)$ is *central*. This completes the proof of assertion (iii). □

Lemma 3.9 (Generalities on Normalizers and Commensurators) *Let G be a profinite group, $N \subseteq G$ a* **normal** *closed subgroup of G, and $H \subseteq G$ a closed subgroup of G. Then the following hold:*

(i) It holds that $C_G(H) \subseteq C_G(H \cap N)$.
(ii) It holds that $C_G(H) \subseteq N_G(Z^{\mathrm{loc}}_G(H))$ [cf. the discussion entitled "Topological groups" in "Notations and Conventions"].
(iii) Suppose that $H \subseteq N$. Then it holds that $C_G(H) \subseteq N_G(C_N(H))$. In particular, if, moreover, H is **commensurably terminal** *in N, then it holds that $C_G(H) = N_G(H)$.*
(iv) Write $\overline{H} \overset{\mathrm{def}}{=} H/(H\cap N) \subseteq \overline{G} \overset{\mathrm{def}}{=} G/N$. If $H\cap N$ is **commensurably terminal** *in N, and the image of $C_G(H) \subseteq G$ in $\overline{G}$ is* **contained** *in $N_{\overline{G}}(\overline{H})$, then $C_G(H) = N_G(H)$.*

Proof Assertion (i) follows immediately from the various definitions involved. Next, we verify assertion (ii). Let $g \in C_G(H)$ and $a \in Z^{\mathrm{loc}}_G(H)$. Since $Z^{\mathrm{loc}}_G(H) = Z^{\mathrm{loc}}_G(H\cap(g^{-1}\cdot H\cdot g)) = Z^{\mathrm{loc}}_G(g^{-1}\cdot H\cdot g)$, there exists an open subgroup $U \subseteq H$ of H such that $a \in Z_G(g^{-1}\cdot U\cdot g)$. But this implies that $gag^{-1} \in Z_G(U) \subseteq Z^{\mathrm{loc}}_G(H)$. This completes the proof of assertion (ii). Next, we verify assertion (iii). Let $g \in C_G(H)$ and $a \in C_N(H)$. Since $C_N(H) \subseteq C_G(H) = C_G(H\cap(g^{-1}\cdot H\cdot g)) = C_G(g^{-1}\cdot H\cdot g)$, we conclude that $ag^{-1}\cdot H\cdot ga^{-1}$ is *commensurate* with $g^{-1}\cdot H\cdot g$. In particular, $gag^{-1}\cdot H\cdot ga^{-1}g^{-1}$ is *commensurate* with H, i.e., $gag^{-1} \in C_G(H)\cap N = C_N(H)$. This completes the proof of assertion (iii). Finally, we verify assertion (iv). First, we observe that since $H \cap N$ is *commensurably terminal* in N, one verifies easily that $H = N_{H\cdot N}(H \cap N)$. Let $g \in C_G(H)$. Then since the image of $C_G(H) \subseteq G$ in $\overline{G}$ is *contained* in $N_{\overline{G}}(\overline{H})$, it is immediate that $g\cdot H\cdot g^{-1} \subseteq H\cdot N$. On the other hand, again by applying the fact that $H \cap N$ is *commensurably terminal* in N, we conclude immediately from assertions (i), (iii) that $C_G(H) \subseteq C_G(H \cap N) = N_G(H \cap N)$. Thus, we obtain that $(g \cdot H \cdot g^{-1}) \cap N = H \cap N$; in particular, $g \cdot H \cdot g^{-1} \subseteq N_{H \cdot N}((g \cdot H \cdot g^{-1}) \cap N) = N_{H \cdot N}(H \cap N) = H$, i.e., $g \in N_G(H)$. This completes the proof of assertion (iv). □

Lemma 3.10 (Restrictions of Outomorphisms) *Let G be a profinite group and $H \subseteq G$ a closed subgroup of G. Write $\mathrm{Out}^H(G) \subseteq \mathrm{Out}(G)$ for the group of outomorphisms of G that* ***preserve*** *the G-conjugacy class of H. Suppose that the homomorphism $N_G(H) \to \mathrm{Aut}(H)$ determined by conjugation* ***factors*** *through $\mathrm{Inn}(H) \subseteq \mathrm{Aut}(H)$. Then the following hold:*

(i) For $\alpha \in \mathrm{Out}^H(G)$, let us write $\alpha|_H$ for the outomorphism of H determined by the restriction to $H \subseteq G$ of a lifting $\widetilde{\alpha} \in \mathrm{Aut}(G)$ of α such that $\widetilde{\alpha}(H) = H$. Then $\alpha|_H$ does ***not depend*** *on the choice of the lifting "$\widetilde{\alpha}$", and the map*

$$\mathrm{Out}^H(G) \longrightarrow \mathrm{Out}(H)$$

given by assigning $\alpha \mapsto \alpha|_H$ is a ***group homomorphism****.*

(ii) The homomorphism

$$\mathrm{Out}^H(G) \longrightarrow \mathrm{Out}(H)$$

of (i) ***depends only*** *on the G-conjugacy class of the closed subgroup $H \subseteq G$, i.e., if we write $H^\gamma \stackrel{\mathrm{def}}{=} \gamma \cdot H \cdot \gamma^{-1}$ for $\gamma \in G$, then the diagram*

$$\begin{array}{ccc} \mathrm{Out}^H(G) & \longrightarrow & \mathrm{Out}(H) \\ \| & & \downarrow \\ \mathrm{Out}^{H^\gamma}(G) & \longrightarrow & \mathrm{Out}(H^\gamma) \end{array}$$

—where the upper (respectively, lower) horizontal arrow is the homomorphism given by mapping $\alpha \mapsto \alpha|_H$ (respectively, $\alpha \mapsto \alpha|_{H^\gamma}$), and the right-hand vertical arrow is the isomorphism obtained by conjugation via the isomorphism $H \stackrel{\sim}{\to} H^\gamma$ determined by conjugation by $\gamma \in G$— ***commutes****.*

Proof Assertion (i) follows immediately from our assumption that the homomorphism $N_G(H) \to \mathrm{Aut}(H)$ determined by conjugation *factors* through $\mathrm{Inn}(H) \subseteq \mathrm{Aut}(H)$, together with the various definitions involved. Assertion (ii) follows immediately from the various definitions involved. This completes the proof of Lemma 3.10. □

Lemma 3.11 (Commensurator of a Tripod Arising from an Edge) *In the notation of Lemma 3.6, suppose that $(j, i) = (1, 2)$; $E = \{i, j\}$; $z_{i,j,x} \in \mathrm{Edge}(\mathcal{G}_{j \in E \setminus \{i\},x})$. [Thus, $\mathcal{G}_{j \in E \setminus \{i\},x} = \mathcal{G}_{i \in E \setminus \{j\},x} = \mathcal{G}$; $\Pi_2 = \Pi_E$; $\Pi_1 = \Pi_{\{j\}} \stackrel{\sim}{\to}$*

$\Pi_{\mathcal{G}_{j\in E\setminus\{i\},x}} = \Pi_{\mathcal{G}}$; $\Pi_{2/1} = \Pi_{E/(E\setminus\{i\})} \overset{\sim}{\to} \Pi_{\mathcal{G}_{i\in E,x}}$.*] Write* $\mathcal{G}_{2/1} \overset{\text{def}}{=} \mathcal{G}_{i\in E,x}$; $\mathcal{G}_{1\setminus 2} \overset{\text{def}}{=} \mathcal{G}_{j\in E,x}$; $p^{\Pi}_{1\setminus 2} \overset{\text{def}}{=} p^{\Pi}_{E/\{2\}}: \Pi_2 \twoheadrightarrow \Pi_{\{2\}}$; $\Pi_{1\setminus 2} \overset{\text{def}}{=} \mathrm{Ker}(p^{\Pi}_{1\setminus 2}) = \Pi_{E/\{2\}} \overset{\sim}{\to} \Pi_{\mathcal{G}_{1\setminus 2}}$; $z_x \overset{\text{def}}{=} z_{i,j,x} \in \mathrm{Edge}(\mathcal{G})$; $c^{\mathrm{diag}} \overset{\text{def}}{=} c^{\mathrm{diag}}_{i,j,x} \in \mathrm{Cusp}(\mathcal{G}_{2/1})$ *[cf. Lemma 3.6, (ii)];* $v^{\mathrm{new}} \overset{\text{def}}{=} v^{\mathrm{new}}_{2/1} \overset{\text{def}}{=} v^{\mathrm{new}}_{i,j,x} \in \mathrm{Vert}(\mathcal{G}_{2/1})$ *[cf. Lemma 3.6, (iv)];* $v^{\mathrm{new}}_{1\setminus 2} \in \mathrm{Vert}(\mathcal{G}_{1\setminus 2})$ *for the vertex that corresponds to* $v^{\mathrm{new}} \in \mathrm{Vert}(\mathcal{G}_{2/1})$ *via the natural bijection* $\mathrm{Vert}(\mathcal{G}_{2/1}) \overset{\sim}{\to} \mathrm{Vert}(\mathcal{G}_{1\setminus 2})$ *induced by the automorphism of* $X^{\log}_E$ *determined by permuting the factors labeled* i, j*;* $Y \to X_E$ *for the base-change—by the morphism* $X_E \to X_{\{1\}} \times_k X_{\{2\}} = X \times_k X$ *determined by* $p^{\log}_{E/\{1\}}$ *and* $p^{\log}_{E/\{2\}}$*—of the geometric point of* $X_{\{1\}} \times_k X_{\{2\}} = X \times_k X$ *determined by the geometric points* $x_{\{1\}}$ *of* $X_{\{1\}} = X$ *and* $x_{\{1\}}$ *of* $X_{\{2\}} = X$ *of Definition 3.1, (i) [i.e., as opposed to the geometric point of* $X_{\{1\}} \times_k X_{\{2\}} = X \times_k X$ *determined by the geometric points* $x_{\{1\}}$ *of* $X_{\{1\}} = X$ *and* $x_{\{2\}}$ *of* $X_{\{2\}} = X$*];* $Y^{\log}$ *for the log scheme obtained by equipping* Y *with the log structure induced by the log structure of* $X^{\log}_E$*;* $U \subseteq Y$ *for the 2-interior of* $Y^{\log}$ *[cf. [MzTa], Definition 5.1, (i)];* $U^{\log}$ *for the log scheme obtained by equipping* U *with the log structure induced by the log structure of* $X^{\log}_E$*;* Π_U *for the maximal pro-*Σ *quotient of the kernel of the natural surjection* $\pi_1(U^{\log}) \twoheadrightarrow \pi_1((\mathrm{Spec}\, k)^{\log})$*. [Thus, one verifies easily that* Y *is* ***isomorphic*** *to* $\mathbb{P}^1_k$*; that the complement* $Y \setminus U$ *consists of* ***three closed points*** *of* Y*; that the vertices* $v^{\mathrm{new}}_{2/1}$ *and* $v^{\mathrm{new}}_{1\setminus 2}$ *correspond to the closed irreducible subscheme* $Y \subseteq X_E$*; and that the point corresponding to the cusp* c^{diag} *is* ***contained*** *in* Y*—cf. Lemma 3.6, (iv).] Let* $\Pi_{z_x} \subseteq \Pi_1$ *be an edge-like subgroup associated to* $z_x \in \mathrm{Edge}(\mathcal{G})$*;* $\Pi_{c^{\mathrm{diag}}} \subseteq \Pi_{2/1} \cap \Pi_{1\setminus 2}$ *a cuspidal subgroup associated to* c^{diag}*;* $\Pi_{v^{\mathrm{new}}} \subseteq \Pi_{2/1}$ *a verticial subgroup associated to* v^{new} *that* ***contains*** $\Pi_{c^{\mathrm{diag}}} \subseteq \Pi_2$*;* $\Pi_{v^{\mathrm{new}}_{2/1}} \overset{\text{def}}{=} \Pi_{v^{\mathrm{new}}}$*;* $\Pi_{v^{\mathrm{new}}_{1\setminus 2}} \subseteq \Pi_{1\setminus 2}$ *a verticial subgroup associated to* $v^{\mathrm{new}}_{1\setminus 2}$ *that* ***contains*** $\Pi_{c^{\mathrm{diag}}} \subseteq \Pi_2$*. Write* $\Pi_2|_{z_x} \overset{\text{def}}{=} \Pi_2 \times_{\Pi_1} \Pi_{z_x} \subseteq \Pi_2$*;* $D_{c^{\mathrm{diag}}} \overset{\text{def}}{=} N_{\Pi_2}(\Pi_{c^{\mathrm{diag}}})$*;* $I_{v^{\mathrm{new}}}|_{z_x} \overset{\text{def}}{=} Z_{\Pi_2|_{z_x}}(\Pi_{v^{\mathrm{new}}}) \subseteq D_{v^{\mathrm{new}}}|_{z_x} \overset{\text{def}}{=} N_{\Pi_2|_{z_x}}(\Pi_{v^{\mathrm{new}}})$*. Then the following hold:*

(i) It holds that $D_{c^{\mathrm{diag}}} \cap \Pi_{2/1} = D_{c^{\mathrm{diag}}} \cap \Pi_{1\setminus 2} = C_{\Pi_2}(\Pi_{c^{\mathrm{diag}}}) \cap \Pi_{2/1} = C_{\Pi_2}(\Pi_{c^{\mathrm{diag}}}) \cap \Pi_{1\setminus 2} = \Pi_{c^{\mathrm{diag}}}$*.*

(ii) It holds that $C_{\Pi_2}(\Pi_{c^{\mathrm{diag}}}) = D_{c^{\mathrm{diag}}}$*.*

(iii) The surjections $p^{\Pi}_{2/1}: \Pi_2 \twoheadrightarrow \Pi_1$*,* $p^{\Pi}_{1\setminus 2}: \Pi_2 \twoheadrightarrow \Pi_{\{2\}}$ *determine* ***isomorphisms*** $D_{c^{\mathrm{diag}}}/\Pi_{c^{\mathrm{diag}}} \overset{\sim}{\to} \Pi_1$*,* $D_{c^{\mathrm{diag}}}/\Pi_{c^{\mathrm{diag}}} \overset{\sim}{\to} \Pi_{\{2\}}$*, respectively, such that the resulting composite outer isomorphism* $\Pi_1 \overset{\sim}{\to} \Pi_{\{2\}}$ *is the* ***identity*** *outer isomorphism.*

(iv) The natural inclusions $\Pi_{v^{\mathrm{new}}}, I_{v^{\mathrm{new}}}|_{z_x} \hookrightarrow D_{v^{\mathrm{new}}}|_{z_x}$ *determine an* ***isomorphism*** $\Pi_{v^{\mathrm{new}}} \times I_{v^{\mathrm{new}}}|_{z_x} \overset{\sim}{\to} D_{v^{\mathrm{new}}}|_{z_x} = C_{\Pi_2|_{z_x}}(\Pi_{v^{\mathrm{new}}})$*. Moreover, the composite* $I_{v^{\mathrm{new}}}|_{z_x} \hookrightarrow D_{v^{\mathrm{new}}}|_{z_x} \to \Pi_{z_x}$ *is an* ***isomorphism****.*

(v) It holds that $C_{\Pi_2}(D_{v^{\mathrm{new}}}|_{z_x}) \subseteq C_{\Pi_2}(\Pi_{v^{\mathrm{new}}})$*.*

(vi) $D_{v^{\mathrm{new}}}|_{z_x}$ *is* ***commensurably terminal*** *in* Π_2*, i.e., it holds that* $D_{v^{\mathrm{new}}}|_{z_x} = C_{\Pi_2}(D_{v^{\mathrm{new}}}|_{z_x})$*.*

(vii) It holds that $Z_{\Pi_2}(\Pi_{v^{\text{new}}}) = Z^{\text{loc}}_{\Pi_2}(\Pi_{v^{\text{new}}}) = I_{v^{\text{new}}}|_{z_x}$. *Moreover, these profinite groups are* ***isomorphic*** *to* $\widehat{\mathbb{Z}}^{\Sigma}$ *[cf. the discussion entitled "Numbers" in [CbTpI], §0].*

(viii) It holds that $C_{\Pi_2}(\Pi_{v^{\text{new}}}) = D_{v^{\text{new}}}|_{z_x} = \Pi_{v^{\text{new}}} \times Z_{\Pi_2}(\Pi_{v^{\text{new}}})$. *In particular, the equality* $C_{\Pi_2}(\Pi_{v^{\text{new}}}) = N_{\Pi_2}(\Pi_{v^{\text{new}}})$ *holds.*

(ix) It holds that $Z(C_{\Pi_2}(\Pi_{v^{\text{new}}})) = Z_{\Pi_2}(\Pi_{v^{\text{new}}})$.

(x) It holds that

$$C_{\Pi_2}(\Pi_{v^{\text{new}}_{2/1}}) \cap \Pi_{2/1} = \Pi_{v^{\text{new}}_{2/1}}, \quad C_{\Pi_2}(\Pi_{v^{\text{new}}_{1\backslash 2}}) \cap \Pi_{1\backslash 2} = \Pi_{v^{\text{new}}_{1\backslash 2}},$$

$$C_{\Pi_2}(\Pi_{v^{\text{new}}_{2/1}}) = C_{\Pi_2}(\Pi_{v^{\text{new}}_{1\backslash 2}}).$$

Moreover, for suitable choices of basepoints of the log schemes $U^{\log}$ *and* $X_E^{\log}$*, the natural morphism* $U^{\log} \to X_E^{\log}$ *induces an* ***isomorphism*** $\Pi_U \xrightarrow{\sim} C_{\Pi_2}(\Pi_{v^{\text{new}}_{2/1}}) = C_{\Pi_2}(\Pi_{v^{\text{new}}_{1\backslash 2}})$.

Proof First, we verify assertion (i). Now it is immediate that we have inclusions $\Pi_{c^{\text{diag}}} \subseteq D_{c^{\text{diag}}} \subseteq C_{\Pi_2}(\Pi_{c^{\text{diag}}})$. In particular, since $\Pi_{c^{\text{diag}}}$ is *commensurably terminal* in $\Pi_{2/1}$ and $\Pi_{1\backslash 2}$ [cf. [CmbGC], Proposition 1.2, (ii)], we obtain that $\Pi_{c^{\text{diag}}} \subseteq D_{c^{\text{diag}}} \cap \Pi_{2/1} \subseteq C_{\Pi_2}(\Pi_{c^{\text{diag}}}) \cap \Pi_{2/1} = C_{\Pi_{2/1}}(\Pi_{c^{\text{diag}}}) = \Pi_{c^{\text{diag}}}$; $\Pi_{c^{\text{diag}}} \subseteq D_{c^{\text{diag}}} \cap \Pi_{1\backslash 2} \subseteq C_{\Pi_2}(\Pi_{c^{\text{diag}}}) \cap \Pi_{1\backslash 2} = C_{\Pi_{1\backslash 2}}(\Pi_{c^{\text{diag}}}) = \Pi_{c^{\text{diag}}}$. This completes the proof of assertion (i). Assertions (ii), (iii) follow immediately from assertion (i), together with the [easily verified] fact that the composites $D_{c^{\text{diag}}} \hookrightarrow \Pi_2 \overset{p^{\Pi}_{2/1}}{\twoheadrightarrow} \Pi_1$ and $D_{c^{\text{diag}}} \hookrightarrow \Pi_2 \overset{p^{\Pi}_{1\backslash 2}}{\twoheadrightarrow} \Pi_{\{2\}}$ are *surjective*.

Next, we verify assertion (iv). It follows immediately from the various definitions involved—by considering a suitable stable log curve of type (g, r) over $(\operatorname{Spec} k)^{\log}$ and applying a suitable *specialization isomorphism* [cf. the discussion preceding [CmbCsp], Definition 2.1, as well as [CbTpI], Remark 5.6.1]—that, to verify assertion (iv), we may assume without loss of generality that $\operatorname{Cusp}(\mathcal{G}) \cup \{z_x\} = \operatorname{Edge}(\mathcal{G})$. Then, in light of the well-known local structure of $X^{\log}$ in a neighborhood of the node or cusp corresponding to z_x, one verifies easily that the outer action $\Pi_{z_x} \to \operatorname{Out}(\Pi_{2/1}) \xrightarrow{\sim} \operatorname{Out}(\Pi_{\mathcal{G}_{2/1}})$ arising from the natural exact sequence

$$1 \longrightarrow \Pi_{2/1} \longrightarrow \Pi_2|_{z_x} \longrightarrow \Pi_{z_x} \longrightarrow 1$$

is of *SNN-type* [cf. [NodNon], Definition 2.4, (iii)], hence, in particular, that the composite $I_{v^{\text{new}}}|_{z_x} \hookrightarrow D_{v^{\text{new}}}|_{z_x} \to \Pi_{z_x}$ is an *isomorphism*. Thus, assertion (iv) follows immediately from [NodNon], Remark 2.7.1, together with the *commensurable terminality* of $\Pi_{v^{\text{new}}}$ in $\Pi_{2/1}$ [cf. [CmbGC], Proposition 1.2, (ii)] and the fact that the composite $D_{v^{\text{new}}}|_{z_x} \hookrightarrow \Pi_2|_{z_x} \twoheadrightarrow \Pi_{z_x}$ is *surjective*. This completes the proof of assertion (iv).

Next, we verify assertion (v). It follows immediately from assertion (iv), together with the *commensurable terminality* of $\Pi_{v^{\mathrm{new}}}$ in $\Pi_{2/1}$ [cf. [CmbGC], Proposition 1.2, (ii)], that $D_{v^{\mathrm{new}}}|_{z_x} \cap \Pi_{2/1} = \Pi_{v^{\mathrm{new}}}$. Thus, since $\Pi_{2/1}$ is *normal* in Π_2, assertion (v) follows immediately from Lemma 3.9, (i). This completes the proof of assertion (v).

Next, we verify assertion (vi). Since the image of the composite $D_{v^{\mathrm{new}}}|_{z_x} \hookrightarrow \Pi_2 \overset{p^{\Pi}_{2/1}}{\twoheadrightarrow} \Pi_1$ coincides with $\Pi_{z_x} \subseteq \Pi_1$ [cf. assertion (iv)], and $\Pi_{z_x} \subseteq \Pi_1$ is *commensurably terminal* in Π_1 [cf. [CmbGC], Proposition 1.2, (ii)], it follows immediately that $C_{\Pi_2}(D_{v^{\mathrm{new}}}|_{z_x}) \subseteq \Pi_2|_{z_x}$. In particular, it follows immediately from assertions (iv), (v) that $D_{v^{\mathrm{new}}}|_{z_x} \subseteq C_{\Pi_2}(D_{v^{\mathrm{new}}}|_{z_x}) \subseteq C_{\Pi_2}(\Pi_{v^{\mathrm{new}}}) \cap \Pi_2|_{z_x} = C_{\Pi_2|_{z_x}}(\Pi_{v^{\mathrm{new}}}) = D_{v^{\mathrm{new}}}|_{z_x}$. This completes the proof of assertion (vi).

Next, we verify assertion (vii). It follows from assertion (iv) and [CmbGC], Remark 1.1.3, that $I_{v^{\mathrm{new}}}|_{z_x}$ is *isomorphic* to $\widehat{\mathbb{Z}}^{\Sigma}$. Moreover, it follows from the various definitions involved that we have inclusions $I_{v^{\mathrm{new}}}|_{z_x} \subseteq Z_{\Pi_2}(\Pi_{v^{\mathrm{new}}}) \subseteq Z^{\mathrm{loc}}_{\Pi_2}(\Pi_{v^{\mathrm{new}}})$. Thus, to verify assertion (vii), it suffices to verify that $Z^{\mathrm{loc}}_{\Pi_2}(\Pi_{v^{\mathrm{new}}}) \subseteq I_{v^{\mathrm{new}}}|_{z_x}$. To this end, let us observe that it follows immediately from the final portion of Lemma 3.6, (iv), that the image $p^{\Pi}_{1\backslash 2}(\Pi_{v^{\mathrm{new}}}) \subseteq \Pi_{\{2\}} \overset{\sim}{\to} \Pi_{\mathcal{G}}$ is an edge-like subgroup of $\Pi_{\{2\}} \overset{\sim}{\to} \Pi_{\mathcal{G}}$ associated to $z_x \in \mathrm{Edge}(\mathcal{G})$. Thus, since every edge-like subgroup is *commensurably terminal* [cf. [CmbGC], Proposition 1.2, (ii)], it follows that the image $p^{\Pi}_{1\backslash 2}(Z^{\mathrm{loc}}_{\Pi_2}(\Pi_{v^{\mathrm{new}}})) \subseteq \Pi_{\{2\}} \overset{\sim}{\to} \Pi_{\mathcal{G}}$ is *contained* in an edge-like subgroup of $\Pi_{\{2\}} \overset{\sim}{\to} \Pi_{\mathcal{G}}$ associated to $z_x \in \mathrm{Edge}(\mathcal{G})$. On the other hand, since $\Pi_{c^{\mathrm{diag}}} \subseteq \Pi_{v^{\mathrm{new}}}$, we have $Z^{\mathrm{loc}}_{\Pi_2}(\Pi_{v^{\mathrm{new}}}) \subseteq Z^{\mathrm{loc}}_{\Pi_2}(\Pi_{c^{\mathrm{diag}}}) \subseteq C_{\Pi_2}(\Pi_{c^{\mathrm{diag}}}) = D_{c^{\mathrm{diag}}}$ [cf. assertion (ii)]. In particular, it follows immediately from assertion (iii), together with the fact [cf. assertion (iv)] that $I_{v^{\mathrm{new}}}|_{z_x} \subseteq Z^{\mathrm{loc}}_{\Pi_2}(\Pi_{v^{\mathrm{new}}})$ *surjects* onto Π_{z_x} [cf. also [NodNon], Lemma 1.5], that $p^{\Pi}_{2/1}(Z^{\mathrm{loc}}_{\Pi_2}(\Pi_{v^{\mathrm{new}}})) \subseteq \Pi_1$ is *contained* in $\Pi_{z_x} \subseteq \Pi_1$, i.e., $Z^{\mathrm{loc}}_{\Pi_2}(\Pi_{v^{\mathrm{new}}}) \subseteq \Pi_2|_{z_x}$. Thus, it follows immediately from assertion (iv), together with the *slimness* of $\Pi_{v^{\mathrm{new}}}$ [cf. [CmbGC], Remark 1.1.3], that $Z^{\mathrm{loc}}_{\Pi_2}(\Pi_{v^{\mathrm{new}}}) \subseteq I_{v^{\mathrm{new}}}|_{z_x}$. This completes the proof of assertion (vii).

Next, we verify assertion (viii). It follows from assertion (vii), together with Lemma 3.9, (ii), that $C_{\Pi_2}(\Pi_{v^{\mathrm{new}}}) \subseteq N_{\Pi_2}(I_{v^{\mathrm{new}}}|_{z_x})$. In particular, since $D_{v^{\mathrm{new}}}|_{z_x}$ is *generated by* $\Pi_{v^{\mathrm{new}}}$, $I_{v^{\mathrm{new}}}|_{z_x}$ [cf. assertion (iv)], it follows immediately that $(D_{v^{\mathrm{new}}}|_{z_x} \subseteq) C_{\Pi_2}(\Pi_{v^{\mathrm{new}}}) \subseteq C_{\Pi_2}(D_{v^{\mathrm{new}}}|_{z_x})$. Thus, the first equality of assertion (viii) follows from assertion (vi); the second equality of assertion (viii) follows immediately from assertions (iv), (vii). This completes the proof of assertion (viii).

Next, we verify assertion (ix). Let us recall from [CmbGC], Remark 1.1.3, that $\Pi_{v^{\mathrm{new}}}$ is *slim*. Thus, assertion (ix) follows from assertion (viii), together with the final portion of assertion (vii). This completes the proof of assertion (ix).

Finally, we verify assertion (x). The first two equalities follow from [CmbGC], Proposition 1.2, (ii). Next, let us observe that since [it is immediate that] the automorphism of $X^{\log}_E$ determined by permuting the factors labeled i, j *stabilizes* U, but *permutes* $v^{\mathrm{new}}_{2/1}$ and $v^{\mathrm{new}}_{1\backslash 2}$, one verifies immediately that, to verify assertion (x), it suffices to verify that, for suitable choices of basepoints of the log schemes

$U^{\log}$ and $X_E^{\log}$, the natural morphism $U^{\log} \to X_E^{\log}$ induces an *isomorphism* $\Pi_U \xrightarrow{\sim} C_{\Pi_2}(\Pi_{v^{\mathrm{new}}})(= C_{\Pi_2}(\Pi_{v^{\mathrm{new}}_{2/1}}))$. To this end, let us observe that since the vertex v^{new} *corresponds* to the closed irreducible subscheme $Y \subseteq X_E$ [cf. the discussion following the definition of Π_U in the statement of Lemma 3.11], it follows immediately from the various definitions involved that, for suitable choices of basepoints of the log schemes $U^{\log}$ and $X_E^{\log}$, the natural morphism $U^{\log} \to X_E^{\log}$ gives rise to a commutative diagram

$$\begin{array}{ccccccccc} 1 & \longrightarrow & \Pi_{U/z_x} & \longrightarrow & \Pi_U & \longrightarrow & \Pi_{z_x} & \longrightarrow & 1 \\ & & \wr\downarrow & & \downarrow & & \parallel & & \\ 1 & \longrightarrow & \Pi_{v^{\mathrm{new}}} & \longrightarrow & D_{v^{\mathrm{new}}}|_{z_x} & \longrightarrow & \Pi_{z_x} & \longrightarrow & 1 \end{array}$$

—where we write Π_{U/z_x} for the kernel of the natural surjection $\Pi_U \twoheadrightarrow \Pi_{z_x}$; the horizontal sequences are *exact*; the exactness of the lower horizontal sequence follows from assertion (iv); the left-hand vertical arrow is an *isomorphism*. Thus, it follows from assertion (viii) that, for suitable choices of basepoints of the log schemes $U^{\log}$ and $X_E^{\log}$, the natural morphism $U^{\log} \to X_E^{\log}$ induces an *isomorphism* $\Pi_U \xrightarrow{\sim} D_{v^{\mathrm{new}}}|_{z_x} = C_{\Pi_2}(\Pi_{v^{\mathrm{new}}})$, as desired. This completes the proof of assertion (x), hence also of Lemma 3.11. □

The first item of the following result [i.e., Lemma 3.12, (i)] is, along with its proof, a routine generalization of [CmbCsp], Corollary 1.10, (ii).

Lemma 3.12 (Commensurator of a Tripod) *Let $E \subseteq \{1, \cdots, n\}$ and $T \subseteq \Pi_E$ an **E-tripod** of Π_n [cf. Definition 3.3, (i)]. Then the following hold:*

*(i) It holds that $C_{\Pi_E}(T) = T \times Z_{\Pi_E}(T)$. Thus, if an outomorphism α of Π_E **preserves** the Π_E-conjugacy class of T, then one may define $\alpha|_T \in \mathrm{Out}(T)$ [cf. Lemma 3.10, (i)].*

*(ii) Suppose that $n = \#E = 3$, and that T is **central** [cf. Definition 3.7, (ii)]. Let $T' \subseteq \Pi_E = \Pi_n$ be a **central E-tripod** of Π_n. Then $C_{\Pi_n}(T)$ (respectively, $N_{\Pi_n}(T)$; $Z_{\Pi_n}(T)$) is a Π_n-conjugate of $C_{\Pi_n}(T')$ (respectively, $N_{\Pi_n}(T')$; $Z_{\Pi_n}(T')$).*

Proof Let $i \in E$; $x \in X_n(k)$; $v \in \mathrm{Vert}(\mathcal{G}_{i\in E,x})$ be such that v is of *type* (0, 3), and, moreover, T is a verticial subgroup of Π_E associated to $v \in \mathrm{Vert}(\mathcal{G}_{i\in E,x})$. [Thus, we have an inclusion $T \subseteq \Pi_{E/(E\setminus\{i\})} \subseteq \Pi_E$—cf. Definition 3.1, (iv).]

First, we verify assertion (i). Since $T \subseteq \Pi_{E/(E\setminus\{i\})} \subseteq \Pi_E$, and T is *commensurably terminal* in $\Pi_{E/(E\setminus\{i\})}$ [cf. [CmbGC], Proposition 1.2, (ii)], it follows from Lemma 3.9, (iii), that $C_{\Pi_E}(T) = N_{\Pi_E}(T)$. Thus, in light of the *slimness* of T [cf. [CmbGC], Remark 1.1.3], to verify assertion (i), it suffices to verify that the natural outer action of $N_{\Pi_E}(T)$ on T is *trivial*. To this end, let $E' \subseteq E$ be such that T is E'-*strict* [cf. Lemma 3.8, (i)]; write $T_{E'} \subseteq \Pi_{E'}$ for the image of T via $p^{\Pi}_{E/E'}\colon \Pi_E \twoheadrightarrow \Pi_{E'}$. Then it is immediate that the image of $N_{\Pi_E}(T)$ via $p^{\Pi}_{E/E'}\colon \Pi_E \twoheadrightarrow \Pi_{E'}$ is *contained* in $N_{\Pi_{E'}}(T_{E'})$, and that the natural surjection

$T \twoheadrightarrow T_{E'}$ is an *isomorphism* [cf. Lemma 3.8, (i)]. Thus, one verifies easily—by replacing E, T by $E', T_{E'}$, respectively—that, to verify that the natural outer action of $N_{\Pi_E}(T)$ on T is *trivial*, we may assume without loss of generality that T is *E-strict*. If T satisfies condition (1) of Lemma 3.8, (ii), then assertion (i) follows from the *commensurable terminality* of T in Π_E [cf. [CmbGC], Proposition 1.2, (ii)]. If T satisfies either condition (2_C) or condition (2_N) of Lemma 3.8, (ii), then assertion (i) follows immediately from Lemma 3.11, (viii). If T satisfies condition (3) of Lemma 3.8, (ii), then one verifies easily from the various definitions involved—by considering a suitable stable log curve of type (g, r) over $(\operatorname{Spec} k)^{\log}$ and applying a suitable *specialization isomorphism* [cf. the discussion preceding [CmbCsp], Definition 2.1, as well as [CbTpI], Remark 5.6.1]—that, to verify assertion (i), we may assume without loss of generality that $\operatorname{Node}(\mathcal{G}) = \emptyset$. Thus, assertion (i) follows immediately from [CmbCsp], Corollary 1.10, (ii). This completes the proof of assertion (i).

Next, we verify assertion (ii). Let us recall from Remark 3.7.1, (iii), that there exist an element $\tau \in \mathfrak{S}_3 \subseteq \operatorname{Out}(\Pi_3)$ [cf. the discussion at the beginning of this chapter] and a lifting $\tilde{\tau} \in \operatorname{Aut}(\Pi_3)$ of τ such that the image of $T \subseteq \Pi_3$ by the automorphism $\tilde{\tau} \in \operatorname{Aut}(\Pi_3)$ *coincides* with $T' \subseteq \Pi_3$. Next, let us observe that one verifies easily that $\tau \in \mathfrak{S}_3$ may be written as a product of *transpositions* in $\mathfrak{S}_3$. Thus, in the remainder of the proof of assertion (ii), we may assume without loss of generality that τ is a *transposition* in $\mathfrak{S}_3$. Moreover, in the remainder of the proof of assertion (ii), we may assume without loss of generality, by conjugating by a suitable element of $\mathfrak{S}_3$, that τ is the *transposition* "(1, 2)" in $\mathfrak{S}_3$. Thus, if, moreover, $i = 3$ [i.e., the E-tripod T is 3-*central*], then it follows from Lemma 3.6, (v), that T is a Π_3-conjugate of T', hence that $C_{\Pi_3}(T)$ (respectively, $N_{\Pi_3}(T)$; $Z_{\Pi_3}(T)$) is a Π_3-*conjugate* of $C_{\Pi_3}(T')$ (respectively, $N_{\Pi_3}(T')$; $Z_{\Pi_3}(T')$). In particular, in the remainder of the proof of assertion (ii), we may assume without loss of generality, by conjugating by $\tau \in \mathfrak{S}_3$ if necessary, that $i = 2$, i.e., that the E-tripods T, T' are 2-*central*, 1-*central*, respectively.

Next, let us observe that, in this situation, one verifies immediately from the various definitions involved that there exists a *natural identification* between $\Pi_{\{1,2,3\}/\{3\}}$ and the "Π_2" that arises in the case where we take "$X^{\log}$" to be the base-change of $p^{\log}_{\{3\}}: X^{\log}_{\{2,3\}} \to X^{\log}_{\{3\}}$ via a suitable morphism of log schemes $(\operatorname{Spec} k)^{\log} \to X^{\log}_{\{3\}}$. Moreover, one also verifies immediately from the various definitions involved [cf. also Lemma 3.6, (v)] that this natural identification maps suitable Π_3-conjugates of T, T', respectively, bijectively onto the closed subgroups "$\Pi_{v^{\mathrm{new}}_{2/1}}$", "$\Pi_{v^{\mathrm{new}}_{1\backslash 2}}$" of the "$\Pi_2$" that appears in the statement of Lemma 3.11. In particular, it follows from Lemma 3.11, (viii), (ix), (x), that the following assertions hold:

(a) The following equalities hold:

$$C_{\Pi_{\{1,2,3\}/\{3\}}}(T) = T \times Z_{\Pi_{\{1,2,3\}/\{3\}}}(T),$$

$$C_{\Pi_{\{1,2,3\}/\{3\}}}(T') = T' \times Z_{\Pi_{\{1,2,3\}/\{3\}}}(T').$$

(b) The following equalities hold:

$$C_{\Pi_{\{1,2,3\}/\{3\}}}(T) \cap \Pi_{\{1,2,3\}/\{1,3\}} = T,$$

$$C_{\Pi_{\{1,2,3\}/\{3\}}}(T') \cap \Pi_{\{1,2,3\}/\{2,3\}} = T'.$$

(c) The subgroup $C_{\Pi_{\{1,2,3\}/\{3\}}}(T)$ (respectively, $Z_{\Pi_{\{1,2,3\}/\{3\}}}(T)$) is a $\Pi_{\{1,2,3\}/\{3\}}$-*conjugate* of the subgroup $C_{\Pi_{\{1,2,3\}/\{3\}}}(T')$ (respectively, $Z_{\Pi_{\{1,2,3\}/\{3\}}}(T')$).

In particular, it follows from (c) that, to verify assertion (ii), it suffices to verify the following assertion:

Claim 3.12.A: The following equalities hold:

$$C_{\Pi_3}(T) = C_{\Pi_3}(C_{\Pi_{\{1,2,3\}/\{3\}}}(T)), \quad C_{\Pi_3}(T') = C_{\Pi_3}(C_{\Pi_{\{1,2,3\}/\{3\}}}(T')),$$

$$N_{\Pi_3}(T) = N_{\Pi_3}(C_{\Pi_{\{1,2,3\}/\{3\}}}(T)), \quad N_{\Pi_3}(T') = N_{\Pi_3}(C_{\Pi_{\{1,2,3\}/\{3\}}}(T')),$$

$$Z_{\Pi_3}(T) = Z_{\Pi_3}(C_{\Pi_{\{1,2,3\}/\{3\}}}(T)), \quad Z_{\Pi_3}(T') = Z_{\Pi_3}(C_{\Pi_{\{1,2,3\}/\{3\}}}(T')).$$

First, we verify the first four equalities of Claim 3.12.A. Observe that since $\Pi_{\{1,2,3\}/\{3\}}$ is a *normal* closed subgroup of Π_3 and *contains* both T and T', it follows from Lemma 3.9, (iii), that the inclusions

$$N_{\Pi_3}(T) \subseteq C_{\Pi_3}(T) \subseteq N_{\Pi_3}(C_{\Pi_{\{1,2,3\}/\{3\}}}(T)) \subseteq C_{\Pi_3}(C_{\Pi_{\{1,2,3\}/\{3\}}}(T)),$$

$$N_{\Pi_3}(T') \subseteq C_{\Pi_3}(T') \subseteq N_{\Pi_3}(C_{\Pi_{\{1,2,3\}/\{3\}}}(T')) \subseteq C_{\Pi_3}(C_{\Pi_{\{1,2,3\}/\{3\}}}(T'))$$

hold. Moreover, by the *normality* of $\Pi_{\{1,2,3\}/\{1,3\}}$ and $\Pi_{\{1,2,3\}/\{2,3\}}$ in Π_3, one verifies easily, by applying (b), that the inclusions

$$N_{\Pi_3}(C_{\Pi_{\{1,2,3\}/\{3\}}}(T)) \subseteq N_{\Pi_3}(T), \quad C_{\Pi_3}(C_{\Pi_{\{1,2,3\}/\{3\}}}(T)) \subseteq C_{\Pi_3}(T),$$

$$N_{\Pi_3}(C_{\Pi_{\{1,2,3\}/\{3\}}}(T')) \subseteq N_{\Pi_3}(T'), \quad C_{\Pi_3}(C_{\Pi_{\{1,2,3\}/\{3\}}}(T')) \subseteq C_{\Pi_3}(T')$$

hold. This completes the proof of the first four equalities of Claim 3.12.A.

Finally, we verify the final two equalities of Claim 3.12.A. Let us first observe that the inclusions $T \subseteq C_{\Pi_{\{1,2,3\}/\{3\}}}(T)$, $T' \subseteq C_{\Pi_{\{1,2,3\}/\{3\}}}(T')$ imply that

$$Z_{\Pi_3}(C_{\Pi_{\{1,2,3\}/\{3\}}}(T)) \subseteq Z_{\Pi_3}(T), \quad Z_{\Pi_3}(C_{\Pi_{\{1,2,3\}/\{3\}}}(T')) \subseteq Z_{\Pi_3}(T').$$

Thus, it follows immediately from (a) that, to verify the final two equalities of Claim 3.12.A, it suffices to verify the following assertion:

Claim 3.12.B: The following inclusions hold:

$$Z_{\Pi_3}(T) \subseteq Z_{\Pi_3}(Z_{\Pi_{\{1,2,3\}/\{3\}}}(T)), \quad Z_{\Pi_3}(T') \subseteq Z_{\Pi_3}(Z_{\Pi_{\{1,2,3\}/\{3\}}}(T')).$$

First, let us observe that one verifies immediately from the various definitions involved that the *natural identification* that appears in the discussion preceding assertion (a) in the present proof of Lemma 3.12, (ii), determines a natural identification between $\Pi_{\{2,3\}/\{3\}}$ and the "$\Pi_1 = \Pi_{\{2\}}$" that arises in the case where we take "$X^{\log}$" to be as in the discussion preceding assertion (a) in the present proof of Lemma 3.12, (ii). Thus, it follows immediately from the final portion of Lemma 3.6, (iv), that the image $J_T \subseteq \Pi_{\{2,3\}/\{3\}}$ of $T \subseteq \Pi_{\{1,2,3\}/\{3\}}$ in $\Pi_{\{2,3\}/\{3\}}$ corresponds, via the natural identification just discussed, to an edge-like subgroup of "$\Pi_1 = \Pi_{\{2\}}$" associated to the edge $z_x \in \mathrm{Edge}(\mathcal{G})$ that appears in the statement of Lemma 3.11. Moreover, it follows immediately from (c) and Lemma 3.11, (iv), (vii), that the surjection $\Pi_{\{1,2,3\}/\{3\}} \twoheadrightarrow \Pi_{\{2,3\}/\{3\}}$ induces an *isomorphism*

$$\Pi_{\{1,2,3\}/\{3\}} \supseteq Z_{\Pi_{\{1,2,3\}/\{3\}}}(T) \xrightarrow{\sim} J_Z \subseteq \Pi_{\{2,3\}/\{3\}}$$

—where the closed subgroup $J_Z \subseteq \Pi_{\{2,3\}/\{3\}}$ corresponds, via the natural identification just discussed, to an edge-like subgroup of "$\Pi_1 = \Pi_{\{2\}}$" associated to the edge $z_x \in \mathrm{Edge}(\mathcal{G})$ that appears in the statement of Lemma 3.11. Thus, we conclude immediately from [CmbGC], Proposition 1.2, (ii), together with the various definitions involved, that $J_T = J_Z$ ($\xleftarrow{\sim} Z_{\Pi_{\{1,2,3\}/\{3\}}}(T)$). In particular, since $Z_{\Pi_3}(T) \subseteq N_{\Pi_3}(Z_{\Pi_{\{1,2,3\}/\{3\}}}(T))$, and the surjection $\Pi_{\{1,2,3\}/\{3\}} \twoheadrightarrow \Pi_{\{2,3\}/\{3\}}$ induces a homomorphism $Z_{\Pi_3}(T) \to Z_{\Pi_{\{2,3\}/\{3\}}}(J_T)$, one verifies easily that the first inclusion of Claim 3.12.B holds. The second inclusion of Claim 3.12.B follows from the first inclusion of Claim 3.12.B by applying $\widetilde{\tau}$. This completes the proof of Claim 3.12.B, hence also of Lemma 3.12. □

Lemma 3.13 (Preservation of Verticial Subgroups) *In the notation of Lemma 3.11, let $\widetilde{\alpha}$ be an F-admissible automorphism of $\Pi_E = \Pi_2$, $v \in \mathrm{Vert}(\mathcal{G})$. Write $v^\circ \in \mathrm{Vert}(\mathcal{G}_{2/1})$ for the vertex of $\mathcal{G}_{2/1}$ that corresponds to $v \in \mathrm{Vert}(\mathcal{G})$ via the bijection of Lemma 3.6, (iv); $\widetilde{\alpha}_1$, $\widetilde{\alpha}_{2/1}$ for the automorphisms of Π_1, $\Pi_{2/1}$ determined by $\widetilde{\alpha}$; α, α_1, $\alpha_{2/1}$ for the outomorphisms of Π_2, Π_1, $\Pi_{2/1}$ determined by $\widetilde{\alpha}$, $\widetilde{\alpha}_1$, $\widetilde{\alpha}_{2/1}$, respectively. Then the following hold:*

(i) Recall the edge-like subgroup $\Pi_{z_x} \subseteq \Pi_1 \xrightarrow{\sim} \Pi_{\mathcal{G}}$ associated to the edge $z_x \in \mathrm{Edge}(\mathcal{G})$. Suppose that

$$\widetilde{\alpha}_1(\Pi_{z_x}) = \Pi_{z_x}.$$

Suppose, moreover, either that

*(a) the outomorphism $\alpha_{2/1}$ of $\Pi_{\mathcal{G}_{2/1}} \xleftarrow{\sim} \Pi_{2/1}$ maps **some** cuspidal inertia subgroup of $\Pi_{\mathcal{G}_{2/1}} \xleftarrow{\sim} \Pi_{2/1}$ to a cuspidal inertia subgroup of $\Pi_{\mathcal{G}_{2/1}} \xleftarrow{\sim} \Pi_{2/1}$, or that*

(b) $z_x \in \mathrm{Cusp}(\mathcal{G})$.

*[For example, condition (a) holds if the outomorphism $\alpha_{2/1}$ of $\Pi_{\mathcal{G}_{2/1}} \stackrel{\sim}{\leftarrow} \Pi_{2/1}$ is **group-theoretically cuspidal**—cf. [CmbGC], Definition 1.4, (iv).] Then $\alpha_{2/1}$ **preserves** the $\Pi_{2/1}$-conjugacy class of the verticial subgroup $\Pi_{v^{\mathrm{new}}} \subseteq \Pi_{2/1} \stackrel{\sim}{\rightarrow} \Pi_{\mathcal{G}_{2/1}}$ associated to the vertex $v^{\mathrm{new}} \in \mathrm{Vert}(\mathcal{G}_{2/1})$. If, moreover, $\alpha_{2/1}$ is **group-theoretically cuspidal**, then the induced outomorphism of $\Pi_{v^{\mathrm{new}}}$ [cf. Lemma 3.12, (i)] is itself **group-theoretically cuspidal**.*

*(ii) In the situation of (i), suppose, moreover, that there exists a verticial subgroup $\Pi_v \subseteq \Pi_{\mathcal{G}} \stackrel{\sim}{\leftarrow} \Pi_1$ of $\Pi_{\mathcal{G}} \stackrel{\sim}{\leftarrow} \Pi_1$ associated to $v \in \mathrm{Vert}(\mathcal{G})$ such that $\widetilde{\alpha}_1$ **preserves** the Π_1-conjugacy class of Π_v. Then $\alpha_{2/1}$ **preserves** the $\Pi_{2/1}$-conjugacy class of a verticial subgroup of $\Pi_{\mathcal{G}_{2/1}} \stackrel{\sim}{\leftarrow} \Pi_{2/1}$ associated to the vertex $v^{\circ} \in \mathrm{Vert}(\mathcal{G}_{2/1})$.*

*(iii) In the situation of (i), suppose, moreover, that $X^{\log}$ is of **type** $(0, 3)$ [which implies that $\Pi_v \stackrel{\mathrm{def}}{=} \Pi_{\mathcal{G}} \stackrel{\sim}{\leftarrow} \Pi_1$ is the unique verticial subgroup of $\Pi_{\mathcal{G}}$ associated to v], and that $\alpha_1 \in \mathrm{Out}^{\mathrm{C}}(\Pi_v)^{\mathrm{cusp}}$ [cf. Definition 3.4, (i)]. Then there exists a **geometric** [cf. Definition 3.4, (ii)] outer isomorphism $\Pi_{v^{\mathrm{new}}} \stackrel{\sim}{\rightarrow} \Pi_v(= \Pi_{\mathcal{G}} \stackrel{\sim}{\leftarrow} \Pi_1)$ which satisfies the following condition:*

> *If either $\alpha_1 \in \mathrm{Out}(\Pi_1) = \mathrm{Out}(\Pi_v)$ is **contained** in $\mathrm{Out}(\Pi_v)^{\Delta}$ [cf. Definition 3.4, (i)] or $\alpha|_{\Pi_{v^{\mathrm{new}}}} \in \mathrm{Out}(\Pi_{v^{\mathrm{new}}})$ [cf. (i); Lemma 3.12, (i)] is **contained** in $\mathrm{Out}(\Pi_{v^{\mathrm{new}}})^{\Delta}$, then the outomorphisms $\alpha|_{\Pi_{v^{\mathrm{new}}}}$, α_1 of $\Pi_{v^{\mathrm{new}}}$, Π_v are **compatible** relative to the outer isomorphism in question $\Pi_{v^{\mathrm{new}}} \stackrel{\sim}{\rightarrow} \Pi_v$.*

Proof First, we verify assertions (i), (ii). Write $S \stackrel{\mathrm{def}}{=} \mathrm{Node}(\mathcal{G}_{2/1}) \setminus \mathcal{N}(v^{\mathrm{new}})$. Then it follows immediately from the well-known local structure of $X^{\log}$ in a neighborhood of the edge corresponding to z_x that if $z_x \in \mathrm{Node}(\mathcal{G})$ (respectively, $z_x \in \mathrm{Cusp}(\mathcal{G})$), then the outer action of Π_{z_x} on $\Pi_{(\mathcal{G}_{2/1})_{\rightsquigarrow S}}$ [cf. [CbTpI], Definition 2.8] obtained by conjugating the natural outer action $\Pi_{z_x} \hookrightarrow \Pi_1 \rightarrow \mathrm{Out}(\Pi_{2/1}) \stackrel{\sim}{\rightarrow} \mathrm{Out}(\Pi_{\mathcal{G}_{2/1}})$ —where the second arrow is the outer action determined by the exact sequence of profinite groups

$$1 \longrightarrow \Pi_{2/1} \longrightarrow \Pi_2 \stackrel{p^{\Pi}_{2/1}}{\longrightarrow} \Pi_1 \longrightarrow 1$$

—by the natural outer isomorphism $\Phi_{(\mathcal{G}_{2/1})_{\rightsquigarrow S}} : \Pi_{(\mathcal{G}_{2/1})_{\rightsquigarrow S}} \stackrel{\sim}{\rightarrow} \Pi_{\mathcal{G}_{2/1}}$ [cf. [CbTpI], Definition 2.10] is of *SNN-type* [cf. [NodNon], Definition 2.4, (iii)] (respectively, *IPSC-type* [cf. [NodNon], Definition 2.4, (i)]). Thus, it follows immediately [in light of the various assumptions made in the statement of assertion (i)!] in the case of condition (a) (respectively, condition (b)) from Theorem 1.9, (i) (respectively, Theorem 1.9, (ii)), that the outomorphism $\alpha_{(\mathcal{G}_{2/1})_{\rightsquigarrow S}}$ of $\Pi_{(\mathcal{G}_{2/1})_{\rightsquigarrow S}}$ obtained by conjugating $\alpha_{2/1}$ by the composite $\Pi_{2/1} \stackrel{\sim}{\rightarrow} \Pi_{\mathcal{G}_{2/1}} \stackrel{\Phi_{(\mathcal{G}_{2/1})_{\rightsquigarrow S}}}{\stackrel{\sim}{\leftarrow}} \Pi_{(\mathcal{G}_{2/1})_{\rightsquigarrow S}}$ is *group-theoretically verticial* [cf. [CmbGC], Definition 1.4, (iv)] and *group-*

theoretically nodal [cf. [NodNon], Definition 1.12]. On the other hand, it follows immediately from condition (3) of [CbTpI], Proposition 2.9, (i), that the image via $\Phi_{(\mathcal{G}_{2/1})_{\rightsquigarrow S}} : \Pi_{(\mathcal{G}_{2/1})_{\rightsquigarrow S}} \xrightarrow{\sim} \Pi_{\mathcal{G}_{2/1}}$ of any verticial subgroup of $\Pi_{(\mathcal{G}_{2/1})_{\rightsquigarrow S}}$ associated to the vertex of $(\mathcal{G}_{2/1})_{\rightsquigarrow S}$ corresponding to v^{new} is a verticial subgroup of $\Pi_{\mathcal{G}_{2/1}}$ associated to v^{new}. Thus, since $\alpha_{(\mathcal{G}_{2/1})_{\rightsquigarrow S}}$ is *group-theoretically verticial*, it follows immediately that $\alpha_{2/1}$ preserves the $\Pi_{2/1}$-conjugacy class of the verticial subgroup $\Pi_{v^{\mathrm{new}}} \subseteq \Pi_{2/1} \xrightarrow{\sim} \Pi_{\mathcal{G}_{2/1}}$ associated to v^{new}. [Here, we observe in passing the following easily verified fact: a vertex of $(\mathcal{G}_{2/1})_{\rightsquigarrow S}$ corresponds to v^{new} if and only if the verticial subgroup of $\Pi_{(\mathcal{G}_{2/1})_{\rightsquigarrow S}}$ associated to this vertex maps, via the composite $\Pi_{(\mathcal{G}_{2/1})_{\rightsquigarrow S}} \xrightarrow{\sim} \Pi_{2/1} \overset{p^{\Pi}_{1\backslash 2}}{\twoheadrightarrow} \Pi_{\{2\}}$, to an *abelian* subgroup of $\Pi_{\{2\}}$.] If, moreover, $\alpha_{2/1}$ is *group-theoretically cuspidal*, then the group-theoretic cuspidality of the resulting outomorphism of $\Pi_{v^{\mathrm{new}}}$ follows immediately from the group-theoretic cuspidality of $\alpha_{2/1}$ and the *group-theoretic nodality* of $\alpha_{(\mathcal{G}_{2/1})_{\rightsquigarrow S}}$. This completes the proof of assertion (i).

To verify assertion (ii), let us first observe that it follows immediately from [CbTpI], Theorem A, (i), that—after possibly replacing $\widetilde{\alpha}$ by the composite of $\widetilde{\alpha}$ with an inner automorphism of Π_2 determined by conjugation by an element of $\Pi_{2/1}$—we may assume without loss of generality that if we write $\widetilde{\alpha}_{\{2\}}$ for the automorphism of $\Pi_{\{2\}}$ determined by $\widetilde{\alpha}$, then

$$\widetilde{\alpha}_{\{2\}}(\Pi_v) = \Pi_v$$

—where, by abuse of notation, we write Π_v for some *fixed* subgroup of $\Pi_{\{2\}}$ whose image in $\Pi_{\mathcal{G}} \xleftarrow{\sim} \Pi_{\{2\}}$ is a verticial subgroup associated to v.

Next, let us *fix* a verticial subgroup $\Pi_{v^\circ} \subseteq \Pi_{2/1} \xrightarrow{\sim} \Pi_{\mathcal{G}_{2/1}}$ of $\Pi_{\mathcal{G}_{2/1}}$ associated to the vertex $v^\circ \in \mathrm{Vert}(\mathcal{G}_{2/1})$ such that the composite $\Pi_{v^\circ} \hookrightarrow \Pi_{2/1} \overset{p^{\Pi}_{1\backslash 2}}{\twoheadrightarrow} \Pi_{\{2\}}$ determines an *isomorphism* $\Pi_{v^\circ} \xrightarrow{\sim} \Pi_v$. Then let us observe that one verifies easily from condition (3) of [CbTpI], Proposition 2.9, (i), together with [NodNon], Lemma 1.9, (ii), that there exists a *unique* vertex $w^\circ \in \mathrm{Vert}((\mathcal{G}_{2/1})_{\rightsquigarrow S})$ such that the image $\Pi_{w^\circ} \subseteq \Pi_{2/1}$ via the composite $\Pi_{(\mathcal{G}_{2/1})_{\rightsquigarrow S}} \overset{\Phi_{(\mathcal{G}_{2/1})_{\rightsquigarrow S}}}{\xrightarrow{\sim}} \Pi_{\mathcal{G}_{2/1}} \xleftarrow{\sim} \Pi_{2/1}$ of some verticial subgroup of $\Pi_{(\mathcal{G}_{2/1})_{\rightsquigarrow S}}$ associated to w° *contains* the verticial subgroup $\Pi_{v^\circ} \subseteq \Pi_{2/1} \xrightarrow{\sim} \Pi_{\mathcal{G}_{2/1}}$. Thus, it follows immediately from the various definitions involved that the composite $\Pi_{w^\circ} \hookrightarrow \Pi_{2/1} \overset{p^{\Pi}_{1\backslash 2}}{\twoheadrightarrow} \Pi_{\{2\}}$ is an *injective* homomorphism whose image $\Pi_w \subseteq \Pi_{\{2\}}$ maps via the composite $\Pi_{\{2\}} \xrightarrow{\sim} \Pi_{\mathcal{G}} \overset{\Phi_{\mathcal{G}_{\rightsquigarrow \overline{S}}}}{\xleftarrow{\sim}} \Pi_{\mathcal{G}_{\rightsquigarrow \overline{S}}}$—where we write $\overline{S} \overset{\mathrm{def}}{=} \mathrm{Node}(\mathcal{G}) \setminus (\mathrm{Node}(\mathcal{G}) \cap \{z_x\})$—to a verticial subgroup of $\Pi_{\mathcal{G}_{\rightsquigarrow \overline{S}}}$

associated to a vertex $w \in \mathrm{Vert}(\mathcal{G}_{\rightsquigarrow \overline{S}})$. Here, we note that the vertex w may also be characterized as the *unique* vertex of $\mathcal{G}_{\rightsquigarrow \overline{S}}$ such that the image via the natural outer isomorphism $\Phi_{\mathcal{G}_{\rightsquigarrow \overline{S}}} : \Pi_{\mathcal{G}_{\rightsquigarrow \overline{S}}} \xrightarrow{\sim} \Pi_{\mathcal{G}}$ of some verticial subgroup associated to w *contains* a verticial subgroup associated to $v \in \mathrm{Vert}(\mathcal{G})$. Thus, we obtain an isomorphism $\Pi_{w^\circ} \xrightarrow{\sim} \Pi_w$, hence also an isomorphism $\widetilde{\alpha}_{2/1}(\Pi_{w^\circ}) \xrightarrow{\sim} \widetilde{\alpha}_{\{2\}}(\Pi_w)$.

Next, let us observe that since $\alpha_{(\mathcal{G}_{2/1})_{\rightsquigarrow S}}$ is *group-theoretically verticial* [cf. the argument given in the proof of assertion (i)], it follows immediately that $\widetilde{\alpha}_{2/1}(\Pi_{w^\circ}) \subseteq \Pi_{2/1} \xrightarrow{\sim} \Pi_{(\mathcal{G}_{2/1})_{\rightsquigarrow S}}$ is a verticial subgroup of $\Pi_{(\mathcal{G}_{2/1})_{\rightsquigarrow S}}$ that maps isomorphically to a verticial subgroup $\widetilde{\alpha}_{\{2\}}(\Pi_w) \subseteq \Pi_{\{2\}} \xrightarrow{\sim} \Pi_{\mathcal{G}_{\rightsquigarrow \overline{S}}}$ of $\Pi_{\mathcal{G}_{\rightsquigarrow \overline{S}}}$ that contains $\widetilde{\alpha}_{\{2\}}(\Pi_v) = \Pi_v$. On the other hand, in light of the *unique* characterization of w given above, this implies that $\widetilde{\alpha}_{\{2\}}(\Pi_w) \subseteq \Pi_{\{2\}} \xrightarrow{\sim} \Pi_{\mathcal{G}_{\rightsquigarrow \overline{S}}}$ is a verticial subgroup associated to w, and hence [as is easily verified] that $\widetilde{\alpha}_{2/1}(\Pi_{w^\circ}) \subseteq \Pi_{2/1} \xrightarrow{\sim} \Pi_{(\mathcal{G}_{2/1})_{\rightsquigarrow S}}$ is a verticial subgroup associated to w°. In particular, one may apply the natural outer isomorphisms $\Pi_{((\mathcal{G}_{2/1})|_{\mathbb{H}_{w^\circ}})_{\succ T_{w^\circ}}} \xrightarrow{\sim} \widetilde{\alpha}_{2/1}(\Pi_{w^\circ})$; $\Pi_{(\mathcal{G}|_{\mathbb{H}_w})_{\succ T_w}} \xrightarrow{\sim} \widetilde{\alpha}_{\{2\}}(\Pi_w)$ [cf. [CbTpI], Definitions 2.2, (ii); 2.5, (ii)] arising from condition (3) of [CbTpI], Proposition 2.9, (i); moreover, one verifies easily that the resulting outer isomorphism $\Pi_{((\mathcal{G}_{2/1})|_{\mathbb{H}_{w^\circ}})_{\succ T_{w^\circ}}} \xrightarrow{\sim} \Pi_{(\mathcal{G}|_{\mathbb{H}_w})_{\succ T_w}}$ [induced by the above isomorphism $\widetilde{\alpha}_{2/1}(\Pi_{w^\circ}) \xrightarrow{\sim} \widetilde{\alpha}_{\{2\}}(\Pi_w)$] arises from *scheme theory*, hence is *graphic* [cf. [CmbGC], Definition 1.4, (i)]. Therefore, we conclude that the closed subgroup $\widetilde{\alpha}_{2/1}(\Pi_{v^\circ}) \subseteq (\widetilde{\alpha}_{2/1}(\Pi_{w^\circ}) \subseteq)\Pi_{2/1} \xrightarrow{\sim} \Pi_{\mathcal{G}_{2/1}}$ is a verticial subgroup of $\Pi_{\mathcal{G}_{2/1}}$ associated to v°. This completes the proof of assertion (ii).

Finally, we verify assertion (iii). First, we recall from [CmbCsp], Corollary 1.14, (ii), that there exists an outer modular symmetry $\sigma \in (\mathfrak{S}_5 \subseteq)\mathrm{Out}(\Pi_2)$ such that the composite $\Pi_{v^{\mathrm{new}}} \hookrightarrow \Pi_2 \overset{\sigma}{\xrightarrow{\sim}} \Pi_2 \overset{p^{\Pi}_{2/1}}{\twoheadrightarrow} \Pi_1 = \Pi_v$ determines a(n) [necessarily *geometric*] outer isomorphism $\Pi_{v^{\mathrm{new}}} \xrightarrow{\sim} \Pi_v$. The remainder of the proof of assertion (iii) is devoted to verifying that this outer isomorphism $\Pi_{v^{\mathrm{new}}} \xrightarrow{\sim} \Pi_v$ satisfies the condition of assertion (iii). First, suppose that $\alpha_1 \in \mathrm{Out}(\Pi_1)^{\Delta}$. Then since $\mathrm{Out}^{\mathrm{F}}(\Pi_2) = \mathrm{Out}^{\mathrm{FC}}(\Pi_2) = \mathrm{Out}^{\mathrm{FCP}}(\Pi_2)$ [cf. [CmbCsp], Definition 1.1, (iv); Theorem 2.3, (ii), (iv), of the present monograph; our assumption that $X^{\log}$ is of *type* $(0, 3)$], it follows from [CmbCsp], Corollary 1.14, (i), together with the *injectivity portion* of [CmbCsp], Theorem A, (i), that α *commutes with every modular outer symmetry* on Π_2; in particular, α *commutes* with σ. Thus, it follows immediately from [CmbCsp], Corollary 1.14, (iii), that the above outer isomorphism $\Pi_{v^{\mathrm{new}}} \xrightarrow{\sim} \Pi_v$ satisfies the condition of assertion (iii).

Next, suppose that $\alpha|_{\Pi_{v^{\mathrm{new}}}} \in \mathrm{Out}(\Pi_{v^{\mathrm{new}}})^{\Delta}$. If we write $\alpha^\sigma \overset{\mathrm{def}}{=} \sigma \circ \alpha \circ \sigma^{-1} (\in \mathrm{Out}^{\mathrm{FC}}(\Pi_2)^{\mathrm{cusp}}$—cf. [CmbCsp], Corollary 1.14, (i); Theorem 2.3, (ii), and Lemma 3.5 of the present monograph) and $(\alpha^\sigma)_1 \in \mathrm{Out}(\Pi_v)$ for the outomorphism of Π_v determined by α^σ, then it follows immediately from [CmbCsp], Corollary

1.14, (iii), that the outomorphisms $\alpha|_{\Pi_{v^{\text{new}}}}$, $(\alpha^\sigma)_1$ of $\Pi_{v^{\text{new}}}$, Π_v are *compatible* relative to the outer isomorphism $\Pi_{v^{\text{new}}} \xrightarrow{\sim} \Pi_v$ discussed above. Thus, since $\alpha|_{\Pi_{v^{\text{new}}}} \in \text{Out}(\Pi_{v^{\text{new}}})^\Delta$, we conclude that $(\alpha^\sigma)_1 \in \text{Out}(\Pi_v)^\Delta$. In particular, [since $\text{Out}^{\text{F}}(\Pi_2) = \text{Out}^{\text{FC}}(\Pi_2) = \text{Out}^{\text{FCP}}(\Pi_2)$—cf. [CmbCsp], Definition 1.1, (iv); Theorem 2.3, (ii), (iv), of the present monograph; our assumption that $X^{\log}$ is of *type* (0, 3)] it follows from [CmbCsp], Corollary 1.14, (i), together with the *injectivity portion* of [CmbCsp], Theorem A, (i), that α^σ *commutes with every modular outer symmetry* on Π_2. Thus, we conclude that α^σ *commutes* with σ^{-1}, which implies that $\alpha = \alpha^\sigma$. This completes the proof of assertion (iii). □

Lemma 3.14 (Commensurator of the Closed Subgroup Arising from a Certain Second Log Configuration Space) *Let $i \in E$, $j \in E$, x, and $z_{i,j,x}$ be as in Lemma 3.6; let $v \in \text{Vert}(\mathcal{G}_{j\in E\setminus\{i\},x})$. Then, by applying a similar argument to the argument used in [CmbCsp], Definition 2.1, (iii), (vi), or [NodNon], Definition 5.1, (ix), (x) [i.e., by considering the portion of the underlying scheme X_E of $X_E^{\log}$ corresponding to the underlying scheme $(X_v)_2$ of the 2-nd log configuration space $(X_v)_2^{\log}$ of the stable log curve $X_v^{\log}$ determined by $\mathcal{G}_{j\in E\setminus\{i\},x}|_v$—cf. [CbTpI], Definition 2.1, (iii)], one obtains a closed subgroup*

$$(\Pi_v)_2 \subseteq \Pi_{E/(E\setminus\{i,j\})}$$

[which is well-defined up to Π_E-conjugation]. Write

$$(\Pi_v)_{2/1} \stackrel{\text{def}}{=} (\Pi_v)_2 \cap \Pi_{E/(E\setminus\{i\})} \subseteq (\Pi_v)_2.$$

[Thus, one verifies easily that there exists a natural commutative diagram

$$\begin{array}{ccccccccc}
1 & \longrightarrow & (\Pi_v)_{2/1} & \longrightarrow & (\Pi_v)_2 & \longrightarrow & \Pi_v & \longrightarrow & 1 \\
 & & \downarrow & & \downarrow & & \downarrow & & \\
1 & \longrightarrow & \Pi_{E/(E\setminus\{i\})} & \longrightarrow & \Pi_{E/(E\setminus\{i,j\})} & \xrightarrow{p^{\Pi}_{E/(E\setminus\{i\})}} & \Pi_{(E\setminus\{i\})/(E\setminus\{i,j\})} & \longrightarrow & 1
\end{array}$$

—where we use the notation Π_v to denote a verticial subgroup of $\Pi_{\mathcal{G}_{j\in E\setminus\{i\},x}} \xleftarrow{\sim} \Pi_{(E\setminus\{i\})/(E\setminus\{i,j\})}$ associated to $v \in \text{Vert}(\mathcal{G}_{j\in E\setminus\{i\},x})$, the horizontal sequences are ***exact****, and the vertical arrows are* ***injective****.] Then the following hold:*

(i) Suppose that $z_{i,j,x} \in \text{VCN}(\mathcal{G}_{j\in E\setminus\{i\},x})$ is ***contained*** *in $\mathcal{E}(v)$. Write $v^\circ \in \text{Vert}(\mathcal{G}_{i\in E,x})$ for the vertex of $\mathcal{G}_{i\in E,x}$ that corresponds to $v \in \text{Vert}(\mathcal{G}_{j\in E\setminus\{i\},x})$ via the bijections of Lemma 3.6, (i), (iv). Let Π_{v°, $\Pi_{v^{\text{new}}_{i,j,x}} \subseteq \Pi_{\mathcal{G}_{i\in E,x}} \xleftarrow{\sim} \Pi_{E/(E\setminus\{i\})}$ be verticial subgroups of $\Pi_{\mathcal{G}_{i\in E,x}} \xleftarrow{\sim} \Pi_{E/(E\setminus\{i\})}$ associated to the vertices v°, $v^{\text{new}}_{i,j,x} \in \text{Vert}(\mathcal{G}_{i\in E,x})$, respectively, such that $\Pi_{v^{\text{new}}_{i,j,x}} \subseteq (\Pi_v)_{2/1}$, and, moreover, $\Pi_{v^\circ} \cap \Pi_{v^{\text{new}}_{i,j,x}} \neq \{1\}$. Let us say that two $\Pi_{E/(E\setminus\{i\})}$-*

conjugates $\Pi^{\gamma}_{v^\circ}$, $\Pi^{\delta}_{v^{\mathrm{new}}_{i,j,x}}$ *[i.e., where* $\gamma, \delta \in \Pi_{E/(E\setminus\{i\})}$*] of* Π_{v°, $\Pi_{v^{\mathrm{new}}_{i,j,x}}$ *are* ***conjugate-adjacent*** *if* $\Pi^{\gamma}_{v^\circ} \cap \Pi^{\delta}_{v^{\mathrm{new}}_{i,j,x}} \neq \{1\}$. *Let us say that a finite sequence of* $\Pi_{E/(E\setminus\{i\})}$*-conjugates of* Π_{v°, $\Pi_{v^{\mathrm{new}}_{i,j,x}}$ *is a* ***conjugate-chain*** *if any two adjacent members of the finite sequence are conjugate-adjacent. Let us say that a subgroup of* $\Pi_{E/(E\setminus\{i\})}$ *is* ***conjugate-tempered*** *if it appears as the* ***first*** *member of a conjugate-chain whose* ***final*** *member is equal to* $\Pi_{v^{\mathrm{new}}_{i,j,x}}$. *Then* $(\Pi_v)_{2/1}$ *is equal to the subgroup of* $\Pi_{E/(E\setminus\{i\})}$ *topologically generated by the conjugate-tempered subgroups and the elements* $\delta \in \Pi_{E/(E\setminus\{i\})}$ *such that* $\Pi^{\delta}_{v^{\mathrm{new}}_{i,j,x}}$ *is conjugate-tempered.*

(ii) If $N_{\Pi_{E\setminus\{i\}}}(\Pi_v) = C_{\Pi_{E\setminus\{i\}}}(\Pi_v)$, *then* $N_{\Pi_E}((\Pi_v)_2) = C_{\Pi_E}((\Pi_v)_2)$.

(iii) If $C_{\Pi_{E\setminus\{i\}}}(\Pi_v) = \Pi_v \times Z_{\Pi_{E\setminus\{i\}}}(\Pi_v)$, *then* $C_{\Pi_E}((\Pi_v)_2) = (\Pi_v)_2 \times Z_{\Pi_E}((\Pi_v)_2)$.

(iv) Suppose that v is of ***type*** $\mathbf{(0,3)}$, *i.e., that* Π_v *is an* $\boldsymbol{(E\setminus\{i\})}$***-tripod*** *of* Π_n *[cf. Definition 3.3, (i)]. Then it holds that* $C_{\Pi_E}((\Pi_v)_2) = (\Pi_v)_2 \times Z_{\Pi_E}((\Pi_v)_2)$. *Thus, if an outomorphism* α *of* Π_E ***preserves*** *the* Π_E*-conjugacy class of* $(\Pi_v)_2$, *then one may define* $\alpha|_{(\Pi_v)_2} \in \mathrm{Out}((\Pi_v)_2)$ *[cf. Lemma 3.10, (i)].*

Proof First, we verify assertion (i). We begin by observing that it follows immediately from [NodNon], Lemma 1.9, (ii), together with the *commensurable terminality* of $\Pi_{v^{\mathrm{new}}_{i,j,x}} \subseteq \Pi_{E/(E\setminus\{i\})}$ [cf. [CmbGC], Proposition 1.2, (ii)], that the subgroup described in the final portion of the statement of assertion (i) is *contained* in $(\Pi_v)_{2/1}$. If $\#(\mathcal{N}(v^\circ) \cap \mathcal{N}(v^{\mathrm{new}}_{i,j,x})) = 1$, then assertion (i) follows immediately from a similar argument to the argument applied in the proof of [CmbCsp], Proposition 1.5, (iii), together with the various definitions involved [cf. also [NodNon], Lemma 1.9, (ii)]. Thus, we may assume without loss of generality that $\#(\mathcal{N}(v^\circ) \cap \mathcal{N}(v^{\mathrm{new}}_{i,j,x})) = 2$.

Write

- $e_1 \in \mathcal{N}(v^\circ) \cap \mathcal{N}(v^{\mathrm{new}}_{i,j,x})$ for the [*uniquely determined*—cf. [NodNon], Lemma 1.5] node such that $\Pi_{v^\circ} \cap \Pi_{v^{\mathrm{new}}_{i,j,x}}$ ($\neq \{1\}$) is a nodal subgroup associated to e_1 [cf. [NodNon], Lemma 1.9, (i)];
- e_2 for the *unique* element of $\mathcal{N}(v^\circ) \cap \mathcal{N}(v^{\mathrm{new}}_{i,j,x})$ such that $e_2 \neq e_1$ [so $\mathcal{N}(v^\circ) \cap \mathcal{N}(v^{\mathrm{new}}_{i,j,x}) = \{e_1, e_2\}$];
- $\mathbb{H}$ for the sub-semi-graph of *PSC-type* [cf. [CbTpI], Definition 2.2, (i)] of the underlying semi-graph of $\mathcal{G}_{i\in E,x}$ whose set of vertices $= \{v^\circ, v^{\mathrm{new}}_{i,j,x}\}$;
- $S \stackrel{\mathrm{def}}{=} \mathrm{Node}(\mathcal{G}_{i\in E,x}|_{\mathbb{H}}) \setminus \{e_1, e_2\}$ [cf. [CbTpI], Definition 2.2, (ii)];
- $\mathcal{H} \stackrel{\mathrm{def}}{=} (\mathcal{G}_{i\in E,x}|_{\mathbb{H}})_{\succ S}$ [which is *well-defined* since, as is easily verified, S is *not of separating type* as a subset of $\mathrm{Node}(\mathcal{G}_{i\in E,x}|_{\mathbb{H}})$—cf. [CbTpI], Definition 2.5, (i), (ii)].

Then it follows immediately from the construction of $\mathcal{H}$ that $\mathcal{H}_{\rightsquigarrow\{e_1\}}$ [cf. [CbTpI], Definition 2.8], where we observe that one verifies easily that the node e_1 of $\mathcal{G}_{i\in E,x}$ may be regarded as a node of $\mathcal{H}$, is *cyclically primitive* [cf. [CbTpI], Definition 4.1]. Moreover, it follows immediately from [NodNon], Lemma 1.9, (ii), together with the various definitions involved, that $(\Pi_v)_{2/1} \subseteq \Pi_{E/(E\setminus\{i\})} \xrightarrow{\sim} \Pi_{\mathcal{G}_{i\in E,x}}$ may be *characterized uniquely* as the closed subgroup of $\Pi_{\mathcal{G}_{i\in E,x}}$ that *contains* $\Pi_{v^{\mathrm{new}}_{i,j,x}} \subseteq \Pi_{\mathcal{G}_{i\in E,x}}$ and, moreover, *belongs* to the $\Pi_{\mathcal{G}_{i\in E,x}}$-conjugacy class of closed subgroups of $\Pi_{\mathcal{G}_{i\in E,x}}$ obtained by forming the image of the composite of outer homomorphisms

$$\Pi_{\mathcal{H}_{\rightsquigarrow\{e_1\}}} \overset{\Phi_{\mathcal{H}_{\rightsquigarrow\{e_1\}}}}{\xrightarrow{\sim}} \Pi_{\mathcal{H}} \hookrightarrow \Pi_{\mathcal{G}_{i\in E,x}}$$

[cf. [CbTpI], Definition 2.10]—where the second arrow is the outer injection discussed in [CbTpI], Proposition 2.11. In particular, it follows from the *commensurable terminality* of $(\Pi_v)_{2/1}$ in $\Pi_{\mathcal{G}_{i\in E,x}}$ [cf. [CmbGC], Proposition 1.2, (ii)] that this characterization of $(\Pi_v)_{2/1}$ determines an outer isomorphism $\Pi_{\mathcal{H}_{\rightsquigarrow\{e_1\}}} \xrightarrow{\sim} (\Pi_v)_{2/1}$.

On the other hand, it follows immediately from a similar argument to the argument applied in the proof of [CmbCsp], Proposition 1.5, (iii), together with the various definitions involved [cf. also [NodNon], Lemma 1.9, (ii)], that the image of the closed subgroup of $(\Pi_v)_{2/1}$ topologically generated by Π_{v° and $\Pi_{v^{\mathrm{new}}_{i,j,x}}$ via the inverse $(\Pi_v)_{2/1} \xrightarrow{\sim} \Pi_{\mathcal{H}_{\rightsquigarrow\{e_1\}}}$ of this outer isomorphism is a verticial subgroup of $\Pi_{\mathcal{H}_{\rightsquigarrow\{e_1\}}}$ associated to the *unique* vertex of $\mathcal{H}_{\rightsquigarrow\{e_1\}}$. Thus, since $\mathcal{H}_{\rightsquigarrow\{e_1\}}$ is *cyclically primitive*, assertion (i) follows immediately from [CmbGC], Proposition 1.2, (ii); [NodNon], Lemma 1.9, (ii), together with the description of the structure of a certain *tempered covering* of $\mathcal{H}_{\rightsquigarrow\{e_1\}}$ given in [CbTpI], Lemma 4.3. This completes the proof of assertion (i).

Next, we verify assertion (ii). Since $(\Pi_v)_{2/1} = (\Pi_v)_2 \cap \Pi_{E/(E\setminus\{i\})}$ is *commensurably terminal* in $\Pi_{E/(E\setminus\{i\})}$ [cf. [CmbGC], Proposition 1.2, (ii)], assertion (ii) follows immediately from Lemma 3.9, (iv). This completes the proof of assertion (ii). Next, we verify assertion (iii). First, let us observe that if $\mathcal{E}(v) = \emptyset$, then one verifies immediately that the vertical arrows of the commutative diagram in the statement of Lemma 3.14 are *isomorphisms*, and hence that assertion (iii) holds. Thus, we may assume that $\mathcal{E}(v) \neq \emptyset$. Next, let us observe that it follows from assertion (ii) that $N_{\Pi_E}((\Pi_v)_2) = C_{\Pi_E}((\Pi_v)_2)$. Thus, in light of the *slimness* of $(\Pi_v)_2$ [cf. [MzTa], Proposition 2.2, (ii)], to verify assertion (iii), it suffices to verify that the natural outer action of $N_{\Pi_E}((\Pi_v)_2)$ on $(\Pi_v)_2$ is *trivial*. On the other hand, since [one verifies easily that] the natural outer action $N_{\Pi_E}((\Pi_v)_2) \to \mathrm{Out}((\Pi_v)_2)$ *factors* through $\mathrm{Out}^{\mathrm{F}}((\Pi_v)_2) \subseteq \mathrm{Out}((\Pi_v)_2)$, it follows from the *injectivity portion* of Theorem 2.3, (i) [cf. our assumption that $\mathcal{E}(v) \neq \emptyset$], that to verify the *triviality* in question, it suffices to verify that the natural outer action of $N_{\Pi_E}((\Pi_v)_2)$ on Π_v is *trivial*. But this follows from the equality $C_{\Pi_{E\setminus\{i\}}}(\Pi_v) = \Pi_v \times Z_{\Pi_{E\setminus\{i\}}}(\Pi_v)$. This completes the proof of assertion (iii). Assertion (iv) follows immediately from assertion (iii), together with Lemma 3.12, (i). This completes the proof of Lemma 3.14. □

Lemma 3.15 (Preservation of Various Subgroups of Geometric Origin) *In the notation of Lemma 3.14, let* $\widetilde{\alpha}$ *be an F-admissible automorphism of* Π_E. *Write* $\widetilde{\alpha}_{E\setminus\{i\}}$, $\widetilde{\alpha}_{E/(E\setminus\{i\})}$ *for the automorphisms of* $\Pi_{E\setminus\{i\}}$, $\Pi_{E/(E\setminus\{i\})}$ *determined by* $\widetilde{\alpha}$; α, $\alpha_{E\setminus\{i\}}$, $\alpha_{E/(E\setminus\{i\})}$ *for the outomorphisms of* Π_E, $\Pi_{E\setminus\{i\}}$, $\Pi_{E/(E\setminus\{i\})}$ *determined by* $\widetilde{\alpha}$, $\widetilde{\alpha}_{E\setminus\{i\}}$, $\widetilde{\alpha}_{E/(E\setminus\{i\})}$, *respectively. Suppose that there exist an edge* $e \in \mathrm{Edge}(\mathcal{G}_{j\in E\setminus\{i\},x})$ *of* $\mathcal{G}_{j\in E\setminus\{i\},x}$ *that belongs to* $\mathcal{E}(v) \subseteq \mathrm{Edge}(\mathcal{G}_{j\in E\setminus\{i\},x})$ *and a pair* $\Pi_e \subseteq \Pi_v \subseteq \Pi_{\mathcal{G}_{j\in E\setminus\{i\},x}} \overset{\sim}{\leftarrow} \Pi_{(E\setminus\{i\})/(E\setminus\{i,j\})}$ *of VCN-subgroups associated to* $e \in \mathrm{Edge}(\mathcal{G}_{j\in E\setminus\{i\},x})$, $v \in \mathrm{Vert}(\mathcal{G}_{j\in E\setminus\{i\},x})$, *respectively, such that*

$$\widetilde{\alpha}_{E\setminus\{i\}}(\Pi_e) = \Pi_e \subseteq \widetilde{\alpha}_{E\setminus\{i\}}(\Pi_v) = \Pi_v.$$

Suppose, moreover, either that

(a) the outomorphism $\alpha_{E/(E\setminus\{i\})}$ *of* $\Pi_{\mathcal{G}_{i\in E,x}} \overset{\sim}{\leftarrow} \Pi_{E/(E\setminus\{i\})}$ *maps* ***some*** *cuspidal inertia subgroup of* $\Pi_{\mathcal{G}_{i\in E,x}} \overset{\sim}{\leftarrow} \Pi_{E/(E\setminus\{i\})}$ *to a cuspidal inertia subgroup of* $\Pi_{\mathcal{G}_{i\in E,x}} \overset{\sim}{\leftarrow} \Pi_{E/(E\setminus\{i\})}$, *or that*
(b) $e \in \mathrm{Cusp}(\mathcal{G}_{j\in E\setminus\{i\},x})$.

[For example, condition (a) holds if the outomorphism $\alpha_{E/(E\setminus\{i\})}$ *of* $\Pi_{\mathcal{G}_{i\in E,x}} \overset{\sim}{\leftarrow} \Pi_{E/(E\setminus\{i\})}$ *is* ***group-theoretically cuspidal****—cf. [CmbGC], Definition 1.4, (iv).] Write* $T \subseteq \Pi_E$ *for the* ***E-tripod*** *of* Π_n *[cf. Definition 3.3, (i)] arising from* $e \in \mathrm{Edge}(\mathcal{G}_{j\in E\setminus\{i\},x})$ *[cf. Definition 3.7, (i)]. Then the following hold:*

(i) The outomorphism α ***preserves*** *the* Π_E*-conjugacy classes of* T, $(\Pi_v)_2 \subseteq \Pi_E$. *If, moreover, the outomorphism* $\alpha_{E/(E\setminus\{i\})}$ *of* $\Pi_{\mathcal{G}_{i\in E,x}} \overset{\sim}{\leftarrow} \Pi_{E/(E\setminus\{i\})}$ *is* ***group-theoretically cuspidal*** *[cf. [CmbGC], Definition 1.4, (iv)], then the outomorphism* $\alpha|_T$ *[cf. Lemma 3.12, (i)] of* T *is* ***contained*** *in* $\mathrm{Out}^{\mathrm{C}}(T)^{\mathrm{cusp}}$ *[cf. Definition 3.4, (i)].*
(ii) Suppose, moreover, that v *is of* ***type*** $\mathbf{(0,3)}$*—i.e., that* Π_v *is an* $\boldsymbol{(E\setminus\{i\})}$***-tripod*** *of* Π_n*—and that* $\alpha_{E\setminus\{i\}}|_{\Pi_v} \in \mathrm{Out}^{\mathrm{C}}(\Pi_v)^{\mathrm{cusp}}$ *[cf. Lemma 3.12, (i)]. Then there exists a* ***geometric*** *[cf. Definition 3.4, (ii)] outer isomorphism* $T \overset{\sim}{\rightarrow} \Pi_v$ *which satisfies the following condition:*

> *If either* $\alpha|_T \in \mathrm{Out}(T)^{\Delta}$ *[cf. (i)] or* $\alpha_{E\setminus\{i\}}|_{\Pi_v} \in \mathrm{Out}(\Pi_v)^{\Delta}$, *then the outomorphisms* $\alpha|_T$, $\alpha_{E\setminus\{i\}}|_{\Pi_v}$ *of* T, Π_v *are* ***compatible*** *relative to the outer isomorphism in question* $T \overset{\sim}{\rightarrow} \Pi_v$.

If, moreover, Π_v *is* $\boldsymbol{(E\setminus\{i\})}$***-strict*** *[cf. Definition 3.3, (iii)], then the following hold:*

(1) If $\#(E\setminus\{i\}) = 1$ *[i.e.,* Π_v *satisfies condition (1) of Lemma 3.8, (ii)], then* T *is* ***E-strict*** *[i.e.,* T *satisfies one of the two conditions* (2_{C}), (2_{N}) *of Lemma 3.8, (ii)].*

(2) *If* $\#(E \setminus \{i\}) = 2$ *[i.e.,* Π_v *satisfies one of the two conditions* (2_{C}), (2_{N}) *of Lemma 3.8, (ii)], and the edge* $e \in \mathrm{Edge}(\mathcal{G}_{j \in E\setminus\{i\},x})$ *is the* ***unique diagonal cusp*** *of* $\mathcal{G}_{j \in E\setminus\{i\},x}$ *[cf. Lemma 3.2, (ii)], then* T *is* ***E-strict*** *[i.e.,* T *satisfies condition (3) of Lemma 3.8, (ii)], hence also* ***central*** *[cf. Definition 3.7, (ii)].*

Proof First, let us observe that one verifies easily—by replacing x by a suitable k-valued geometric point of $X_n(k)$ that *lifts* $x_{E\setminus\{i,j\}} \in X_{E\setminus\{i,j\}}(k)$ [note that this does *not affect* "$\mathcal{G}_{j \in E\setminus\{i\},x}$"!]—that, to verify Lemma 3.15, we may assume without loss of generality that $z_{i,j,x} = e \in \mathrm{Edge}(\mathcal{G}_{j \in E\setminus\{i\},x})$.

Now we verify assertion (i). First, let us observe that one verifies easily—by replacing $X_E^{\log}$ by the base-change of $p_{E\setminus\{i,j\}}^{\log} : X_E^{\log} \to X_{E\setminus\{i,j\}}^{\log}$ by a suitable morphism of log schemes $(\mathrm{Spec}\, k)^{\log} \to X_{E\setminus\{i,j\}}^{\log}$ that lies over $x_{E\setminus\{i,j\}} \in X_{E\setminus\{i,j\}}(k)$ [cf. Definition 3.1, (i)]—that, to verify assertion (i), we may assume without loss of generality that $\#E = 2$. Then it follows immediately from Lemma 3.13, (i), that $\alpha_{E/(E\setminus\{i\})}$ *preserves* the $\Pi_{E/(E\setminus\{i\})}$-conjugacy class of $T (= \Pi_{v^{\mathrm{new}}_{i,j,x}}) \subseteq \Pi_{E/(E\setminus\{i\})}$. Moreover, it follows immediately from Lemma 3.13, (i), (ii), together with Lemma 3.14, (i), that $\alpha_{E/(E\setminus\{i\})}$ *preserves* the $\Pi_{E/(E\setminus\{i\})}$-conjugacy classes of the *normally terminal* closed subgroups $\Pi_{v^\circ} \subseteq (\Pi_v)_{2/1} \subseteq \Pi_{E/(E\setminus\{i\})}$ [cf. [CmbGC], Proposition 1.2, (ii)]. In particular, since $\widetilde{\alpha}_{E\setminus\{i\}}(\Pi_v) = \Pi_v$, by considering the natural isomorphism $(\Pi_v)_2 \xrightarrow{\sim} (\Pi_v)_{2/1} \overset{\mathrm{out}}{\rtimes} \Pi_v$ [cf. the upper exact sequence of the commutative diagram in the statement of Lemma 3.14; the discussion entitled "*Topological groups*" in [CbTpI], §0], we conclude that α_E *preserves* the Π_E-conjugacy class of $(\Pi_v)_2 \subseteq \Pi_E$.

Next, suppose that the outomorphism $\alpha_{E/(E\setminus\{i\})}$ of $\Pi_{\mathcal{G}_{i\in E,x}} \xleftarrow{\sim} \Pi_{E/(E\setminus\{i\})}$ is *group-theoretically cuspidal*. Then it follows from Lemma 3.13, (i), that $\alpha|_T \in \mathrm{Out}^{\mathrm{C}}(T)$. Moreover, since $\alpha_{E/(E\setminus\{i\})}$ is *group-theoretically cuspidal*, it follows immediately from Lemma 3.2, (iv), that $\alpha_{E/(E\setminus\{i\})}$ *fixes* the $\Pi_{E/(E\setminus\{i\})}$-conjugacy class of cuspidal inertia subgroups associated to each element $\in C(v^{\mathrm{new}}_{i,j,x}) (\ni c^{\mathrm{diag}}_{i,j,x})$. Thus, to verify that $\alpha|_T \in \mathrm{Out}^{\mathrm{C}}(T)^{\mathrm{cusp}}$, it suffices to verify that $\alpha_{E/(E\setminus\{i\})}$ *fixes* the $\Pi_{E/(E\setminus\{i\})}$-conjugacy class of nodal subgroups of $\Pi_{\mathcal{G}_{i\in E,x}} \xleftarrow{\sim} \Pi_{E/(E\setminus\{i\})}$ associated to each element of $\mathcal{N}(v^{\mathrm{new}}_{i,j,x}) \cap \mathcal{N}(v^\circ)$. To this end, let $e^\circ \in \mathcal{N}(v^{\mathrm{new}}_{i,j,x}) \cap \mathcal{N}(v^\circ)$ and $\Pi_{e^\circ} \subseteq \Pi_{\mathcal{G}_{i\in E,x}} \xleftarrow{\sim} \Pi_{E/(E\setminus\{i\})}$ a nodal subgroup associated to the node e° such that $\Pi_{e^\circ} \subseteq \Pi_{v^\circ}$. Now let us observe that one verifies easily that the closed subgroups $\Pi_{e^\circ} \subseteq \Pi_{v^\circ} \subseteq \Pi_{\mathcal{G}_{i\in E,x}} \xleftarrow{\sim} \Pi_{E/(E\setminus\{i\})}$ map *bijectively onto* VCN-subgroups of $\Pi_{\mathcal{G}_{i\in E\setminus\{j\},x}} \xleftarrow{\sim} \Pi_{(E\setminus\{j\})/(E\setminus\{i,j\})}$ associated, respectively, to the edge and vertex of $\mathcal{G}_{i\in E\setminus\{j\},x}$ that correspond, via the bijections of Lemma 3.6, (i), to e, $v \in \mathrm{VCN}(\mathcal{G}_{j\in E\setminus\{i\},x})$. In particular, if $\widetilde{\beta}$ is the composite of $\widetilde{\alpha}$ with some $\Pi_{E/(E\setminus\{i\})}$-inner automorphism such that $\widetilde{\beta}(\Pi_{v^\circ}) = \Pi_{v^\circ}$ [cf. the preceding paragraph], then it follows immediately from our assumption that $\widetilde{\alpha}_{E\setminus\{i\}}(\Pi_e) = \Pi_e \subseteq \widetilde{\alpha}_{E\setminus\{i\}}(\Pi_v) = \Pi_v$, together with [CbTpI], Theorem A, (i), and [CmbGC], Proposition 1.2, (ii), that the automorphism of Π_{v° determined by $\widetilde{\beta}$ *preserves* the Π_{v°-conjugacy class of Π_{e°. Thus, $\alpha_{E/(E\setminus\{i\})}$ *fixes* the $\Pi_{E/(E\setminus\{i\})}$-conjugacy class of Π_{e°, as desired. This completes the proof of assertion (i).

Next, we verify assertion (ii). Since v is of *type* $(0, 3)$, it follows from assertion (i), together with Lemma 3.14, (iv), that one may define $\alpha|_{(\Pi_v)_2} \in \mathrm{Out}((\Pi_v)_2)$. Thus, by applying Lemma 3.13, (iii), to $\alpha|_{(\Pi_v)_2} \in \mathrm{Out}((\Pi_v)_2)$, one verifies easily that the first portion of assertion (ii) holds. The final portion of assertion (ii) follows immediately from the descriptions given in the four conditions of Lemma 3.8, (ii), together with the various definitions involved. This completes the proof of assertion (ii). □

Theorem 3.16 (Outomorphisms Preserving Tripods) *In the notation of the beginning of this chapter, let* $E \subseteq \{1, \cdots, n\}$ *and* $T \subseteq \Pi_E$ *an* ***E-tripod*** *of* Π_n *[cf. Definition 3.3, (i)]. Let us write*

$$\mathrm{Out}^{\mathrm{F}}(\Pi_n)[T] \subseteq \mathrm{Out}^{\mathrm{F}}(\Pi_n)$$

for the [closed] subgroup of $\mathrm{Out}^{\mathrm{F}}(\Pi_n)$ *[cf. [CmbCsp], Definition 1.1, (ii)] consisting of F-admissible outomorphisms* α *of* Π_n *such that the outomorphism of* Π_E *determined by* α *preserves the* Π_E*-conjugacy class of* $T \subseteq \Pi_E$*. Then the following hold:*

(i) It holds that

$$C_{\Pi_E}(T) = T \times Z_{\Pi_E}(T).$$

Thus, by applying Lemma 3.10, (i), to outomorphisms of Π_E *determined by elements of* $\mathrm{Out}^{\mathrm{F}}(\Pi_n)[T]$*, one obtains a natural homomorphism*

$$\mathfrak{T}_T : \mathrm{Out}^{\mathrm{F}}(\Pi_n)[T] \longrightarrow \mathrm{Out}(T).$$

Let us write

$$\mathrm{Out}^{\mathrm{F}}(\Pi_n)[T : \{C\}], \ \mathrm{Out}^{\mathrm{F}}(\Pi_n)[T : \{|C|\}], \ \mathrm{Out}^{\mathrm{F}}(\Pi_n)[T : \{\Delta\}],$$
$$\mathrm{Out}^{\mathrm{F}}(\Pi_n)[T : \{+\}] \subseteq \mathrm{Out}^{\mathrm{F}}(\Pi_n)[T]$$

for the [closed] subgroups of $\mathrm{Out}^{\mathrm{F}}(\Pi_n)[T]$ *obtained by forming the respective inverse images via* $\mathfrak{T}_T$ *of the closed subgroups* $\mathrm{Out}^{\mathrm{C}}(T)$, $\mathrm{Out}^{\mathrm{C}}(T)^{\mathrm{cusp}}$, $\mathrm{Out}(T)^{\Delta}$, $\mathrm{Out}(T)^{+} \subseteq \mathrm{Out}(T)$ *[cf. Definition 3.4, (i)]. For each subset* $S \subseteq \{C, |C|, \Delta, +\}$*, let us write*

$$\mathrm{Out}^{\mathrm{F}}(\Pi_n)[T : S] \stackrel{\mathrm{def}}{=} \bigcap_{\square \in S} \mathrm{Out}^{\mathrm{F}}(\Pi_n)[T : \{\square\}] \subseteq \mathrm{Out}^{\mathrm{F}}(\Pi_n)[T];$$

$$\mathrm{Out}^{\mathrm{FC}}(\Pi_n)[T : S] \stackrel{\mathrm{def}}{=} \mathrm{Out}^{\mathrm{F}}(\Pi_n)[T : S] \cap \mathrm{Out}^{\mathrm{FC}}(\Pi_n) \subseteq \mathrm{Out}^{\mathrm{FC}}(\Pi_n)$$

[cf. [CmbCsp], Definition 1.1, (ii)]. Suppose, moreover, that we are given an element $\sigma \in \mathfrak{S}_n \subseteq \mathrm{Out}(\Pi_n)$ *[cf. the discussion at the beginning of this chapter] and a lifting* $\widetilde{\sigma} \in \mathrm{Aut}(\Pi_n)$ *of* $\sigma \in \mathfrak{S}_n \subseteq \mathrm{Out}(\Pi_n)$*. Write*

$$T^{\widetilde{\sigma}} \subseteq \Pi_{\sigma(E)}$$

for the image of $T \subseteq \Pi_E$ *by the isomorphism* $\Pi_E \xrightarrow{\sim} \Pi_{\sigma(E)}$ *determined by* $\widetilde{\sigma} \in \mathrm{Aut}(\Pi_n)$ *[which thus implies that* $T^{\widetilde{\sigma}} \subseteq \Pi_{\sigma(E)}$ *is a* $\sigma(E)$*-tripod of* Π_n*— cf. Remark 3.7.1] and*

$$\mathrm{Out}^{\mathrm{F}}(\Pi_n)[T, \widetilde{\sigma}] \stackrel{\mathrm{def}}{=} \mathrm{Out}^{\mathrm{F}}(\Pi_n)[T] \cap \mathrm{Out}^{\mathrm{F}}(\Pi_n)[T^{\widetilde{\sigma}}] \subseteq \mathrm{Out}^{\mathrm{F}}(\Pi_n),$$

$$\mathrm{Out}^{\mathrm{FC}}(\Pi_n)[T, \widetilde{\sigma}] \stackrel{\mathrm{def}}{=} \mathrm{Out}^{\mathrm{F}}(\Pi_n)[T, \widetilde{\sigma}] \cap \mathrm{Out}^{\mathrm{FC}}(\Pi_n) \subseteq \mathrm{Out}^{\mathrm{FC}}(\Pi_n).$$

Then the resulting isomorphism $T \xrightarrow{\sim} T^{\widetilde{\sigma}}$ *is* ***geometric*** *[cf. Definition 3.4, (ii)]. Moreover, we have a commutative diagram*

$$\begin{array}{ccc} \mathrm{Out}^{\mathrm{F}}(\Pi_n)[T, \widetilde{\sigma}] & = & \mathrm{Out}^{\mathrm{F}}(\Pi_n)[T, \widetilde{\sigma}] \\ {\scriptstyle \mathfrak{T}_T}\downarrow & & \downarrow{\scriptstyle \mathfrak{T}_{T^{\widetilde{\sigma}}}} \\ \mathrm{Out}(T) & \xrightarrow{\sim} & \mathrm{Out}(T^{\widetilde{\sigma}}) \end{array}$$

—where the upper horizontal equality is an equality of subgroups of the group $\mathrm{Out}^{\mathrm{F}}(\Pi_n)$*, and the lower horizontal arrow is the isomorphism obtained by conjugating by the above* ***geometric*** *isomorphism* $T \xrightarrow{\sim} T^{\widetilde{\sigma}}$ *[i.e., induced by* $\widetilde{\sigma} \in \mathrm{Aut}(\Pi_n)$*]. Finally, the equalities*

$$\mathrm{Out}^{\mathrm{FC}}(\Pi_n)[T, \widetilde{\sigma}] = \mathrm{Out}^{\mathrm{FC}}(\Pi_n)[T] = \mathrm{Out}^{\mathrm{FC}}(\Pi_n)[T^{\widetilde{\sigma}}]$$

hold; if, moreover, one of the following conditions is satisfied, then the equalities

$$\mathrm{Out}^{\mathrm{F}}(\Pi_n)[T, \widetilde{\sigma}] = \mathrm{Out}^{\mathrm{F}}(\Pi_n)[T] = \mathrm{Out}^{\mathrm{F}}(\Pi_n)[T^{\widetilde{\sigma}}]$$

hold:

(i-1) $(r, n) \neq (0, 2)$.
(i-2) T *is* ***E-strict*** *[cf. Definition 3.3, (iii)].*

(ii) It holds that

$$\mathrm{Out}^{\mathrm{F}}(\Pi_n)[T : \{C, \Delta\}] = \mathrm{Out}^{\mathrm{F}}(\Pi_n)[T : \{|C|, \Delta\}].$$

*(iii) Suppose that T is **1-descendable** [cf. Definition 3.3, (iv)]. Then it holds that*

$$\mathrm{Out}^{\mathrm{FC}}(\Pi_n)[T : \{|C|\}] = \mathrm{Out}^{\mathrm{FC}}(\Pi_n)[T : \{|C|, +\}].$$

If, moreover, one of the following conditions is satisfied, then it holds that

$$\mathrm{Out}^{\mathrm{F}}(\Pi_n)[T : \{|C|\}] = \mathrm{Out}^{\mathrm{F}}(\Pi_n)[T : \{|C|, +\}].$$

*(iii-1) T is **2-descendable** [cf. Definition 3.3, (iv)].*
(iii-2) There exists a subset $E' \subseteq E$ such that:

(iii-2-a) $E' \neq \{1, \cdots, n\}$;
*(iii-2-b) the image $p^{\Pi}_{E/E'}(T) \subseteq \Pi_{E'}$ is a **cusp-supporting E'-tripod** of Π_n [cf. Definition 3.3, (i)].*

*(iv) Let $i, j \in E$ be two **distinct** elements of E; $e \in \mathrm{Edge}(\mathcal{G}_{j \in E \setminus \{i\}, x})$ [cf. Definition 3.1, (iii)]; $\alpha \in \mathrm{Out}^{\mathrm{F}}(\Pi_n)$. Suppose that T **arises** from $e \in \mathrm{Edge}(\mathcal{G}_{j \in E \setminus \{i\}, x})$ [cf. Definition 3.7, (i)], and that the outomorphism of $\Pi_{E \setminus \{i\}}$ determined by α **preserves** the $\Pi_{E \setminus \{i\}}$-conjugacy class of an edge-like subgroup of $\Pi_{E \setminus \{i\}}$ associated to $e \in \mathrm{Edge}(\mathcal{G}_{j \in E \setminus \{i\}, x})$ [cf. Definition 3.1, (iv)]. Suppose, moreover, that one of the following conditions is satisfied:*

(iv-1) $\alpha \in \mathrm{Out}^{\mathrm{FC}}(\Pi_n)$.
(iv-2) $\#E \leq n - 1$.
(iv-3) $e \in \mathrm{Cusp}(\mathcal{G}_{j \in E \setminus \{i\}, x})$.

Then $\alpha \in \mathrm{Out}^{\mathrm{F}}(\Pi_n)[T]$. Suppose, further, that either condition (iv-1) or condition (iv-2) is satisfied. Then $\alpha \in \mathrm{Out}^{\mathrm{F}}(\Pi_n)[T : \{C\}]$; if, in addition, condition (iv-3) is satisfied, then $\alpha \in \mathrm{Out}^{\mathrm{F}}(\Pi_n)[T : \{|C|\}]$.

*(v) Suppose that T is **central** [cf. Definition 3.7, (ii)]. If $n \geq 4$ [i.e., T is **1-descendable**], then it holds that*

$$\mathrm{Out}^{\mathrm{F}}(\Pi_n) = \mathrm{Out}^{\mathrm{FC}}(\Pi_n)[T : \{|C|, \Delta, +\}].$$

*If $n = 3$ [i.e., T is **not 1-descendable**], then it holds that*

$$\mathrm{Out}^{\mathrm{FC}}(\Pi_n) = \mathrm{Out}^{\mathrm{FC}}(\Pi_n)[T : \{|C|, \Delta\}]$$
$$\subseteq \mathrm{Out}^{\mathrm{F}}(\Pi_n) = \mathrm{Out}^{\mathrm{F}}(\Pi_n)[T : \{\Delta\}];$$

if, moreover, $r \neq 0$, then

$$\mathrm{Out}^{\mathrm{F}}(\Pi_n) = \mathrm{Out}^{\mathrm{FC}}(\Pi_n)[T : \{|C|, \Delta, +\}].$$

Proof We begin the proof of Theorem 3.16 with the following claim:

Claim 3.16.A: Let $E' \subseteq E$ be a subset such that the image $T_{E'}$ of T via $p^{\Pi}_{E/E'}\colon \Pi_E \twoheadrightarrow \Pi_{E'}$ is an E'-*tripod*. Thus, one verifies easily that one obtains a(n) [necessarily *geometric*] outer isomorphism $T \xrightarrow{\sim} T_{E'}$ [induced by $p^{\Pi}_{E/E'}$]. Then we have an *inclusion* $\mathrm{Out}^{\mathrm{F}}(\Pi_n)[T] \subseteq \mathrm{Out}^{\mathrm{F}}(\Pi_n)[T_{E'}]$, and, moreover, the diagram

$$\begin{array}{ccc}
\mathrm{Out}^{\mathrm{F}}(\Pi_n)[T] & \subseteq & \mathrm{Out}^{\mathrm{F}}(\Pi_n)[T_{E'}] \\
\mathfrak{T}_T \downarrow & & \downarrow \mathfrak{T}_{T_{E'}} \\
\mathrm{Out}(T) & \xrightarrow{\sim} & \mathrm{Out}(T_{E'})
\end{array}$$

—where the lower horizontal arrow is the isomorphism determined by the isomorphism $T \xrightarrow{\sim} T_{E'}$ induced by $p^{\Pi}_{E/E'}$—*commutes*.

Indeed, this follows immediately from the various definitions involved. This completes the proof of Claim 3.16.A.

Next, we verify assertion (i). The equality $C_{\Pi_E}(T) = T \times Z_{\Pi_E}(T)$ of the first display in assertion (i) follows from Lemma 3.12, (i). Moreover, the *geometricity* of the isomorphism $T \xrightarrow{\sim} T^{\tilde{\sigma}}$ follows immediately from the various definitions involved. Next, let us observe that if $(r, n) \neq (0, 2)$, then the *commutativity* of the displayed diagram in assertion (i) and the equalities

$$\mathrm{Out}^{\mathrm{F}}(\Pi_n)[T, \tilde{\sigma}] = \mathrm{Out}^{\mathrm{F}}(\Pi_n)[T] = \mathrm{Out}^{\mathrm{F}}(\Pi_n)[T^{\tilde{\sigma}}]$$

in assertion (i) may be easily derived from the fact that the closed subgroup $\mathrm{Out}^{\mathrm{F}}(\Pi_n) \subseteq \mathrm{Out}(\Pi_n)$ *centralizes* the closed subgroup $\mathfrak{S}_n \subseteq \mathrm{Out}^{\mathrm{F}}(\Pi_n)$ [cf. Theorem 2.3, (iv)]. Moreover, the equalities

$$\mathrm{Out}^{\mathrm{FC}}(\Pi_n)[T, \tilde{\sigma}] = \mathrm{Out}^{\mathrm{FC}}(\Pi_n)[T] = \mathrm{Out}^{\mathrm{FC}}(\Pi_n)[T^{\tilde{\sigma}}]$$

in assertion (i) may be easily derived from the fact that the closed subgroup $\mathrm{Out}^{\mathrm{FC}}(\Pi_n) \subseteq \mathrm{Out}(\Pi_n)$ *centralizes* the closed subgroup $\mathfrak{S}_n \subseteq \mathrm{Out}^{\mathrm{F}}(\Pi_n)$ [cf. [NodNon], Theorem B].

Next, let us observe that if T is E'-*strict* for some subset $E' \subseteq E$ of *cardinality one*, then the *commutativity* of the displayed diagram in assertion (i) follows immediately from Claim 3.16.A and [CbTpI], Theorem A, (i). Thus, it follows from Lemma 3.8, (ii), that, to complete the verification of assertion (i), it suffices to verify, under the assumption that $\sigma \neq \mathrm{id}$,

(a) the *commutativity* of the displayed diagram in assertion (i) in the case where $(r, n) = (0, 2)$, and T is $\{1, 2\}$-*strict*, and
(b) the equalities

$$\mathrm{Out}^{\mathrm{F}}(\Pi_n)[T, \tilde{\sigma}] = \mathrm{Out}^{\mathrm{F}}(\Pi_n)[T] = \mathrm{Out}^{\mathrm{F}}(\Pi_n)[T^{\tilde{\sigma}}]$$

in assertion (i) in the case where $(r, n) = (0, 2)$, and T is $\{1, 2\}$-*strict*.

In particular, to verify assertion (i), we may assume without loss of generality [cf. conditions (2_{C}) and (2_{N}) of Lemma 3.8, (ii)] that we are in the situation of Lemma 3.11 in the case where we take the "n", "E" of Lemma 3.11 to be 2, $\{1, 2\}$, respectively. Moreover, it follows immediately from Lemma 3.8, (ii), that the Π_n-conjugacy classes of T, $T^{\widetilde{\sigma}}$ *coincide* with the Π_n-conjugacy classes of the closed subgroups $\Pi_{v^{\mathrm{new}}_{2/1}}$, $\Pi_{v^{\mathrm{new}}_{1\backslash 2}}$ of Π_n that appear in the statement of Lemma 3.11, respectively. Then the above equalities in (b) follows immediately from Lemma 3.11, (x). Moreover, it follows from Lemma 3.11, (viii), (ix), that the composites

$$T \hookrightarrow C_{\Pi_n}(T) \twoheadrightarrow C_{\Pi_n}(T)/Z(C_{\Pi_n}(T)),$$

$$T^{\widetilde{\sigma}} \hookrightarrow C_{\Pi_n}(T^{\widetilde{\sigma}}) \twoheadrightarrow C_{\Pi_n}(T^{\widetilde{\sigma}})/Z(C_{\Pi_n}(T^{\widetilde{\sigma}}))$$

are *isomorphisms*. Thus, the *commutativity* in (a) follows immediately from Lemma 3.11, (x). This completes the proof of assertion (i). Assertion (ii) follows from Lemma 3.5.

Next, we verify assertion (iii). First, to verify the first displayed equality of assertion (iii), let us observe that since T is 1-*descendable*, there exists a subset $E' \subseteq E$ such that the image of $T \subseteq \Pi_E$ via $p^{\Pi}_{E/E'}\colon \Pi_E \twoheadrightarrow \Pi_{E'}$ is an E'-*tripod*, and, moreover, $\#E' \leq n - 1$. Thus, it follows immediately from Claim 3.16.A, together with Remark 3.4.1—by replacing T, E, by $p^{\Pi}_{E/E'}(T)$, E', respectively—that, to verify the first displayed equality of assertion (iii), we may assume without loss of generality that $E \neq \{1, \cdots, n\}$. Then the first displayed equality of assertion (iii) follows immediately from Lemma 3.14, (iv); the portion of Lemma 3.15, (i) [where we observe that the "T" of Lemma 3.15 *differs* from the T of the present discussion!], concerning "$(\Pi_v)_2$" [cf. condition (a) of Lemma 3.15]. This completes the proof of the first displayed equality of assertion (iii).

Next, suppose that condition (iii-1) is satisfied; thus, there exists a subset $E' \subseteq E$ such that the image $p^{\Pi}_{E/E'}(T) \subseteq \Pi_E$ is an E'-*tripod*, and, moreover, $\#E' \leq n - 2$. Then—by replacing T, E by $p^{\Pi}_{E/E'}(T)$, E', respectively [and applying Claim 3.16.A]— we may assume without loss of generality that $\#E \leq n - 2$. Thus, by applying [CbTpI], Theorem A, (ii), we conclude that the second displayed equality of assertion (iii) follows immediately from the first displayed equality of assertion (iii).

Next, suppose that condition (iii-2) is satisfied. Then—by replacing T, E by the $p^{\Pi}_{E/E'}(T)$, E' in condition (iii-2) [and applying Claim 3.16.A]—we may assume without loss of generality that $E \neq \{1, \cdots, n\}$, and, moreover, that T is a *cusp-supporting* E-*tripod*. Then it follows immediately from Lemma 3.14, (iv); the portion of Lemma 3.15, (i), concerning $(\Pi_v)_2$ [cf. condition (b) of Lemma 3.15], that the second displayed equality of assertion (iii) holds. This completes the proof of assertion (iii).

Next, we verify assertion (iv). If either condition (iv-1) or condition (iv-3) is satisfied, then one reduces immediately to the case where $n = 2$, in which case it follows immediately from Lemma 3.13, (i), that $\alpha \in \mathrm{Out}^{\mathrm{F}}(\Pi_n)[T]$. If condition (iv-1) is satisfied, then one reduces immediately to the case where $n = 2$, in which case it follows immediately from Lemma 3.13, (i), that $\alpha \in \mathrm{Out}^{\mathrm{F}}(\Pi_n)[T : \{C\}]$. If both condition (iv-1) and condition (iv-3) are satisfied, then—by applying a suitable *specialization isomorphism* [cf. the discussion preceding [CmbCsp], Definition 2.1, as well as [CbTpI], Remark 5.6.1]—one reduces immediately to the case where $n = 2$ and $\mathrm{Node}(\mathcal{G}) = \emptyset$, in which case it follows immediately from Lemma 3.15, (i), that $\alpha \in \mathrm{Out}^{\mathrm{F}}(\Pi_n)[T : \{|C|\}]$. Finally, if condition (iv-2) is satisfied, then, by applying [CbTpI], Theorem A, (ii), one reduces immediately to the case where "n" is taken to be $n - 1$, and condition (iv-1) is satisfied. This completes the proof of assertion (iv).

Finally, we verify assertion (v). First, we claim that the following assertion holds:

Claim 3.16.B: $\mathrm{Out}^{\mathrm{F}}(\Pi_n) = \mathrm{Out}^{\mathrm{F}}(\Pi_n)[T]$.

Indeed, to verify Claim 3.16.B, by reordering the factors of X_n, we may assume without loss of generality that $E = \{1, 2, 3\}$. Let $\widetilde{\alpha} \in \mathrm{Aut}^{\mathrm{F}}(\Pi_n)$. Then since $n \geq 3$, it follows immediately from [CbTpI], Theorem A, (ii), together with Lemma 3.2, (iv), that the outomorphism of $\Pi_{2/1}$ determined by $\widetilde{\alpha}$ *preserves* the $\Pi_{2/1}$-conjugacy class of cuspidal subgroups of $\Pi_{2/1}$ associated to the [*unique*—cf. Lemma 3.2, (ii)] diagonal cusp. Thus, it follows immediately from assertion (iv) in the case where condition (iv-3) is satisfied that the outomorphism of Π_3 determined by $\widetilde{\alpha}$ *preserves* the Π_3-conjugacy class of $T \subseteq \Pi_3$. This completes the proof of Claim 3.16.B.

Next, we claim that the following assertion holds:

Claim 3.16.C: $\mathrm{Out}^{\mathrm{F}}(\Pi_n)[T] = \mathrm{Out}^{\mathrm{F}}(\Pi_n)[T : \{\Delta\}]$.

Indeed, since $n \geq 3$, this follows immediately from Theorem 2.3, (iv), together with a similar argument to the argument used in the proof of [CmbCsp], Corollary 3.4, (i). This completes the proof of Claim 3.16.C.

Now it follows immediately from Claims 3.16.B, 3.16.C that we have an equality $\mathrm{Out}^{\mathrm{F}}(\Pi_n) = \mathrm{Out}^{\mathrm{F}}(\Pi_n)[T : \{\Delta\}]$. Thus, it follows from assertion (ii) and the first displayed equality of assertion (iii), together with Theorem 2.3, (ii), that, to complete the proof of the content of the first two displays of assertion (v), it suffices to verify the equality $\mathrm{Out}^{\mathrm{FC}}(\Pi_n) = \mathrm{Out}^{\mathrm{FC}}(\Pi_n)[T : \{C\}]$. On the other hand, this follows immediately from the portion of Lemma 3.15, (i), concerning $\alpha|_T$. [Note that one verifies easily that every *central* tripod *arises* from a *cusp*.]

Thus, it remains to verify the equality of the final display of assertion (v). In light of what has already been verified [cf. also Theorem 2.3, (ii)], to verify the final equality of assertion (v), it suffices to verify the condition "+" on the right-hand side of this equality. On the other hand, it follows immediately—by replacing an element of the left-hand side of the equality under consideration by a composite of the element with a suitable outomorphism arising from an element of $\mathrm{Out}^{\mathrm{FC}}(\Pi_4)$ [cf. the equality of the first display of assertion (v)]—from [CmbCsp], Lemma 2.4, that it suffices to verify the condition "+" on an element of the left-hand side of the equality under consideration that induces the *identity automorphism* on $\mathrm{Cusp}(\mathcal{G})$. Then the equality under consideration follows immediately, in light of the assumption that $r \neq 0$, by first applying Lemma 3.15, (i) [in the case where we take the "E" of *loc. cit.* to be a subset of E of cardinality two, and we apply the argument involving *specialization isomorphisms* applied in the proof of assertion (iv)], and then applying Lemma 3.15, (i), (ii) [in the case where we take the "E" of *loc. cit.* to be E]. This completes the proof of assertion (v). □

Remark 3.16.1 Theorem 3.16, (i), may be regarded as a *generalization* of [CmbCsp], Corollary 1.10, (ii). On the other hand, Theorem 3.16, (v), may be regarded as a *more precise version* of [CmbCsp], Corollary 3.4.

Theorem 3.17 (Synchronization of Tripods in Two Dimensions) *In the notation of Theorem 3.16, suppose that $n = 2$, and that $\#E = 1$; thus, one may regard the E-tripod T of Π_n as a verticial subgroup of $\Pi_E \xrightarrow{\sim} \Pi_{\mathcal{G}}$ associated to a vertex $v_T \in \mathrm{Vert}(\mathcal{G})$ of type $(0, 3)$ [cf. Definition 3.1, (ii)]. Let $E' \subseteq \{1, \cdots, n\}$ and $T' \subseteq \Pi_{E'}$ an **E'-tripod** of Π_n. Then the following hold:*

*(i) Suppose that there exists an edge $e \in \mathcal{E}(v_T)$ from which T' **arises** [cf. Definition 3.7, (i)]. [Thus, it holds that $E' = \{1, 2\}$.] Then it holds that*

$$\mathrm{Out}^{\mathrm{FC}}(\Pi_n)[T : \{|C|, \Delta\}] \subseteq \mathrm{Out}^{\mathrm{FC}}(\Pi_n)[T' : \{|C|, \Delta, +\}]$$

*[cf. the notational conventions of Theorem 3.16, (i)]. Moreover, there exists a **geometric** [cf. Definition 3.4, (ii)] outer isomorphism $T \xrightarrow{\sim} T'$ such that the diagram*

$$\begin{array}{ccc} \mathrm{Out}^{\mathrm{FC}}(\Pi_n)[T : \{|C|, \Delta\}] & \subseteq & \mathrm{Out}^{\mathrm{FC}}(\Pi_n)[T' : \{|C|, \Delta, +\}] \\ \mathfrak{T}_T \downarrow & & \downarrow \mathfrak{T}_{T'} \\ \mathrm{Out}(T) & \xrightarrow{\sim} & \mathrm{Out}(T') \end{array}$$

*[cf. the notation of Theorem 3.16, (i)]—where the lower horizontal arrow is the isomorphism induced by the outer isomorphism in question $T \xrightarrow{\sim} T'$—**commutes**.*

*(ii) Suppose that $\#E' = 1$. Thus, one may regard the E'-tripod T' of Π_n as a verticial subgroup of $\Pi_{E'} \xrightarrow{\sim} \Pi_{\mathcal{G}}$ associated to a vertex $v_{T'} \in \mathrm{Vert}(\mathcal{G})$ of type $(0, 3)$. Suppose, moreover, that $\mathcal{N}(v_T) \cap \mathcal{N}(v_{T'}) \neq \emptyset$. Then there exists a **geometric** [cf. Definition 3.4, (ii)] outer isomorphism $T \xrightarrow{\sim} T'$ such that if we write*

$$\mathrm{Out}^{\mathrm{FC}}(\Pi_n)[T, T' : \{|C|, \Delta\}]$$

$$\stackrel{\mathrm{def}}{=} \mathrm{Out}^{\mathrm{FC}}(\Pi_n)[T : \{|C|, \Delta\}] \cap \mathrm{Out}^{\mathrm{FC}}(\Pi_n)[T' : \{|C|, \Delta\}],$$

then the diagram

$$\begin{array}{ccc} \mathrm{Out}^{\mathrm{FC}}(\Pi_n)[T, T' : \{|C|, \Delta\}] & = & \mathrm{Out}^{\mathrm{FC}}(\Pi_n)[T, T' : \{|C|, \Delta\}] \\ \mathfrak{T}_T \downarrow & & \downarrow \mathfrak{T}_{T'} \\ \mathrm{Out}(T) & \xrightarrow{\sim} & \mathrm{Out}(T') \end{array}$$

*—where the lower horizontal arrow is the isomorphism induced by the outer isomorphism in question $T \xrightarrow{\sim} T'$—**commutes**.*

Proof First, we verify assertion (i). Let us observe that the inclusion $\mathrm{Out}^{\mathrm{FC}}(\Pi_n)[T : \{|C|\}] \subseteq \mathrm{Out}^{\mathrm{FC}}(\Pi_n)[T']$, hence also the inclusion $\mathrm{Out}^{\mathrm{FC}}(\Pi_n)[T : \{|C|, \Delta\}] \subseteq \mathrm{Out}^{\mathrm{FC}}(\Pi_n)[T']$, follows immediately from Theorem 3.16, (iv), in the case where condition (iv-1) is satisfied. Thus, one verifies easily from Lemma 3.15, (i), (ii) [cf. also Lemma 3.14, (iv)], that the remainder of assertion (i) holds. This completes the proof of assertion (i). Next, we verify assertion (ii). It follows immediately from [CmbCsp], Proposition 1.2, (iii), that we may assume without loss of generality that $E' = E$. Write $T'' \subseteq \Pi_n$ for the $\{1, 2\}$-tripod of Π_n arising from $e \in \mathcal{N}(v_T) \cap \mathcal{N}(v_{T'})$. Then it follows from assertion (i) that there exist *geometric* outer isomorphisms $T \xrightarrow{\sim} T''$, $T' \xrightarrow{\sim} T''$ that satisfy the condition of assertion (i) [i.e., for the pairs (T, T'') and (T', T'')]. Thus, one verifies easily that the [necessarily *geometric*] outer isomorphism $T \xrightarrow{\sim} T'' \xleftarrow{\sim} T'$ obtained by forming the composite of these two outer isomorphisms satisfies the condition of assertion (ii). This completes the proof of assertion (ii). □

Theorem 3.18 (Synchronization of Tripods in Three or More Dimensions) *In the notation of Theorem 3.16, suppose that $n \geq 3$. Then the following hold:*

(i) It holds that

$$\mathrm{Out}^{\mathrm{FC}}(\Pi_n)[T : \{|C|\}] = \mathrm{Out}^{\mathrm{FC}}(\Pi_n)[T : \{|C|, \Delta\}]$$

[cf. the notational conventions of Theorem 3.16, (i)]. If, moreover, $n \geq 4$ or $r \neq 0$, then it holds that

$$\mathrm{Out}^{\mathrm{FC}}(\Pi_n)[T : \{|C|\}] = \mathrm{Out}^{\mathrm{FC}}(\Pi_n)[T : \{|C|, \Delta, +\}]$$

[cf. the notational conventions of Theorem 3.16, (i)].

*(ii) Let $E' \subseteq \{1, \cdots, n\}$ and $T' \subseteq \Pi_{E'}$ an **E'-tripod** of Π_n. Then there exists a **geometric** [cf. Definition 3.4, (ii)] outer isomorphism $T \xrightarrow{\sim} T'$ such that if we write*

$$\mathrm{Out}^{\mathrm{FC}}(\Pi_n)[T, T' : \{|C|\}]$$

$$\stackrel{\mathrm{def}}{=} \mathrm{Out}^{\mathrm{FC}}(\Pi_n)[T : \{|C|\}] \cap \mathrm{Out}^{\mathrm{FC}}(\Pi_n)[T' : \{|C|\}],$$

then the diagram

$$\begin{array}{ccc} \mathrm{Out}^{\mathrm{FC}}(\Pi_n)[T, T' : \{|C|\}] & = & \mathrm{Out}^{\mathrm{FC}}(\Pi_n)[T, T' : \{|C|\}] \\ \mathfrak{T}_T \downarrow & & \downarrow \mathfrak{T}_{T'} \\ \mathrm{Out}(T) & \xrightarrow{\sim} & \mathrm{Out}(T') \end{array}$$

[cf. the notation of Theorem 3.16, (i)]—where the lower horizontal arrow is the isomorphism induced by the outer isomorphism in question $T \xrightarrow{\sim} T'$— ***commutes****.*

Proof First, we verify the first displayed equality of assertion (i). Observe that it follows immediately from Lemma 3.8, (i), together with a similar argument to the argument applied in the proof of the first displayed equality of Theorem 3.16, (iii), that we may assume without loss of generality that T is *E-strict*, which thus implies that $\#E \in \{1, 2, 3\}$ [cf. Lemma 3.8, (ii)]. Now we apply *induction on* $3 - \#E \in \{0, 1, 2\}$. If $3 - \#E = 0$, i.e., T is *central* [cf. Lemma 3.8, (ii)], then the first displayed equality of assertion (i) follows immediately from Theorem 3.16, (v). Now suppose that $3 - \#E > 0$, and that the *induction hypothesis* is in force. Let $\alpha \in \mathrm{Out}^{\mathrm{FC}}(\Pi_n)[T : \{|C|\}]$. Then it follows immediately from Lemma 3.15, (i), (ii) [cf. also conditions (1), (2) of Lemma 3.15, (ii), where we note that the E, E', T, T' of the present discussion correspond, respectively, to the "$E \setminus \{i\}$", "E", "Π_v", "T" of Lemma 3.15], that there exist a subset $E \subsetneq E' \subseteq \{1, \cdots, n\}$ and an E'-tripod $T' \subseteq \Pi_{E'}$ such that $3 - \#E' < 3 - \#E$, $T' \subseteq \Pi_{E'}$ is *E'-strict*, and $\alpha \in \mathrm{Out}^{\mathrm{FC}}(\Pi_n)[T' : \{|C|\}]$ [cf. Lemma 3.15, (i)]. Thus, it follows immediately from the *induction hypothesis* that $\alpha \in \mathrm{Out}^{\mathrm{FC}}(\Pi_n)[T' : \{|C|, \Delta\}]$. In particular, it follows immediately from Lemma 3.15, (ii), that—for a suitable choice of the pair (E', T') [cf. the statement of Lemma 3.15, (ii)]—the actions of α on T and T' may be related by means of a *geometric* outer isomorphism, which thus implies that $\alpha \in \mathrm{Out}^{\mathrm{FC}}(\Pi_n)[T : \{|C|, \Delta\}]$ [cf. Remark 3.4.1]. This completes the proof of the first displayed equality of assertion (i).

Next, we verify assertion (ii). First, we claim that the following assertion holds:

> Claim 3.18.A: If both T and T' are *central*, then the pair (T, T') satisfies the property stated in assertion (ii).

Indeed, this assertion follows immediately from the commutativity of the displayed diagram of Theorem 3.16, (i).

Next, we claim that the following assertion holds:

> Claim 3.18.B: Suppose that T is *E-strict*, and that $\#E \neq 3$ [i.e., $\#E \in \{1, 2\}$—cf. Lemma 3.8, (ii)]. Then there exist a subset $E \subsetneq E'' \subseteq \{1, \cdots, n\}$ and an E''-tripod $T'' \subseteq \Pi_{E''}$ such that T'' is *E''-strict*, $\mathrm{Out}^{\mathrm{FC}}(\Pi_n)[T : \{|C|\}] \subseteq \mathrm{Out}^{\mathrm{FC}}(\Pi_n)[T'' : \{|C|\}]$, and, moreover, the pair (T, T'') satisfies the property stated in assertion (ii) [i.e., where one takes "T'" to be T''].

Indeed, this follows immediately from Lemma 3.15, (i), (ii) [cf. also conditions (1), (2) of Lemma 3.15, (ii), where we note that the E, E'', T, T'' of the present discussion correspond, respectively, to the "$E \setminus \{i\}$", "E", "Π_v", "T"

of Lemma 3.15], together with the first displayed equality of assertion (i). This completes the proof of Claim 3.18.B.

To verify assertion (ii), let us observe that it follows immediately from Lemma 3.8, (i), together with a similar argument to the argument applied in the proof of the first displayed equality of Theorem 3.16, (iii), that we may assume without loss of generality that T is E-*strict*; in particular, $\#E \in \{1, 2, 3\}$ [cf. Lemma 3.8, (ii)]. Next, let us observe that, by comparing two *arbitrary* tripods of Π_n to a *fixed central* tripod of Π_n [and applying Theorem 3.16, (v)], one may reduce immediately to the case where T' is *central*. Moreover, by successive application of Claim 3.18.B, one reduces immediately to the case where both T and T' are *central*, which was verified in Claim 3.18.A. This completes the proof of assertion (ii). Finally, the second displayed equality of assertion (i) follows immediately from assertion (ii), together with Theorem 3.16, (v). This completes the proof of Theorem 3.18. □

Definition 3.19 Suppose that $n \geq 3$. Let us write

$$\Pi^{\text{tpd}}$$

for the i-central E-tripod of Π_n [cf. Definitions 3.3, (i); 3.7, (ii)], where $E \subseteq \{1, \ldots, n\}$ is a subset of cardinality 3, and $i \in E$. Then it follows from Theorem 3.16, (i), (v), that one has a natural homomorphism

$$\mathfrak{T}_{\Pi^{\text{tpd}}} : \text{Out}^{\text{FC}}(\Pi_n) = \text{Out}^{\text{FC}}(\Pi_n)[\Pi^{\text{tpd}} : \{|C|, \Delta\}] \longrightarrow \text{Out}^{\text{C}}(\Pi^{\text{tpd}})^{\Delta}$$

[cf. Definition 3.4, (i)], which is in fact *independent* of E and i [cf. Theorem 3.16, (i)]. We shall refer to this homomorphism as the *tripod homomorphism* associated to Π_n and write

$$\text{Out}^{\text{FC}}(\Pi_n)^{\text{geo}} \subseteq \text{Out}^{\text{FC}}(\Pi_n)$$

for the kernel of this homomorphism [cf. Remark 3.19.1 below]. Note that it follows from Theorem 3.16, (v), that if $n \geq 4$ or $r \neq 0$, then the image of the tripod homomorphism is contained in $\text{Out}^{\text{C}}(\Pi^{\text{tpd}})^{\Delta+} \subseteq \text{Out}^{\text{C}}(\Pi^{\text{tpd}})^{\Delta}$ [cf. Definition 3.4, (i)]. If $n \geq 4$ or $r \neq 0$, then $\mathfrak{T}_{\Pi^{\text{tpd}}}$ may also be regarded as a homomorphism defined on $\text{Out}^{\text{F}}(\Pi_n)(= \text{Out}^{\text{FC}}(\Pi_n)$—cf. Theorem 2.3, (ii)); in this case, we shall write $\text{Out}^{\text{F}}(\Pi_n)^{\text{geo}} \stackrel{\text{def}}{=} \text{Out}^{\text{FC}}(\Pi_n)^{\text{geo}}$.

Remark 3.19.1 Let us recall that if we write $\pi_1((\mathcal{M}_{g,[r]})_{\mathbb{Q}})$ for the étale fundamental group of the moduli stack $(\mathcal{M}_{g,[r]})_{\mathbb{Q}}$ of hyperbolic curves of type (g, r) over $\mathbb{Q}$ [cf. the discussion entitled "*Curves*" in "Notations and Conventions"], then we have a natural outer homomorphism

$$\pi_1((\mathcal{M}_{g,[r]})_{\mathbb{Q}}) \longrightarrow \text{Out}^{\text{FC}}(\Pi_n).$$

Suppose that $n \geq 4$. Then $\mathrm{Out}^{\mathrm{FC}}(\Pi_n) = \mathrm{Out}^{\mathrm{F}}(\Pi_n)$ does *not depend* on n [cf. Theorem 2.3, (ii); [NodNon], Theorem B]. Moreover, one verifies easily that the image of the geometric fundamental group $\pi_1((\mathcal{M}_{g,[r]})_{\overline{\mathbb{Q}}}) \subseteq \pi_1((\mathcal{M}_{g,[r]})_{\mathbb{Q}})$—where we use the notation $\overline{\mathbb{Q}}$ to denote an algebraic closure of $\mathbb{Q}$—via the above displayed outer homomorphism is contained in the *kernel* $\mathrm{Out}^{\mathrm{FC}}(\Pi_n)^{\mathrm{geo}} \subseteq \mathrm{Out}^{\mathrm{FC}}(\Pi_n)$ of the *tripod homomorphism* associated to Π_n [cf. Definition 3.19]. Thus, the outer homomorphism of the above display fits into a commutative diagram of profinite groups

$$\begin{array}{ccccccccc} 1 & \longrightarrow & \pi_1((\mathcal{M}_{g,[r]})_{\overline{\mathbb{Q}}}) & \longrightarrow & \pi_1((\mathcal{M}_{g,[r]})_{\mathbb{Q}}) & \longrightarrow & \mathrm{Gal}(\overline{\mathbb{Q}}/\mathbb{Q}) & \longrightarrow & 1 \\ & & \downarrow & & \downarrow & & \downarrow & & \\ 1 & \longrightarrow & \mathrm{Out}^{\mathrm{F}}(\Pi_n)^{\mathrm{geo}} & \longrightarrow & \mathrm{Out}^{\mathrm{F}}(\Pi_n) & \xrightarrow{\mathfrak{T}_{\Pi^{\mathrm{tpd}}}} & \mathrm{Out}^{\mathrm{C}}(\Pi^{\mathrm{tpd}})^{\Delta+} & & \end{array}$$

—where the horizontal sequences are *exact*. In Chap. 4 below, we shall verify that the lower right-hand horizontal arrow $\mathfrak{T}_{\Pi^{\mathrm{tpd}}}$ is *surjective* [cf. Corollary 4.15 below]. On the other hand, if Σ is the set of all prime numbers, then it follows from *Belyi's Theorem* that the right-hand vertical arrow is *injective*; moreover, the *surjectivity* of the right-hand vertical arrow has been conjectured in the theory of the *Grothendieck-Teichmüller group*. From this point of view, one may regard the quotient $\mathrm{Out}^{\mathrm{F}}(\Pi_n) \overset{\mathfrak{T}_{\Pi^{\mathrm{tpd}}}}{\twoheadrightarrow} \mathrm{Out}^{\mathrm{C}}(\Pi^{\mathrm{tpd}})^{\Delta+}$ as a sort of *arithmetic quotient* of $\mathrm{Out}^{\mathrm{F}}(\Pi_n)$ and the subgroup $\mathrm{Out}^{\mathrm{F}}(\Pi_n)^{\mathrm{geo}} \subseteq \mathrm{Out}^{\mathrm{F}}(\Pi_n)$ as a sort of *geometric portion* of $\mathrm{Out}^{\mathrm{F}}(\Pi_n)$.

Definition 3.20 Let m be a positive integer and $Y^{\log}$ a stable log curve over $(\mathrm{Spec}\, k)^{\log}$. For each nonnegative integer i, write ${}^{Y}\Pi_i$ for the "Π_i" that occurs in the case where we take "$X^{\log}$" to be $Y^{\log}$. Then we shall say that an isomorphism (respectively, outer isomorphism) $\Pi_1 \xrightarrow{\sim} {}^{Y}\Pi_1$ is *m-cuspidalizable* if it arises from a [*necessarily unique*, up to a permutation of the m factors, by [NodNon], Theorem B] PFC-admissible [cf. [CbTpI], Definition 1.4, (iii)] isomorphism $\Pi_m \xrightarrow{\sim} {}^{Y}\Pi_m$.

Proposition 3.21 (Tripod Homomorphisms and Finite Étale Coverings) *Let $Y^{\log}$ be a* ***stable log curve*** *over* $(\mathrm{Spec}\, k)^{\log}$ *and $Y^{\log} \to X^{\log}$ a finite log étale covering over* $(\mathrm{Spec}\, k)^{\log}$*. For each positive integer i, write $Y_i^{\log}$ (respectively, ${}^{Y}\Pi_i$) for the "$X_i^{\log}$" (respectively, "Π_i") that occurs in the case where we take "$X^{\log}$" to be $Y^{\log}$. Suppose that $Y^{\log} \to X^{\log}$ is* ***geometrically pro-Σ*** *and* ***geometrically Galois****, i.e., $Y^{\log} \to X^{\log}$ determines an* ***injection*** *${}^{Y}\Pi_1 \hookrightarrow \Pi_1$ [that is well-defined up to Π_1-conjugation] whose image is* ***normal****. Let $\widetilde{\alpha}$ be an automorphism of Π_1 that preserves ${}^{Y}\Pi_1 \subseteq \Pi_1$. Suppose, moreover, that the outomorphism α of Π_1 determined by $\widetilde{\alpha}$ is* ***n-cuspidalizable*** *[cf. Definition 3.20]. Then the following hold:*

(i) *The outomorphism ${}^{Y}\alpha$ of ${}^{Y}\Pi_1$ determined by $\widetilde{\alpha}$ is* ***n-cuspidalizable*** *[cf. Definition 3.20].*

(ii) *Suppose that* $n \geq 3$. *Let* $\Pi^{\mathrm{tpd}} \subseteq \Pi_3$, ${}^Y\Pi^{\mathrm{tpd}} \subseteq {}^Y\Pi_3$ *be* **1-central** *[*$\{1, 2, 3\}$*-]tripods [cf. Definitions 3.3, (i); 3.7, (ii)] of* Π_n, ${}^Y\Pi_n$, *respectively. Write* α_n, ${}^Y\!\alpha_n$ *for the respective FC-admissible outomorphisms of* Π_n, ${}^Y\Pi_n$ *determined by the* n*-cuspidalizable outomorphisms* α, ${}^Y\!\alpha$ *[cf. (i)]. Then there exists a* **geometric** *[cf. Definition 3.4, (ii)] outer isomorphism* $\phi^{\mathrm{tpd}}\colon \Pi^{\mathrm{tpd}} \xrightarrow{\sim} {}^Y\Pi^{\mathrm{tpd}}$ *such that the outomorphism* $\mathfrak{T}_{\Pi^{\mathrm{tpd}}}(\alpha_n)$ *[cf. Definition 3.19] of* Π^{tpd} *is* **compatible** *with the outomorphism* $\mathfrak{T}_{{}^Y\Pi^{\mathrm{tpd}}}({}^Y\!\alpha_n)$ *[cf. Definition 3.19] of* ${}^Y\Pi^{\mathrm{tpd}}$ *relative to* ϕ^{tpd}.

Proof First, let us observe that, to verify Proposition 3.21—by applying a suitable *specialization isomorphism* [cf. the discussion preceding [CmbCsp], Definition 2.1, as well as [CbTpI], Remark 5.6.1]—we may assume without loss of generality that $X^{\log}$ and $Y^{\log}$ are *smooth log curves* over $(\mathrm{Spec}\, k)^{\log}$. Write $(U_X)_n$, $(U_Y)_n$ for the [open subschemes of X_n, Y_n determined by the] 1-*interiors* [cf. [MzTa], Definition 5.1, (i)] of $X_n^{\log}$, $Y_n^{\log}$, respectively. [Here, we note that in the present situation, the 0-*interior* of $(\mathrm{Spec}\, k)^{\log}$, hence also of $X_n^{\log}$, $Y_n^{\log}$, is *empty*!] Thus, one verifies easily that $U_X \overset{\mathrm{def}}{=} (U_X)_1$, $U_Y \overset{\mathrm{def}}{=} (U_Y)_1$ are *hyperbolic curves* over k, and that $(U_X)_n$, $(U_Y)_n$ are naturally isomorphic to the n-th *configuration spaces* of U_X, U_Y, respectively. Write $U_X^{\times n}$, $U_Y^{\times n}$ for the respective fiber products of n copies of U_X, U_Y over k; $\Pi_1^{\times n}$, ${}^Y\Pi_1^{\times n}$ for the respective direct products of n copies of Π_1, ${}^Y\Pi_1$; V_n for the fiber product of the natural open immersion $(U_X)_n \hookrightarrow U_X^{\times n}$ and the natural finite étale covering $U_Y^{\times n} \to U_X^{\times n}$. Then one verifies easily that the resulting open immersion $V_n \hookrightarrow U_Y^{\times n}$ factors through the natural open immersion $(U_Y)_n \hookrightarrow U_Y^{\times n}$, i.e., we obtain an open immersion $V_n \hookrightarrow (U_Y)_n$. That is to say, whereas $(U_Y)_n$ is the open subscheme of $U_Y^{\times n}$ obtained by removing the various *diagonals* of $U_Y^{\times n}$, the scheme V_n may be thought of as the open subscheme of $U_Y^{\times n}$ obtained by removing the various *Galois conjugates of these diagonals*, relative to the action of the Galois group $\mathrm{Gal}(U_Y^{\times n}/U_X^{\times n}) = \mathrm{Gal}(U_Y/U_X)^{\times n}$. In particular, we obtain a natural outer isomorphism and outer surjection

$$\Pi_n \times_{\Pi_1^{\times n}} {}^Y\Pi_1^{\times n} \xleftarrow{\sim} \Pi_{V_n} \twoheadrightarrow {}^Y\Pi_n$$

—where we write Π_{V_n} for the maximal pro-Σ quotient of the étale fundamental group of V_n.

Now we verify assertion (i). Let $\widetilde{\alpha}_n$ be an FC-admissible automorphism of Π_n that lies over the automorphism $\widetilde{\alpha}$ of Π_1 with respect to each of the n natural projections $\Pi_n \twoheadrightarrow \Pi_1$. Then since $\widetilde{\alpha}_n$ is *FC-admissible* and *commutes* with the image of the natural inclusion $\mathfrak{S}_n \hookrightarrow \mathrm{Out}(\Pi_n)$ [cf. [NodNon], Theorem B], one verifies easily, in light of the description given above of V_n, that the outomorphism of $\Pi_n \times_{\Pi_1^{\times n}} {}^Y\Pi_1^{\times n}$ induced by $\widetilde{\alpha}_n$ and ${}^Y\!\alpha$ *preserves* the inertia subgroups associated to each irreducible component of the complement $U_Y^{\times n} \setminus V_n$. Thus, since [by the *Zariski-Nagata purity theorem*] the inertia subgroups of the irreducible components of the complement $(U_Y)_n \setminus V_n$ normally topologically generate the kernel of the above outer surjection $\Pi_{V_n} \twoheadrightarrow {}^Y\Pi_n$, we conclude, by applying the morphisms of the above display, that the outomorphism of $\Pi_n \times_{\Pi_1^{\times n}} {}^Y\Pi_1^{\times n}$ induced by $\widetilde{\alpha}_n$ and ${}^Y\alpha$ determines an FC-admissible outomorphism of ${}^Y\Pi_n$. Moreover, one verifies easily that the resulting outomorphism of ${}^Y\Pi_n$ lies over the outomorphism ${}^Y\alpha$ of ${}^Y\Pi_1$. This completes the proof of assertion (i).

Next, we verify assertion (ii). First, let us observe that the natural inclusion $\Pi^{\text{tpd}} \hookrightarrow \Pi_3$, together with the trivial homomorphism $\Pi^{\text{tpd}} \to (\{1\} \hookrightarrow){}^Y\Pi_1^{\times 3}$ [cf. Definition 3.3, (ii); Lemma 3.6, (v); Definition 3.7, (ii)], determines an injection $\Pi^{\text{tpd}} \hookrightarrow \Pi_3 \times_{\Pi_1^{\times 3}} {}^Y\Pi_1^{\times 3} \overset{\sim}{\leftarrow} \Pi_{V_3}$. Moreover, it follows immediately from the fact that the *blow-up* operation that gives rise to a central tripod is *compatible* with *étale localization* [cf. the discussion of [CmbCsp], Definition 1.8] that —after possibly replacing ${}^Y\Pi^{\text{tpd}} \subseteq {}^Y\Pi_3$ by a suitable ${}^Y\Pi_3$-conjugate of ${}^Y\Pi^{\text{tpd}}$—the composite of this injection $\Pi^{\text{tpd}} \hookrightarrow \Pi_{V_3}$ with the natural outer surjection $\Pi_{V_3} \twoheadrightarrow {}^Y\Pi_3$ of the above display determines a *geometric* outer [cf. Lemma 3.12, (i)] isomorphism $\phi^{\text{tpd}}\colon \Pi^{\text{tpd}} \overset{\sim}{\to} {}^Y\Pi^{\text{tpd}} \subseteq {}^Y\Pi_3$. On the other hand, one verifies easily [cf. the construction of ${}^Y\alpha_n$ given in the proof of assertion (i)] that this outer isomorphism ϕ^{tpd} satisfies the property stated in assertion (ii). This completes the proof of assertion (ii). □

Corollary 3.22 (Non-surjectivity Result) *In the notation of Theorem 3.16, suppose that* $(g, r) \notin \{(0, 3); (1, 1)\}$. *Then the natural* ***injection***

$$\text{Out}^{\text{FC}}(\Pi_2) \hookrightarrow \text{Out}^{\text{FC}}(\Pi_1)$$

of [NodNon], Theorem B, is ***not surjective***.

Proof First, let us observe—by considering a suitable stable log curve of type (g, r) over $(\text{Spec}\, k)^{\log}$ and applying a suitable *specialization isomorphism* [cf. the discussion preceding [CmbCsp], Definition 2.1, as well as [CbTpI], Remark 5.6.1]—that, to verify Corollary 3.22, we may assume without loss of generality that $\mathcal{G}$ is *totally degenerate* [cf. [CbTpI], Definition 2.3, (iv)], i.e., that every vertex of $\mathcal{G}$ is a tripod of $X_n^{\log}$ [cf. Definition 3.1, (v)]. Note that [since $(g, r) \notin \{(0, 3); (1, 1)\}$] this implies that $\#\text{Vert}(\mathcal{G}) \geq 2$. Let us fix a vertex $v_0 \in \text{Vert}(\mathcal{G})$ and write $\alpha_{v_0} \overset{\text{def}}{=} \text{id}_{\mathcal{G}|_{v_0}} \in \text{Aut}^{|\text{grph}|}(\mathcal{G}|_{v_0})$ [cf. [CbTpI], Definitions 2.1, (iii), and 2.6, (i); Remark 4.1.2 of the present monograph]. For each $v \in \text{Vert}(\mathcal{G}) \setminus \{v_0\}$, let $\alpha_v \in \text{Aut}^{|\text{grph}|}(\mathcal{G}|_v)$ be a *nontrivial* automorphism of $\mathcal{G}|_v$ such that $\alpha_v \in \text{Out}^{\text{C}}(\Pi_{\mathcal{G}|_v})^{\Delta}$, and, moreover, $\chi_{\mathcal{G}|_v}(\alpha_v) = 1$ [cf. [CbTpI], Definition 3.8, (ii)]. Here, we note that since the image of the natural outer Galois representation of the absolute Galois group of $\mathbb{Q}$ associated to $\mathbb{P}^1_{\mathbb{Q}} \setminus \{0, 1, \infty\}$ is contained in "$\text{Out}^{\text{C}}(-)^{\Delta}$", by considering a *nontrivial* element of this image whose image via the cyclotomic character is *trivial*, one verifies immediately [e.g., by applying [LocAn], Theorem A] that such an automorphism $\alpha_v \in \text{Aut}^{|\text{grph}|}(\mathcal{G}|_v)$ always *exists*. Then it follows immediately from [CbTpI], Theorem B, (iii), that there exists an automorphism $\alpha \in \text{Aut}^{|\text{grph}|}(\mathcal{G})$ such that $\rho_{\mathcal{G}}^{\text{Vert}}(\alpha) = (\alpha_v)_{v \in \text{Vert}(\mathcal{G})}$. Now assume that there exists an outomorphism

$\alpha_2 \in \mathrm{Out}^{\mathrm{FC}}(\Pi_2)$ such that $\alpha \in \mathrm{Aut}^{|\mathrm{grph}|}(\mathcal{G})(\subseteq \mathrm{Out}(\Pi_{\mathcal{G}}) \stackrel{\sim}{\leftarrow} \mathrm{Out}(\Pi_1))$ is equal to the image of α_2 via the injection in question $\mathrm{Out}^{\mathrm{FC}}(\Pi_2) \hookrightarrow \mathrm{Out}^{\mathrm{FC}}(\Pi_1)$. Then, for each $v \in \mathrm{Vert}(\mathcal{G})$, since $\alpha_v \in \mathrm{Out}^{\mathrm{C}}(\Pi_{\mathcal{G}|_v})^{\Delta}$, and $\alpha \in \mathrm{Aut}^{|\mathrm{grph}|}(\mathcal{G})$, it follows immediately from the various definitions involved that $\alpha_2 \in \mathrm{Out}^{\mathrm{FC}}(\Pi_2)[\Pi_v : \{|C|, \Delta\}]$—where we use the notation Π_v to denote a verticial subgroup of $\Pi_{\mathcal{G}} \stackrel{\sim}{\leftarrow} \Pi_1$ associated to $v \in \mathrm{Vert}(\mathcal{G})$. Thus, since $\alpha_{v_0} \stackrel{\mathrm{def}}{=} \mathrm{id}_{\mathcal{G}|_{v_0}}$, it follows from Theorem 3.17, (ii), that $\alpha_v = \mathrm{id}_{\mathcal{G}|_v}$ for every $v \in \mathrm{Vert}(\mathcal{G})$, in contradiction to the fact that for $v \in \mathrm{Vert}(\mathcal{G}) \setminus \{v_0\}(\neq \emptyset)$, the automorphism $\alpha_v \in \mathrm{Aut}^{|\mathrm{grph}|}(\mathcal{G}|_v)$ is *nontrivial*. This completes the proof of Corollary 3.22. □

Remark 3.22.1

(i) Let us recall from [NodNon], Corollary 6.6, that, in the *discrete case*, the homomorphism that corresponds to the homomorphism discussed in Corollary 3.22 is, in fact, *surjective*; moreover, this *surjectivity* may be regarded as an immediate consequence of the *Dehn-Nielsen-Baer theorem*—cf. the proof of [CmbCsp], Theorem 5.1, (ii). This phenomenon illustrates that, in general, analogous constructions in the *discrete* and *profinite* cases may in fact exhibit quite *different behavior*.

(ii) In the context of (i), we recall another famous example of substantially different behavior in the *discrete* and *profinite* cases: As is well-known, in classical algebraic topology, *singular cohomology* with coefficients in $\mathbb{Z}$ yields a "good" cohomology theory with coefficients in $\mathbb{Z}$. On the other hand, in the 1960s, Serre gave an argument involving supersingular elliptic curves in characteristic $p > 0$ which shows that such a "good" cohomology theory with coefficients in $\mathbb{Z}$ [or even in $\mathbb{Z}_p$!] *cannot exist* for smooth varieties of positive characteristic.

(iii) In [Lch], various conjectures concerning [in the notation of the present monograph] the profinite group "$\mathrm{Out}(\Pi_1)$" were introduced. However, at the time of writing, the authors of the present monograph were unable to find any justification for the validity of these conjectures that goes beyond the observation that the *discrete* analogues of these conjectures are indeed valid. That is to say, there does not appear to exist any justification for excluding the possibility that—just as in the case of the examples discussed in (i), (ii), i.e., the *Dehn-Nielsen-Baer theorem* and *singular cohomology with coefficients in* $\mathbb{Z}$—the *discrete* and *profinite* cases exhibit substantially different behavior. In particular, it appears to the authors that it is desirable that this issue be addressed in a satisfactory fashion in the context of these conjectures.

Remark 3.22.2 As discussed in Remark 3.22.1, (i), in the *discrete case*, the homomorphism that corresponds to the homomorphism discussed in Corollary 3.22 is, in fact, *bijective*. The proof of Corollary 3.22 *fails* in the *discrete case* for the following reason: The *pro-*Σ "Π_1" of a tripod admits *nontrivial C-admissible* outomorphisms that *commute* with the outer modular symmetries and, moreover, lie in the *kernel* of the cyclotomic character [cf. the proof of Corollary 3.22]. By contrast, the discrete "Π_1" of a tripod does *not admit* such outomorphisms. Indeed, it follows from a classical result of *Nielsen* [cf. [CmbCsp], Remark 5.3.1] that the discrete "$\mathrm{Out}^{\mathrm{C}}(\Pi_1)^{\mathrm{cusp}}$" in the case of a tripod is a *finite group of order* 2 whose *unique nontrivial element* arises from *complex conjugation*.

Remark 3.22.3 It follows from [NodNon], Theorem B, together with Corollary 3.22, that if $(g, r) \notin \{(0, 3); (1, 1)\}$, then the homomorphism $\mathrm{Out}^{\mathrm{FC}}(\Pi_{n+1}) \to \mathrm{Out}^{\mathrm{FC}}(\Pi_n)$ of [NodNon], Theorem B, fits into the following sequences of homomorphisms of profinite groups: If $r \neq 0$, then for any $n \geq 3$,

$$\mathrm{Out}^{\mathrm{FC}}(\Pi_n) \overset{\sim}{\to} \mathrm{Out}^{\mathrm{FC}}(\Pi_3) \overset{\simeq ?}{\hookrightarrow} \mathrm{Out}^{\mathrm{FC}}(\Pi_2) \overset{\not\simeq}{\hookrightarrow} \mathrm{Out}^{\mathrm{FC}}(\Pi_1).$$

If $r = 0$, then for any $n \geq 4$,

$$\mathrm{Out}^{\mathrm{FC}}(\Pi_n) \overset{\sim}{\to} \mathrm{Out}^{\mathrm{FC}}(\Pi_4) \overset{\simeq ?}{\hookrightarrow} \mathrm{Out}^{\mathrm{FC}}(\Pi_3) \overset{\simeq ?}{\hookrightarrow} \mathrm{Out}^{\mathrm{FC}}(\Pi_2) \overset{\not\simeq}{\hookrightarrow} \mathrm{Out}^{\mathrm{FC}}(\Pi_1).$$

Definition 3.23 Let Σ_0 be a nonempty set of prime numbers and $\mathcal{G}_0$ a semi-graph of anabelioids of pro-Σ_0 PSC-type. Write $\Pi_{\mathcal{G}_0}$ for the [pro-Σ_0] fundamental group of $\mathcal{G}_0$.

(i) Let $\mathcal{H}$ be a semi-graph of anabelioids of pro-Σ_0 PSC-type, $S \subseteq \mathrm{Node}(\mathcal{H})$, and $\phi\colon \mathcal{H}_{\rightsquigarrow S} \overset{\sim}{\to} \mathcal{G}_0$ [cf. [CbTpI], Definition 2.8, for more on this notation] an isomorphism [of semi-graphs of anabelioids of PSC-type]. Then we shall refer to the triple $(\mathcal{H}, S, \phi)$ as a *degeneration structure* on $\mathcal{G}_0$.

(ii) Let $(\mathcal{H}_1, S_1, \phi_1)$, $(\mathcal{H}_2, S_2, \phi_2)$ be two degeneration structures on $\mathcal{G}_0$ [cf. (i)]. Then we shall write

$$(\mathcal{H}_2, S_2, \phi_2) \preceq (\mathcal{H}_1, S_1, \phi_1)$$

if there exist a subset $S_{2,1} \subseteq S_2$ of S_2 and a(n) [*uniquely determined*, by ϕ_1 and ϕ_2!—cf. [CmbGC], Proposition 1.5, (ii)] isomorphism $\phi_{2,1}\colon (\mathcal{H}_2)_{\rightsquigarrow S_{2,1}} \overset{\sim}{\to} \mathcal{H}_1$ [i.e., a degeneration structure $(\mathcal{H}_2, S_{2,1}, \phi_{2,1})$ on $\mathcal{H}_1$] such that $\phi_{2,1}$ maps $S_2 \setminus S_{2,1}$ bijectively onto S_1, and the diagram

$$\begin{array}{ccc} ((\mathcal{H}_2)_{\rightsquigarrow S_{2,1}})_{\rightsquigarrow S_2 \setminus S_{2,1}} & \overset{\sim}{\longrightarrow} & (\mathcal{H}_1)_{\rightsquigarrow S_1} \\ \wr\downarrow & & \wr\downarrow \phi_1 \\ (\mathcal{H}_2)_{\rightsquigarrow S_2} & \underset{\sim}{\overset{\phi_2}{\longrightarrow}} & \mathcal{G}_0 \end{array}$$

—where the upper horizontal arrow is the isomorphism induced by $\phi_{2,1}$, and the left-hand vertical arrow is the natural isomorphism—*commutes*. [Here, we note that the subset $S_{2,1}$ is also *uniquely determined* by ϕ_1 and ϕ_2—cf. [CmbGC], Proposition 1.2, (i).]

(iii) Let $(\mathcal{H}_1, S_1, \phi_1)$, $(\mathcal{H}_2, S_2, \phi_2)$ be two degeneration structures on $\mathcal{G}_0$ [cf. (i)]. Then we shall say that $(\mathcal{H}_1, S_1, \phi_1)$ is *co-Dehn* to $(\mathcal{H}_2, S_2, \phi_2)$ if there exists a degeneration structure $(\mathcal{H}_3, S_3, \phi_3)$ on $\mathcal{G}_0$ such that

$$(\mathcal{H}_3, S_3, \phi_3) \preceq (\mathcal{H}_1, S_1, \phi_1); \quad (\mathcal{H}_3, S_3, \phi_3) \preceq (\mathcal{H}_2, S_2, \phi_2)$$

[cf. (ii)].

(iv) Let $(\mathcal{H}, S, \phi)$ be a degeneration structure on $\mathcal{G}_0$ [cf. (i)] and $\alpha \in \mathrm{Out}(\Pi_{\mathcal{G}_0})$. Then we shall say that α is an *$(\mathcal{H}, S, \phi)$-Dehn multi-twist* of $\mathcal{G}_0$ if α is contained in the image of the composite

$$\mathrm{Dehn}(\mathcal{H}) \hookrightarrow \mathrm{Out}(\Pi_{\mathcal{H}}) \stackrel{\sim}{\leftarrow} \mathrm{Out}(\Pi_{\mathcal{H}_{\leadsto S}}) \stackrel{\sim}{\rightarrow} \mathrm{Out}(\Pi_{\mathcal{G}_0})$$

—where the first arrow is the natural inclusion [cf. [CbTpI], Definition 4.4], the second arrow is the isomorphism determined by $\Phi_{\mathcal{H}_{\leadsto S}}$ [cf. [CbTpI], Definition 2.10], and the third arrow is the isomorphism determined by ϕ. We shall say that α is a *nondegenerate* (respectively, *positive definite*) *$(\mathcal{H}, S, \phi)$-Dehn multi-twist* of $\mathcal{G}_0$ if α is the image of a nondegenerate [cf. [CbTpI], Definition 5.8, (ii)] (respectively, positive definite [cf. [CbTpI], Definition 5.8, (iii)]) profinite Dehn multi-twist of $\mathcal{H}$ via the above composite.

(v) Let m be a positive integer and $Y^{\log}$ a stable log curve over $(\mathrm{Spec}\, k)^{\log}$. If $m \geq 2$, then suppose that Σ_0 is either equal to $\mathfrak{Primes}$ or of cardinality one. For each nonnegative integer i, write ${}^{Y}\Pi_i$ (respectively, $\mathcal{H}$) for the "Π_i" (respectively, "$\mathcal{G}$") that occurs in the case where we take "$X^{\log}$" to be $Y^{\log}$. Then we shall say that a degeneration structure $(\mathcal{H}, S, \phi)$ on $\mathcal{G}$ [cf. (i)] is *m-cuspidalizable* if the composite

$${}^{Y}\Pi_1 \stackrel{\sim}{\longrightarrow} \Pi_{\mathcal{H}} \stackrel{\Phi_{\mathcal{H}_{\leadsto S}}}{\stackrel{\sim}{\longleftarrow}} \Pi_{\mathcal{H}_{\leadsto S}} \stackrel{\phi}{\stackrel{\sim}{\longrightarrow}} \Pi_{\mathcal{G}} \stackrel{\sim}{\longleftarrow} \Pi_1$$

—where the first and fourth arrows are the natural outer isomorphisms [cf. Definition 3.1, (ii)], and the second arrow $\Phi_{\mathcal{H}_{\leadsto S}}$ is the natural outer isomorphism of [CbTpI], Definition 2.10— is m-cuspidalizable [cf. Definition 3.20].

Remark 3.23.1 One interesting open problem in the theory of *profinite Dehn multi-twists* developed in [CbTpI], §4, is the following: In the notation of Definition 3.23, for $i = 1, 2$, let $(\mathcal{H}_i, S_i, \phi_i)$ be a degeneration structure on $\mathcal{G}_0$ [cf. Definition 3.23, (i)]; $\alpha_i \in \mathrm{Out}(\Pi_{\mathcal{G}_0})$ a *nondegenerate* $(\mathcal{H}_i, S_i, \phi_i)$-Dehn multi-twist [cf. Definition 3.23, (iv)]. Then:

> Suppose that α_1 *commutes* with α_2. Then is $(\mathcal{H}_1, S_1, \phi_1)$ *co-Dehn* to $(\mathcal{H}_2, S_2, \phi_2)$ [cf. Definition 3.23, (iii)]?

It is not clear to the authors at the time of writing whether or not this question may be answered in the affirmative. Nevertheless, we are able to obtain a *partial result* in this direction [cf. Corollary 3.25 below].

Proposition 3.24 (Compatibility of Tripod Homomorphisms) *Suppose that* $n \geq 3$. *Then the following hold:*

(i) *Let* $Y^{\log}$ *be a stable log curve over* $(\operatorname{Spec} k)^{\log}$. *For each nonnegative integer* i, *write* ${}^{Y}\Pi_i$ *(respectively,* $\mathcal{H}$*) for the "*Π_i*" (respectively, "*$\mathcal{G}$*") that occurs in the case where we take "*$X^{\log}$*" to be* $Y^{\log}$. *Let* $(\mathcal{H}, S, \phi)$ *be an* ***n-cuspidalizable degeneration structure*** *on* $\mathcal{G}$ *[cf. Definition 3.23, (i), (v)];* $\phi_n : {}^{Y}\Pi_n \xrightarrow{\sim} \Pi_n$ *a PFC-admissible outer isomorphism [cf. [CbTpI], Definition 1.4, (iii)] that lies over the displayed composite isomorphism of Definition 3.23, (v);* $\Pi^{\mathrm{tpd}} \subseteq \Pi_3$, ${}^{Y}\Pi^{\mathrm{tpd}} \subseteq {}^{Y}\Pi_3$ ***1-central*** *[*$\{1,2,3\}$*-]tripods [cf. Definitions 3.3, (i); 3.7, (ii)] of* Π_n, ${}^{Y}\Pi_n$, *respectively. Then there exists an outer isomorphism* $\phi^{\mathrm{tpd}} : {}^{Y}\Pi^{\mathrm{tpd}} \xrightarrow{\sim} \Pi^{\mathrm{tpd}}$ *such that the diagram*

$$\begin{array}{ccc} \mathrm{Out}^{\mathrm{FC}}({}^{Y}\Pi_n) & \xrightarrow{\sim} & \mathrm{Out}^{\mathrm{FC}}(\Pi_n) \\ \mathfrak{T}_{{}^{Y}\Pi^{\mathrm{tpd}}} \downarrow & & \downarrow \mathfrak{T}_{\Pi^{\mathrm{tpd}}} \\ \mathrm{Out}({}^{Y}\Pi^{\mathrm{tpd}}) & \xrightarrow{\sim} & \mathrm{Out}(\Pi^{\mathrm{tpd}}) \end{array}$$

[cf. Definition 3.19]—where the upper and lower horizontal arrows are the isomorphisms induced by ϕ_n, ϕ^{tpd}, *respectively—****commutes****, up to inner automorphisms of* $\mathrm{Out}(\Pi^{\mathrm{tpd}})$. *In particular,* ϕ_n *determines an isomorphism*

$$\mathrm{Out}^{\mathrm{FC}}({}^{Y}\Pi_n)^{\mathrm{geo}} \xrightarrow{\sim} \mathrm{Out}^{\mathrm{FC}}(\Pi_n)^{\mathrm{geo}}$$

[cf. Definition 3.19].

(ii) *If we regard* $\mathrm{Out}^{\mathrm{FC}}(\Pi_n)$ *as a closed subgroup of* $\mathrm{Out}^{\mathrm{FC}}(\Pi_1)$ *by means of the* ***natural injection*** $\mathrm{Out}^{\mathrm{FC}}(\Pi_n) \hookrightarrow \mathrm{Out}^{\mathrm{FC}}(\Pi_1)$ *of [NodNon], Theorem B, then the closed subgroup* $\mathrm{Dehn}(\mathcal{G}) \subseteq (\mathrm{Aut}(\mathcal{G}) \subseteq)\mathrm{Out}(\Pi_{\mathcal{G}}) \xleftarrow{\sim} \mathrm{Out}(\Pi_1)$ *[cf. [CbTpI], Definition 4.4] is* ***contained*** *in* $\mathrm{Out}^{\mathrm{FC}}(\Pi_n)^{\mathrm{geo}} \subseteq \mathrm{Out}^{\mathrm{FC}}(\Pi_n)$, *i.e.,*

$$\mathrm{Dehn}(\mathcal{G}) \subseteq \mathrm{Out}^{\mathrm{FC}}(\Pi_n)^{\mathrm{geo}}.$$

Proof First, we verify assertion (i). Let us observe that if the outer isomorphism ϕ_n arises *scheme-theoretically* as a *specialization isomorphism*—cf. the discussion preceding [CmbCsp], Definition 2.1, as well as [CbTpI], Remark 5.6.1—then the *commutativity* in question follows immediately from the various definitions involved [cf. also the discussion preceding [CmbCsp], Definition 2.1]. Now the general case follows from the observation that the *scheme-theoretic* case treated above allows one to reduce to the case where $Y^{\log} = X^{\log}$, and ϕ_n is an FC-admissible outomorphism, in which case the *commutativity* in question is a *tautological consequence* of the fact that $\mathfrak{T}_{\Pi^{\mathrm{tpd}}}$ is a *group homomorphism*. This completes the proof of assertion (i).

Next, we verify assertion (ii). The inclusion $\mathrm{Dehn}(\mathcal{G}) \subseteq \mathrm{Out}^{\mathrm{FC}}(\Pi_n)$ follows immediately from the fact that every profinite Dehn multi-twist arises *scheme-theoretically*. Next, we observe that the inclusion $\mathrm{Dehn}(\mathcal{G}) \subseteq \mathrm{Out}^{\mathrm{FC}}(\Pi_n)^{\mathrm{geo}}$ may be regarded *either* as a consequence of the fact that every profinite Dehn multi-twist arises "$\overline{\mathbb{Q}}$-*scheme-theoretically*", i.e., from scheme theory over $\overline{\mathbb{Q}}$ [cf. the commutative diagram of Remark 3.19.1], *or* as a consequence of the following argument: Observe that it follows immediately from assertion (i), together with [CbTpI], Theorem 4.8, (ii), (iv), that, by applying a suitable *specialization isomorphism*—cf. the discussion preceding [CmbCsp], Definition 2.1, as well as [CbTpI], Remark 5.6.1—we may assume without loss of generality that $\mathcal{G}$ is *totally degenerate*. Then the inclusion $\mathrm{Dehn}(\mathcal{G}) \subseteq \mathrm{Out}^{\mathrm{FC}}(\Pi_n)^{\mathrm{geo}}$ follows immediately from Theorem 3.18, (ii) [cf. also Theorem 3.16, (v); [CbTpI], Definition 4.4!]. This completes the proof of assertion (ii). □

Corollary 3.25 (Co-Dehn-ness of Degeneration Structures in the Totally Degenerate Case) *In the notation of Theorem 3.16, for $i = 1, 2$, let $Y_i^{\log}$ be a stable log curve over $(\mathrm{Spec}\, k)^{\log}$; $\mathcal{H}_i$ the "$\mathcal{G}$" that occurs in the case where we take "$X^{\log}$" to be $Y_i^{\log}$; $(\mathcal{H}_i, S_i, \phi_i)$ a* ***3-cuspidalizable degeneration structure*** *on $\mathcal{G}$ [cf. Definition 3.23, (i), (v)]; $\alpha_i \in \mathrm{Out}(\Pi_{\mathcal{G}})$ a* ***nondegenerate*** *$(\mathcal{H}_i, S_i, \phi_i)$-Dehn multi-twist of $\mathcal{G}$ [cf. Definition 3.23, (iv)]. Suppose that α_1* ***commutes*** *with α_2, and that $\mathcal{H}_2$ is* ***totally degenerate*** *[cf. [CbTpI], Definition 2.3, (iv)]. Suppose, moreover, that one of the following conditions is satisfied:*

(a) $r \neq 0$.
(b) α_1 and α_2 are ***positive definite*** *[cf. Definition 3.23, (iv)].*

Then $(\mathcal{H}_1, S_1, \phi_1)$ is ***co-Dehn*** *to $(\mathcal{H}_2, S_2, \phi_2)$ [cf. Definition 3.23, (iii)], or, equivalently [since $\mathcal{H}_2$ is* ***totally degenerate****], $(\mathcal{H}_2, S_2, \phi_2) \preceq (\mathcal{H}_1, S_1, \phi_1)$ [cf. Definition 3.23, (ii)].*

Proof For $i = 1, 2$, write $\psi_i : \Pi_{\mathcal{G}} \xrightarrow{\sim} \Pi_{\mathcal{H}_i}$ for the composite outer isomorphism

$$\psi_i : \Pi_{\mathcal{G}} \overset{\phi_i}{\underset{\sim}{\leftarrow}} \Pi_{(\mathcal{H}_i)_{\rightsquigarrow S_i}} \overset{\Phi_{(\mathcal{H}_i)_{\rightsquigarrow S_i}}}{\underset{\sim}{\rightarrow}} \Pi_{\mathcal{H}_i}$$

and $\psi \overset{\mathrm{def}}{=} \psi_1 \circ \psi_2^{-1}$. Write $\alpha_1[\mathcal{H}_2] \in \mathrm{Out}(\Pi_{\mathcal{H}_2})$ for the outomorphism obtained by conjugating α_1 by ψ_2. First, we claim that the following assertion holds:

Claim 3.25.A: There exists a positive integer a such that $\beta \overset{\mathrm{def}}{=} \alpha_1[\mathcal{H}_2]^a \in \mathrm{Dehn}(\mathcal{H}_2)$.

Indeed, since α_1 is an $(\mathcal{H}_1, S_1, \phi_1)$-*Dehn multi-twist* of $\mathcal{G}$, the outomorphism $\alpha_1[\mathcal{H}_2]$ of $\Pi_{\mathcal{H}_2}$ is *group-theoretically cuspidal*. Thus, since α_1 *commutes* with α_2, it follows, in the case of condition (a) (respectively, (b)), from Theorem 1.9, (i) (respectively Theorem 1.9, (ii)), which may be applied in light of [CbTpI], Corollary

5.9, (ii) (respectively, [CbTpI], Corollary 5.9, (iii)), that $\alpha_1[\mathcal{H}_2] \in \mathrm{Aut}(\mathcal{H}_2)$. In particular, since the underlying semi-graph of $\mathcal{H}_2$ is *finite*, there exists a positive integer a such that $\alpha_1[\mathcal{H}_2]^a \in \mathrm{Aut}^{|\mathrm{grph}|}(\mathcal{H}_2)$ [cf. [CbTpI], Definition 2.6, (i); Remark 4.1.2 of the present monograph]. On the other hand, since α_1 is an $(\mathcal{H}_1, S_1, \phi_1)$-*Dehn multi-twist* of $\mathcal{G}$, it follows immediately from Proposition 3.24, (i), (ii), that the image of α_1 via the tripod homomorphism associated to Π_3 [cf. Definition 3.19] is *trivial*. Thus, since $\mathcal{H}_2$ is *totally degenerate*, and $\alpha_1[\mathcal{H}_2]^a \in \mathrm{Aut}^{|\mathrm{grph}|}(\mathcal{H}_2)$, by applying Theorem 3.18, (ii), together with Proposition 3.24, (i), we conclude that $\beta = \alpha_1[\mathcal{H}_2]^a \in \mathrm{Dehn}(\mathcal{H}_2)$. This completes the proof of Claim 3.25.A.

Next, let us fix an element $l \in \Sigma$. For $i \in \{1, 2\}$, write $\mathcal{H}_i^{\{l\}}$ for the semi-graph of anabelioids of pro-l PSC-type obtained by forming the pro-l completion of $\mathcal{H}_i$ [cf. [SemiAn], Definition 2.9, (ii)]. Then it follows immediately from Claim 3.25.A, together with [CbTpI], Theorem 4.8, (ii), (iv), that there exists a subset $S \subseteq \mathrm{Node}(\mathcal{H}_2)$ [which may *depend* on l!] such that the automorphism $\beta^{\{l\}} \in \mathrm{Aut}(\mathcal{H}_2^{\{l\}})$ induced by β is contained in $\mathrm{Dehn}((\mathcal{H}_2^{\{l\}})_{\rightsquigarrow S}) \subseteq \mathrm{Dehn}(\mathcal{H}_2^{\{l\}}) \subseteq \mathrm{Aut}(\mathcal{H}_2^{\{l\}})$ [i.e., $\beta^{\{l\}}$ is a *profinite Dehn multi-twist* of $(\mathcal{H}_2^{\{l\}})_{\rightsquigarrow S}$], and, moreover, $\beta^{\{l\}}$ is *nondegenerate* as a *profinite Dehn multi-twist of* $(\mathcal{H}_2^{\{l\}})_{\rightsquigarrow S}$. Write $\alpha_1^{\{l\}}$ for the outomorphism of the pro-l group $\Pi_{\mathcal{H}_1^{\{l\}}}$ [which is naturally isomorphic to the maximal pro-l quotient of $\Pi_{\mathcal{H}_1}$] obtained by conjugating α_1 by ψ_1 and $\psi^{\{l\}}\colon \Pi_{\mathcal{H}_2^{\{l\}}} \xrightarrow{\sim} \Pi_{\mathcal{H}_1^{\{l\}}}$ for the outer isomorphism induced by ψ [cf. the discussion preceding Claim 3.25.A].

Next, we claim that the following assertion holds:

Claim 3.25.B: The composite outer isomorphism

$$\psi_S\colon \Pi_{(\mathcal{H}_2)_{\rightsquigarrow S}} \overset{\Phi_{(\mathcal{H}_2)_{\rightsquigarrow S}}}{\xrightarrow{\sim}} \Pi_{\mathcal{H}_2} \overset{\psi}{\xrightarrow{\sim}} \Pi_{\mathcal{H}_1}$$

is *graphic*, i.e., arises from an isomorphism $(\mathcal{H}_2)_{\rightsquigarrow S} \xrightarrow{\sim} \mathcal{H}_1$.

Indeed, let $\widetilde{\psi}_S\colon \Pi_{(\mathcal{H}_2)_{\rightsquigarrow S}} \xrightarrow{\sim} \Pi_{\mathcal{H}_1}$ be an isomorphism that *lifts* ψ_S. Then it follows immediately from [CmbGC], Proposition 1.5, (ii)—by considering the *functorial bijections* between the sets "VCN" [cf. [NodNon], Definition 1.1, (iii)] of various connected finite étale coverings of $\mathcal{H}_1$, $(\mathcal{H}_2)_{\rightsquigarrow S}$—that, to verify Claim 3.25.B, it suffices to verify the following:

Let $\mathcal{I}_2 \to (\mathcal{H}_2)_{\rightsquigarrow S}$ be a connected finite étale covering of $(\mathcal{H}_2)_{\rightsquigarrow S}$ that corresponds to a *characteristic* open subgroup $\Pi_{\mathcal{I}_2} \subseteq \Pi_{(\mathcal{H}_2)_{\rightsquigarrow S}}$. Write $\mathcal{I}_1 \to \mathcal{H}_1$ for the connected finite étale covering of $\mathcal{H}_1$ that corresponds to the [necessarily *characteristic*] open subgroup $\Pi_{\mathcal{I}_1} \overset{\mathrm{def}}{=} \widetilde{\psi}_S(\Pi_{\mathcal{I}_2}) \subseteq \Pi_{\mathcal{H}_1}$ and $\mathcal{I}_1^{\{l\}}$, $\mathcal{I}_2^{\{l\}}$ for the semi-graphs of anabelioids of pro-l PSC-type obtained by forming the pro-l completions of $\mathcal{I}_1$, $\mathcal{I}_2$, respectively. Then the outer isomorphism $\Pi_{\mathcal{I}_2^{\{l\}}} \xrightarrow{\sim} \Pi_{\mathcal{I}_1^{\{l\}}}$ determined by $\widetilde{\psi}_S$ is *graphic*.

To verify this *graphicity*, let us first recall that the automorphisms $\beta^{\{l\}} \in \mathrm{Aut}((\mathcal{H}_2^{\{l\}})_{\rightsquigarrow S})$ and $\alpha_1 \in \mathrm{Aut}(\mathcal{H}_1)$ are *nondegenerate profinite Dehn multi-*

twists. Thus, it follows immediately from Lemma 3.26, (i), (ii), below [cf. also Claim 3.25.A], that there exist liftings $\widetilde{\beta} \in \mathrm{Aut}(\Pi_{(\mathcal{H}_2)_{\rightsquigarrow S}})$, $\widetilde{\alpha}_1 \in \mathrm{Aut}(\Pi_{\mathcal{H}_1})$ of β, α_1, respectively, and a positive integer b such that the outomorphisms γ_2, γ_1 of $\Pi_{\mathcal{I}_2^{\{l\}}}$, $\Pi_{\mathcal{I}_1^{\{l\}}}$ determined by $\widetilde{\beta}^b$, $\widetilde{\alpha}_1^b$ are *nondegenerate profinite Dehn multi-twists* of $\mathcal{I}_2^{\{l\}}$, $\mathcal{I}_1^{\{l\}}$, respectively, and, moreover, γ_2 and γ_1^a are *compatible* relative to the outer isomorphism in question $\Pi_{\mathcal{I}_2^{\{l\}}} \xrightarrow{\sim} \Pi_{\mathcal{I}_1^{\{l\}}}$. Moreover, if condition (b) is satisfied, then γ_1 is a *positive definite profinite Dehn multi-twist* of $\mathcal{I}_1^{\{l\}}$ [cf. Lemma 3.26, (ii), below]. Thus, it follows, in the case of condition (a) (respectively, (b)), from Theorem 1.9, (i) (respectively Theorem 1.9, (ii)), which may be applied in light of [CbTpI], Corollary 5.9, (ii) (respectively, [CbTpI], Corollary 5.9, (iii)), that the outer isomorphism in question $\Pi_{\mathcal{I}_2^{\{l\}}} \xrightarrow{\sim} \Pi_{\mathcal{I}_1^{\{l\}}}$ is *graphic*. This completes the proof of Claim 3.25.B. On the other hand, one verifies easily from the various definitions involved that Claim 3.25.B implies that $(\mathcal{H}_2, S_2, \phi_2) \preceq (\mathcal{H}_1, S_1, \phi_1)$. This completes the proof of Corollary 3.25. □

Lemma 3.26 (Profinite Dehn Multi-twists and Pro-Σ Completions of Finite Étale Coverings) *Let $\Sigma_1 \subseteq \Sigma_0$ be nonempty sets of prime numbers, $\mathcal{G}_0$ a semi-graph of anabelioids of pro-Σ_0 PSC-type, $\mathcal{H}_0 \to \mathcal{G}_0$ a connected finite étale Galois covering that arises from a normal open subgroup $\Pi_{\mathcal{H}_0} \subseteq \Pi_{\mathcal{G}_0}$ of $\Pi_{\mathcal{G}_0}$, and $\widetilde{\alpha} \in \mathrm{Aut}(\Pi_{\mathcal{G}_0})$. Write $\mathcal{G}_1$, $\mathcal{H}_1$ for the semi-graphs of anabelioids of pro-Σ_1 PSC-type obtained by forming the pro-Σ_1 completions of $\mathcal{G}_0$, $\mathcal{H}_0$, respectively [cf. [SemiAn], Definition 2.9, (ii)]. Suppose that $\widetilde{\alpha} \in \mathrm{Aut}(\Pi_{\mathcal{G}_0})$ preserves the normal open subgroup $\Pi_{\mathcal{H}_0} \subseteq \Pi_{\mathcal{G}_0}$ corresponding to $\mathcal{H}_0 \to \mathcal{G}_0$. Write $\alpha_{\mathcal{G}_0}$, $\alpha_{\mathcal{H}_0}$, $\alpha_{\mathcal{G}_1}$, $\alpha_{\mathcal{H}_1}$ for the respective outomorphisms of $\Pi_{\mathcal{G}_0}$, $\Pi_{\mathcal{H}_0}$, $\Pi_{\mathcal{G}_1}$, $\Pi_{\mathcal{H}_1}$ induced by $\widetilde{\alpha}$. Suppose, moreover, that $\alpha_{\mathcal{G}_0} \in \mathrm{Dehn}(\mathcal{G}_0)$ [cf. [CbTpI], Definition 4.4]. Then the following hold:*

(i) It holds that $\alpha_{\mathcal{G}_1} \in \mathrm{Dehn}(\mathcal{G}_1)$. Moreover, there exists a positive integer a such that

$$\alpha_{\mathcal{H}_0}^a \in \mathrm{Dehn}(\mathcal{H}_0), \quad \alpha_{\mathcal{H}_1}^a \in \mathrm{Dehn}(\mathcal{H}_1).$$

*(ii) If, moreover, $\alpha_{\mathcal{G}_1} \in \mathrm{Dehn}(\mathcal{G}_1)$ [cf. (i)] is **nondegenerate** (respectively, **positive definite**) [cf. [CbTpI], Definition 5.8, (ii), (iii)], then $\alpha_{\mathcal{H}_1}^a \in \mathrm{Dehn}(\mathcal{H}_1)$ [cf. (i)] is **nondegenerate** (respectively, **positive definite**).*

Proof First, we verify assertion (i). One verifies easily from [NodNon], Lemma 2.6, (i), together with [CbTpI], Corollary 5.9, (i), that there exists a positive integer a such that $\alpha_{\mathcal{H}_0}^a \in \mathrm{Dehn}(\mathcal{H}_0)$. Now since $\alpha_{\mathcal{G}_0} \in \mathrm{Dehn}(\mathcal{G}_0)$, $\alpha_{\mathcal{H}_0}^a \in \mathrm{Dehn}(\mathcal{H}_0)$, it follows immediately from the various definitions involved that $\alpha_{\mathcal{G}_1} \in \mathrm{Dehn}(\mathcal{G}_1)$, $\alpha_{\mathcal{H}_1}^a \in \mathrm{Dehn}(\mathcal{H}_1)$. This completes the proof of assertion (i). Assertion (ii) follows immediately, in the *nondegenerate* (respectively, *positive definite*) case, from

[NodNon], Lemma 2.6, (i), together with [CbTpI], Corollary 5.9, (ii) (respectively, from Corollary 5.9, (iii), (v)). This completes the proof of Lemma 3.26. □

Corollary 3.27 (Commensurator of Profinite Dehn Multi-twists in the Totally Degenerate Case) *In the notation of Theorem 3.16, Definition 3.19 [so $n \geq 3$], suppose further that $\mathcal{G}$ is* ***totally degenerate*** *[cf. [CbTpI], Definition 2.3, (iv)]. Write $s\colon \mathrm{Spec}\, k \to (\overline{\mathcal{M}}_{g,[r]})_k \stackrel{\mathrm{def}}{=} (\overline{\mathcal{M}}_{g,[r]})_{\mathrm{Spec}\, k}$ [cf. the discussion entitled "Curves" in "Notations and Conventions"] for the underlying (1-)morphism of algebraic stacks of the classifying (1-)morphism $(\mathrm{Spec}\, k)^{\log} \to (\overline{\mathcal{M}}^{\log}_{g,[r]})_k \stackrel{\mathrm{def}}{=} (\overline{\mathcal{M}}^{\log}_{g,[r]})_{\mathrm{Spec}\, k}$ [cf. the discussion entitled "Curves" in "Notations and Conventions"] of the stable log curve $X^{\log}$ over $(\mathrm{Spec}\, k)^{\log}$; $\widetilde{\mathcal{N}}^{\log}_s$ for the log scheme obtained by equipping $\widetilde{\mathcal{N}}_s \stackrel{\mathrm{def}}{=} \mathrm{Spec}\, k$ with the log structure induced, via s, by the log structure of $(\overline{\mathcal{M}}^{\log}_{g,[r]})_k$; $\mathcal{N}^{\log}_s$ for the log stack obtained by forming the [stack-theoretic] quotient of the log scheme $\widetilde{\mathcal{N}}^{\log}_s$ by the natural action of the finite k-group "$s \times_{(\overline{\mathcal{M}}_{g,[r]})_k} s$", i.e., the fiber product over $(\overline{\mathcal{M}}_{g,[r]})_k$ of two copies of s; $\mathcal{N}_s$ for the underlying stack of the log stack $\mathcal{N}^{\log}_s$; $I_{\mathcal{N}_s} \subseteq \pi_1(\mathcal{N}^{\log}_s)$ for the closed subgroup of the log fundamental group $\pi_1(\mathcal{N}^{\log}_s)$ of $\mathcal{N}^{\log}_s$ given by the kernel of the natural surjection $\pi_1(\mathcal{N}^{\log}_s) \twoheadrightarrow \pi_1(\mathcal{N}_s)$ [induced by the (1-)morphism $\mathcal{N}^{\log}_s \to \mathcal{N}_s$ obtained by forgetting the log structure]; $\pi_1^{(\Sigma)}(\mathcal{N}^{\log}_s)$ for the quotient of $\pi_1(\mathcal{N}^{\log}_s)$ by the kernel of the natural surjection from $I_{\mathcal{N}_s}$ to its maximal pro-Σ quotient $I^{\Sigma}_{\mathcal{N}_s}$. Then the following hold:*

(i) The natural homomorphism $\pi_1(\mathcal{N}^{\log}_s) \to \mathrm{Out}(\Pi_1)$ [cf. the natural outer homomorphism of the first display of Remark 3.19.1] ***factors*** *through the quotient $\pi_1(\mathcal{N}^{\log}_s) \twoheadrightarrow \pi_1^{(\Sigma)}(\mathcal{N}^{\log}_s)$ and the natural inclusion $N_{\mathrm{Out}^{\mathrm{FC}}(\Pi_n)^{\mathrm{geo}}}(\mathrm{Dehn}(\mathcal{G})) \hookrightarrow \mathrm{Out}(\Pi_1)$ [cf. Proposition 3.24, (ii)]. In particular, we obtain a homomorphism*

$$\pi_1^{(\Sigma)}(\mathcal{N}^{\log}_s) \longrightarrow N_{\mathrm{Out}^{\mathrm{FC}}(\Pi_n)^{\mathrm{geo}}}(\mathrm{Dehn}(\mathcal{G})),$$

hence also a homomorphism

$$\pi_1^{(\Sigma)}(\mathcal{N}^{\log}_s) \longrightarrow C_{\mathrm{Out}^{\mathrm{FC}}(\Pi_n)^{\mathrm{geo}}}(\mathrm{Dehn}(\mathcal{G})).$$

(ii) The second displayed homomorphism of (i) fits into a natural commutative diagram of profinite groups

$$\begin{array}{ccccccccc}
1 & \longrightarrow & I^{\Sigma}_{\mathcal{N}_s} & \longrightarrow & \pi_1^{(\Sigma)}(\mathcal{N}^{\log}_s) & \longrightarrow & \pi_1(\mathcal{N}_s) & \longrightarrow & 1 \\
 & & \downarrow & & \downarrow & & \downarrow & & \\
1 & \longrightarrow & \mathrm{Dehn}(\mathcal{G}) & \longrightarrow & C_{\mathrm{Out}^{\mathrm{FC}}(\Pi_n)^{\mathrm{geo}}}(\mathrm{Dehn}(\mathcal{G})) & \longrightarrow & \mathrm{Aut}(\mathbb{G}) & \longrightarrow & 1
\end{array}$$

[cf. Definition 3.1, (ii), concerning the notation "$\mathbb{G}$"]—where the horizontal sequences are ***exact****, and the vertical arrows are* ***isomorphisms****.*

(iii) $\mathrm{Dehn}(\mathcal{G})$ *is* ***open*** *in* $C_{\mathrm{Out}^{\mathrm{FC}}(\Pi_n)^{\mathrm{geo}}}(\mathrm{Dehn}(\mathcal{G}))$.

(iv) We have an equality

$$N_{\mathrm{Out}^{\mathrm{FC}}(\Pi_n)^{\mathrm{geo}}}(\mathrm{Dehn}(\mathcal{G})) = C_{\mathrm{Out}^{\mathrm{FC}}(\Pi_n)^{\mathrm{geo}}}(\mathrm{Dehn}(\mathcal{G})).$$

Proof First, we verify assertion (i). The fact that the image of the homomorphism in question is *contained* in $\mathrm{Out}^{\mathrm{FC}}(\Pi_n)^{\mathrm{geo}}$ follows immediately from the [*tautological!*] fact that this image arises "$\overline{\mathbb{Q}}$*-scheme-theoretically*", i.e., from scheme theory over $\overline{\mathbb{Q}}$ [cf. the discussion of Remark 3.19.1]. Thus, assertion (i) follows immediately from the fact that the natural homomorphism $\pi_1(\mathcal{N}_s^{\log}) \to \mathrm{Out}(\Pi_1)$ determines an *isomorphism* $I^{\Sigma}_{\mathcal{N}_s} \xrightarrow{\sim} \mathrm{Dehn}(\mathcal{G})$ [cf. [CbTpI], Proposition 5.6, (ii)]. This completes the proof of assertion (i).

Next, we verify assertion (ii). First, let us observe that it follows from [CbTpI], Theorem 5.14, (iii), that $C_{\mathrm{Out}^{\mathrm{FC}}(\Pi_n)^{\mathrm{geo}}}(\mathrm{Dehn}(\mathcal{G})) \subseteq \mathrm{Aut}(\mathcal{G})$. Thus, we obtain a natural homomorphism $C_{\mathrm{Out}^{\mathrm{FC}}(\Pi_n)^{\mathrm{geo}}}(\mathrm{Dehn}(\mathcal{G})) \to \mathrm{Aut}(\mathbb{G})$, whose kernel contains $\mathrm{Dehn}(\mathcal{G})$ [cf. the definition of a profinite Dehn multi-twist given in [CbTpI], Definition 4.4]. On the other hand, if an element $\alpha \in C_{\mathrm{Out}^{\mathrm{FC}}(\Pi_n)^{\mathrm{geo}}}(\mathrm{Dehn}(\mathcal{G}))$ acts *trivially* on $\mathbb{G}$, then, since $\mathcal{G}$ is *totally degenerate*, it follows immediately from Theorem 3.18, (ii), that $\alpha \in \mathrm{Dehn}(\mathcal{G})$. This completes the proof of the existence of the lower exact sequence in the diagram of assertion (ii), except for the *surjectivity* of the third arrow of this sequence. Thus, it follows immediately from the proof of assertion (i) that, to complete the proof of assertion (ii), it suffices to verify that the right-hand vertical arrow $\pi_1(\mathcal{N}_s) \to \mathrm{Aut}(\mathbb{G})$ of the diagram is an *isomorphism*. Write $X^{\log}_{\widetilde{\mathcal{N}}_s}$ for the stable log curve over $\widetilde{\mathcal{N}}_s^{\log}$ whose classifying (1-)morphism is given by the natural (1-)morphism $\widetilde{\mathcal{N}}_s^{\log} \to (\overline{\mathcal{M}}^{\log}_{g,[r]})_k$ and $\mathrm{Aut}_{\widetilde{\mathcal{N}}_s^{\log}}(X^{\log}_{\widetilde{\mathcal{N}}_s})$ for the group of automorphisms of $X^{\log}_{\widetilde{\mathcal{N}}_s}$ over $\widetilde{\mathcal{N}}_s^{\log}$. Then since $X^{\log}$, hence also $X^{\log}_{\widetilde{\mathcal{N}}_s}$, is *totally degenerate*, one verifies easily that the natural homomorphism $\mathrm{Aut}_{\widetilde{\mathcal{N}}_s^{\log}}(X^{\log}_{\widetilde{\mathcal{N}}_s}) \to \mathrm{Aut}(\mathbb{G})$ is an *isomorphism*. Thus, it follows immediately from the various definitions involved that the right-hand vertical arrow $\pi_1(\mathcal{N}_s) \to \mathrm{Aut}(\mathbb{G})$ of the diagram is an *isomorphism*. This completes the proof of assertion (ii).

Assertion (iii) follows immediately from the *exactness* of the lower sequence of the diagram of assertion (ii), together with the *finiteness* of $\mathbb{G}$. Assertion (iv) follows immediately from the fact that the middle vertical arrow of the diagram of assertion (ii) is an *isomorphism* which *factors* through $N_{\mathrm{Out}^{\mathrm{FC}}(\Pi_n)^{\mathrm{geo}}}(\mathrm{Dehn}(\mathcal{G})) \subseteq C_{\mathrm{Out}^{\mathrm{FC}}(\Pi_n)^{\mathrm{geo}}}(\mathrm{Dehn}(\mathcal{G}))$ [cf. assertion (i)]. This completes the proof of Corollary 3.27. □

Remark 3.27.1 One interesting consequence of Corollary 3.27 is the following: The profinite group $\mathrm{Out}^{\mathrm{FC}}(\Pi_n)^{\mathrm{geo}}$ [which, as discussed in Remark 3.19.1, may be regarded as the *geometric portion* of the group of FC-admissible outomorphisms of the configuration space group Π_n], hence also the commensurator $C_{\mathrm{Out}^{\mathrm{FC}}(\Pi_n)^{\mathrm{geo}}}(\mathrm{Dehn}(\mathcal{G}))$, is defined in a *purely combinatorial/group-theoretic* fashion. In particular, it follows from the commutative diagram of Corollary 3.27, (ii), that this commensurator $C_{\mathrm{Out}^{\mathrm{FC}}(\Pi_n)^{\mathrm{geo}}}(\mathrm{Dehn}(\mathcal{G}))$ yields a *purely combinatorial/group-theoretic algorithm* for reconstructing the profinite groups of *scheme-theoretic* origin that appear in the upper sequence of this diagram.

Chapter 4
Glueability of Combinatorial Cuspidalizations

In this chapter, we discuss the *glueability of combinatorial cuspidalizations*. The resulting theory may be regarded as a higher-dimensional analogue of the displayed exact sequence of [CbTpI], Theorem B, (iii) [cf. Theorem 4.14, (iii), below, of the present monograph]. This theory implies a certain key *surjectivity* property of the *tripod homomorphism* [cf. Corollary 4.15 below]. Finally, we apply this result to construct *cuspidalizations* of the log fundamental group of a stable log curve over a finite field [cf. Corollary 4.16 below] and to compute certain *commensurators* of the corresponding Galois image in the *totally degenerate case* [cf. Corollary 4.17 below].

In this chapter, we maintain the notation of the preceding Chap. 3 [cf. also Definition 3.1]. In addition, let Σ_0 be a nonempty set of prime numbers and $\mathcal{G}_0$ a semi-graph of anabelioids of pro-Σ_0 PSC-type. Write $\mathbb{G}_0$ for the underlying semi-graph of $\mathcal{G}_0$ and $\Pi_{\mathcal{G}_0}$ for the [pro-Σ_0] fundamental group of $\mathcal{G}_0$.

Definition 4.1

(i) We shall write

$$\mathrm{Aut}^{|\mathrm{Brch}(\mathcal{G}_0)|}(\mathcal{G}_0) \subseteq (\mathrm{Aut}^{|\mathrm{Vert}(\mathcal{G}_0)|}(\mathcal{G}_0) \cap \mathrm{Aut}^{|\mathrm{Node}(\mathcal{G}_0)|}(\mathcal{G}_0) \subseteq) \mathrm{Aut}(\mathcal{G}_0)$$

[cf. [CbTpI], Definition 2.6, (i)] for the [*closed*] subgroup of $\mathrm{Aut}(\mathcal{G}_0)$ consisting of automorphisms α of $\mathcal{G}_0$ that induce the identity automorphism of $\mathrm{Vert}(\mathcal{G}_0)$, $\mathrm{Node}(\mathcal{G}_0)$ and, moreover, fix each of the branches of every node of $\mathcal{G}_0$. Thus, we have a *natural exact sequence* of profinite groups

$$1 \longrightarrow \mathrm{Aut}^{|\mathrm{grph}|}(\mathcal{G}_0) \longrightarrow \mathrm{Aut}^{|\mathrm{Brch}(\mathcal{G}_0)|}(\mathcal{G}_0) \longrightarrow \mathrm{Aut}(\mathrm{Cusp}(\mathcal{G}_0))$$

[cf. [CbTpI], Definition 2.6, (i); Remark 4.1.2 of the present monograph].

Y. Hoshi, S. Mochizuki, *Topics Surrounding the Combinatorial Anabelian Geometry of Hyperbolic Curves II*, Lecture Notes in Mathematics 2299,
https://doi.org/10.1007/978-981-19-1096-8_4

(ii) Let $v \in \mathrm{Vert}(\mathcal{G}_0)$. Then we shall write

$$\mathcal{E}(\mathcal{G}_0|_v : \mathcal{G}_0) \subseteq \mathrm{Edge}(\mathcal{G}_0|_v) \; (= \mathrm{Cusp}(\mathcal{G}_0|_v))$$

[cf. [CbTpI], Definition 2.1, (iii)] for the subset of $\mathrm{Edge}(\mathcal{G}_0|_v)$ $(= \mathrm{Cusp}(\mathcal{G}_0|_v))$ consisting of cusps of $\mathcal{G}_0|_v$ that arise from nodes of $\mathcal{G}_0$.

(iii) We shall write

$$\mathrm{Glu}^{\mathrm{brch}}(\mathcal{G}_0) \subseteq \prod_{v \in \mathrm{Vert}(\mathcal{G}_0)} \mathrm{Aut}^{|\mathcal{E}(\mathcal{G}|_v:\mathcal{G})|}(\mathcal{G}_0|_v)$$

[cf. (ii); [CbTpI], Definition 2.6, (i)] for the [*closed*] subgroup of $\prod_{v \in \mathrm{Vert}(\mathcal{G}_0)} \mathrm{Aut}^{|\mathcal{E}(\mathcal{G}|_v:\mathcal{G})|}(\mathcal{G}_0|_v)$ consisting of "*glueable*" collections of automorphisms of the various $\mathcal{G}_0|_v$, i.e., the subgroup consisting of $(\alpha_v)_{v \in \mathrm{Vert}(\mathcal{G}_0)}$ such that, for every $v, w \in \mathrm{Vert}(\mathcal{G}_0)$, it holds that $\chi_v(\alpha_v) = \chi_w(\alpha_w)$ [cf. [CbTpI], Definition 3.8, (ii)].

Remark 4.1.1 In the notation of Definition 4.1, one verifies easily from the various definitions involved that

$$\mathrm{Glu}(\mathcal{G}_0) = \mathrm{Glu}^{\mathrm{brch}}(\mathcal{G}_0) \cap \Big(\prod_{v \in \mathrm{Vert}(\mathcal{G}_0)} \mathrm{Aut}^{|\mathrm{grph}|}(\mathcal{G}_0|_v) \Big)$$

[cf. [CbTpI], Definitions 2.6, (i), and 4.9; Remark 4.1.2 of the present monograph].

Remark 4.1.2 Here, we take the opportunity to correct a *minor error* in the exposition of [CbTpI]. In [CbTpI], Definition 2.6, (i), "$\mathrm{Aut}^{|\mathrm{grph}|}(\mathcal{G})$" should be defined as the subgroup of $\mathrm{Aut}(\mathcal{G})$ of automorphisms of $\mathcal{G}$ which induce the identity automorphism on the underlying semi-graph of $\mathcal{G}$ [cf. the definition given in [CbTpI], Theorem B]. In a similar vein, in [CbTpI], Definition 2.6, (iii), "$\mathrm{Aut}^{|\mathbb{H}|}(\mathcal{G})$" should be defined as the subgroup of $\mathrm{Aut}(\mathcal{G})$ of automorphisms of $\mathcal{G}$ which preserve the sub-semi-graph $\mathbb{H}$ of the underlying semi-graph of $\mathcal{G}$ and, moreover, induce the identity automorphism of $\mathbb{H}$. Since the correct definitions are applied throughout the exposition of [CbTpI], these errors in the statement of the definitions have *no substantive effect* on the exposition of [CbTpI], except for the following two instances [which themselves do not have any substantive effect on the exposition of [CbTpI]]:

(i) In [CbTpI], Proposition 2.7, (ii), "$\mathrm{Aut}^{|\mathrm{grph}|}(\mathcal{G})$" should be replaced by "$\mathrm{Aut}^{|\mathrm{VCN}(\mathcal{G})|}(\mathcal{G})$".
(ii) In [CbTpI], Proposition 2.7, (iii), the phrase "In particular" should be replaced by the word "Finally".

Theorem 4.2 (Glueability of Combinatorial Cuspidalizations in the One-Dimensional Case) *Let Σ_0 be a nonempty set of prime numbers and $\mathcal{G}_0$ a* *semi-graph of anabelioids of pro-*Σ_0 *PSC-type. Write* $\Pi_{\mathcal{G}_0}$ *for the [pro-*Σ_0*] fundamental group of* $\mathcal{G}_0$*. Then the following hold:*

(i) The closed subgroup $\mathrm{Dehn}(\mathcal{G}_0) \subseteq \mathrm{Aut}(\mathcal{G}_0)$ *[cf. [CbTpI], Definition 4.4] is* **contained** *in* $\mathrm{Aut}^{|\mathrm{Brch}(\mathcal{G}_0)|}(\mathcal{G}_0) \subseteq \mathrm{Aut}(\mathcal{G}_0)$ *[cf. Definition 4.1, (i)], i.e.,* $\mathrm{Dehn}(\mathcal{G}_0) \subseteq \mathrm{Aut}^{|\mathrm{Brch}(\mathcal{G}_0)|}(\mathcal{G}_0)$.

(ii) The natural homomorphism

$$\begin{array}{ccc}\mathrm{Aut}^{|\mathrm{Brch}(\mathcal{G}_0)|}(\mathcal{G}_0) & \longrightarrow & \prod_{v\in \mathrm{Vert}(\mathcal{G}_0)} \mathrm{Aut}(\mathcal{G}_0|_v) \\ \alpha & \mapsto & (\alpha_{\mathcal{G}_0|_v})_{v\in\mathrm{Vert}(\mathcal{G}_0)}\end{array}$$

[cf. [CbTpI], Definition 2.14, (ii); [CbTpI], Remark 2.5.1, (ii)] **factors** *through*

$$\mathrm{Glu}^{\mathrm{brch}}(\mathcal{G}_0) \subseteq \prod_{v\in\mathrm{Vert}(\mathcal{G}_0)} \mathrm{Aut}(\mathcal{G}_0|_v)$$

[cf. Definition 4.1, (iii)].

(iii) The natural inclusion $\mathrm{Dehn}(\mathcal{G}_0) \hookrightarrow \mathrm{Aut}^{|\mathrm{Brch}(\mathcal{G}_0)|}(\mathcal{G}_0)$ *of (i) and the natural homomorphism* $\rho^{\mathrm{brch}}_{\mathcal{G}_0} \colon \mathrm{Aut}^{|\mathrm{Brch}(\mathcal{G}_0)|}(\mathcal{G}_0) \to \mathrm{Glu}^{\mathrm{brch}}(\mathcal{G}_0)$ *[cf. (ii)] fit into an* **exact sequence** *of profinite groups*

$$1 \longrightarrow \mathrm{Dehn}(\mathcal{G}_0) \longrightarrow \mathrm{Aut}^{|\mathrm{Brch}(\mathcal{G}_0)|}(\mathcal{G}_0) \overset{\rho^{\mathrm{brch}}_{\mathcal{G}_0}}{\longrightarrow} \mathrm{Glu}^{\mathrm{brch}}(\mathcal{G}_0) \longrightarrow 1.$$

Proof Assertion (i) follows immediately from the various definitions involved. Assertion (ii) follows immediately from [CbTpI], Corollary 3.9, (iv). Assertion (iii) follows, in light of Remark 4.1.1, from the exact sequence of [CbTpI], Theorem B, (iii), together with the existence of automorphisms of $\mathcal{G}_0$ that induce *arbitrary permutations of the cusps on each vertex of* $\mathcal{G}_0$ and, moreover, restrict to automorphisms of each $\mathcal{G}_0|_v$ that lie in the *kernel* of χ_v [cf. the automorphisms constructed in the proof of [CmbCsp], Lemma 2.4]. □

Definition 4.3 Let $\mathbb{H}$ be a sub-semi-graph of *PSC-type* [cf. [CbTpI], Definition 2.2, (i)] of $\mathbb{G}$ [cf. Definition 3.1, (ii)] and $S \subseteq \mathrm{Node}(\mathcal{G}|_{\mathbb{H}})$ [cf. [CbTpI], Definition 2.2, (ii)] a subset of $\mathrm{Node}(\mathcal{G}|_{\mathbb{H}})$ that is *not of separating type* [cf. [CbTpI], Definition 2.5, (i)]. Then, by applying a similar argument to the argument applied in [CmbCsp], Definition 2.1, (iii), (vi), or [NodNon], Definition 5.1, (ix), (x) [i.e., by considering the portion of the underlying scheme X_n of $X_n^{\log}$ corresponding to the underlying scheme $(X_{\mathbb{H},S})_n$ of the n-th log configuration space $(X_{\mathbb{H},S})_n^{\log}$ of the stable log curve $X^{\log}_{\mathbb{H},S}$ determined by $(\mathcal{G}|_{\mathbb{H}})_{\succ S}$—cf. [CbTpI], Definition 2.5, (ii)], one obtains a closed subgroup

$$(\Pi_{\mathbb{H},S})_n \subseteq \Pi_n$$

[which is well-defined up to Π_n-conjugation]. We shall refer to $(\Pi_{\mathbb{H},S})_n \subseteq \Pi_n$ as a *configuration space subgroup* [*associated to* $(\mathbb{H}, S)$]. For each $0 \leq i \leq j \leq n$, we shall write

$$(\Pi_{\mathbb{H},S})_{n/i} \stackrel{\text{def}}{=} (\Pi_{\mathbb{H},S})_n \cap \Pi_{n/i} \subseteq \Pi_{n/i}$$

[which is well-defined up to Π_n-conjugation];

$$(\Pi_{\mathbb{H},S})_{j/i} \stackrel{\text{def}}{=} (\Pi_{\mathbb{H},S})_{n/i}/(\Pi_{\mathbb{H},S})_{n/j} \subseteq \Pi_{j/i}$$

[which is well-defined up to Π_j-conjugation]. In particular,

$$(\Pi_{\mathbb{H},S})_j = (\Pi_{\mathbb{H},S})_{j/0} \subseteq \Pi_j$$

[where we recall that, in fact, the subgroups on either side of the "=" are only well-defined up to Π_j-conjugation]. Thus, by applying [CbTpI], Proposition 2.11, inductively, we conclude that each $(\Pi_{\mathbb{H},S})_{j/i}$ is a *pro-*Σ *configuration space group* [cf. [MzTa], Definition 2.3, (i)], and that we have a *natural exact sequence* of profinite groups

$$1 \longrightarrow (\Pi_{\mathbb{H},S})_{j/i} \longrightarrow (\Pi_{\mathbb{H},S})_j \longrightarrow (\Pi_{\mathbb{H},S})_i \longrightarrow 1.$$

Finally, let $v \in \text{Vert}(\mathcal{G})$. Then the semi-graph of anabelioids of PSC-type $\mathcal{G}|_v$ [cf. [CbTpI], Definition 2.1, (iii)] may be naturally identified with $(\mathcal{G}|_{\mathbb{H}_v})_{\succ S_v}$ for suitable choices of $\mathbb{H}_v$, S_v [cf. [CbTpI], Remark 2.5.1, (ii)]. We shall refer to

$$(\Pi_v)_n \stackrel{\text{def}}{=} (\Pi_{\mathbb{H}_v,S_v})_n \subseteq \Pi_n$$

as a *configuration space subgroup associated to* v. Thus, $(\Pi_v)_1 \subseteq \Pi_1$ is a verticial subgroup associated to $v \in \text{Vert}(\mathcal{G})$, i.e., a subgroup that is typically denoted "Π_v". We shall write

$$(\Pi_v)_{j/i} \stackrel{\text{def}}{=} (\Pi_{\mathbb{H}_v,S_v})_{j/i} \subseteq \Pi_{j/i}.$$

Remark 4.3.1 In the notation of Definition 4.3, one verifies easily—by applying a suitable *specialization isomorphism* [cf. the discussion preceding [CmbCsp], Definition 2.1, as well as [CbTpI], Remark 5.6.1]—that there exist a stable log curve $Y^{\log}$ over $(\text{Spec}\, k)^{\log}$ and an *n-cuspidalizable degeneration structure* $(\mathcal{G}, S, \phi)$ on ${}^Y\mathcal{G}$ [cf. Definition 3.23, (i), (v)]—where we write ${}^Y\mathcal{G}$ for the "$\mathcal{G}$" that occurs in the case where we take "$X^{\log}$" to be $Y^{\log}$—which satisfy the following: Write ${}^Y\Pi_n$ for the "Π_n" that occurs in the case where we take "$X^{\log}$" to be $Y^{\log}$. Then:

The image of a configuration space subgroup of Π_n associated to $(\mathbb{H}, S)$ [cf. Definition 4.3] via a PFC-admissible outer isomorphism $\Pi_n \xrightarrow{\sim} {}^Y\Pi_n$ that lies over the displayed composite isomorphism of Definition 3.23, (v) [where we note that, in *loc. cit.*, the roles of "${}^{Y}\Pi_n$" and "Π_n" are *reversed*!], is a *configuration space subgroup of* ${}^{Y}\Pi_n$ *associated to a vertex of* ${}^{Y}\mathcal{G}$.

Lemma 4.4 (Commensurable Terminality and Slimness) *Every* **configuration space subgroup** *[cf. Definition 4.3] of* Π_n *is* **topologically finitely generated, slim,** *and* **commensurably terminal** *in* Π_n.

Proof Since any configuration space subgroup is, in particular, a configuration space group, the fact that such a subgroup is topologically finitely generated and slim follows from [MzTa], Proposition 2.2, (ii). Thus, it remains to verify *commensurable terminality*. By applying the *observation* of Remark 4.3.1, we reduce immediately to the case of a configuration space subgroup *associated to a vertex*. But then the desired commensurable terminality follows, in light of Lemma 4.5 below, by induction on n, together with the corresponding fact for $n = 1$ [cf. [CmbGC], Proposition 1.2, (ii)]. This completes the proof of Lemma 4.4. □

Lemma 4.5 (Extensions and Commensurable Terminality) *Let*

$$\begin{array}{ccccccccc} 1 & \longrightarrow & N_H & \longrightarrow & H & \longrightarrow & Q_H & \longrightarrow & 1 \\ & & \downarrow & & \downarrow & & \downarrow & & \\ 1 & \longrightarrow & N & \longrightarrow & G & \longrightarrow & Q & \longrightarrow & 1 \end{array}$$

be a commutative diagram of profinite groups, where the horizontal sequences are **exact**, *and the vertical arrows are* **injective**. *Suppose that* $N_H \subseteq N$, $Q_H \subseteq Q$ *are* **commensurably terminal** *in* N, Q, *respectively. Then* $H \subseteq G$ *is* **commensurably terminal** *in* G.

Proof This follows immediately from Lemma 3.9, (i). □

Definition 4.6

(i) We shall write

$$\mathrm{Out}^{\mathrm{FC}}(\Pi_n)^{\mathrm{brch}} \subseteq \mathrm{Out}^{\mathrm{FC}}(\Pi_n)$$

for the closed subgroup of $\mathrm{Out}^{\mathrm{FC}}(\Pi_n)$ given by the inverse image of

$$\mathrm{Aut}^{|\mathrm{Brch}(\mathcal{G})|}(\mathcal{G}) \subseteq \ (\mathrm{Aut}(\mathcal{G}) \subseteq)\ \mathrm{Out}(\Pi_{\mathcal{G}}) \stackrel{\sim}{\leftarrow} \mathrm{Out}(\Pi_1)$$

[cf. Definition 4.1, (i)] via the natural injection $\mathrm{Out}^{\mathrm{FC}}(\Pi_n) \hookrightarrow \mathrm{Out}^{\mathrm{FC}}(\Pi_1) \subseteq \mathrm{Out}(\Pi_1)$ of [NodNon], Theorem B.

(ii) Let $v \in \mathrm{Vert}(\mathcal{G})$; write $\Pi_v \stackrel{\mathrm{def}}{=} (\Pi_v)_1$ [cf. Definition 4.3]. Then we shall write

$$\mathrm{Out}^{\mathrm{FC}}((\Pi_v)_n)^{\mathcal{G}\text{-node}} \subseteq \mathrm{Out}^{\mathrm{FC}}((\Pi_v)_n)$$

for the [*closed*] subgroup of $\mathrm{Out}^{\mathrm{FC}}((\Pi_v)_n)$ given by the inverse image of

$$\mathrm{Aut}^{|\mathcal{E}(\mathcal{G}|_v:\mathcal{G})|}(\mathcal{G}|_v) \subseteq (\mathrm{Aut}(\mathcal{G}|_v) \subseteq) \ \mathrm{Out}(\Pi_v)$$

[cf. Definition 4.1, (ii); [CbTpI], Definition 2.6, (i)] via the natural injection $\mathrm{Out}^{\mathrm{FC}}((\Pi_v)_n) \hookrightarrow \mathrm{Out}^{\mathrm{FC}}(\Pi_v) \subseteq \mathrm{Out}(\Pi_v)$ of [NodNon], Theorem B.

Theorem 4.7 (Graphicity of Outomorphisms of Certain Subquotients) *In the notation of the preceding Chap. 3 [cf. also Definition 3.1], let $x \in X_n(k)$. Write*

$$C_x \subseteq \mathrm{Cusp}(\mathcal{G})$$

for the [possibly empty] set consisting of cusps c of $\mathcal{G}$ such that, for some $i \in \{1, \cdots, n\}$, $x_{\{i\}} \in X_{\{i\}}(k) = X(k)$ [cf. Definition 3.1, (i)] **lies** *on the cusp of $X^{\log}$ corresponding to $c \in \mathrm{Cusp}(\mathcal{G})$. For each $i \in \{1, \cdots, n\}$, write*

$$\mathcal{G}_{i/i-1,x} \stackrel{\mathrm{def}}{=} \mathcal{G}_{i\in\{1,\cdots,i\},x}$$

[cf. Definition 3.1, (iii)] and

$$z_{i/i-1,x} \in \mathrm{VCN}(\mathcal{G}_{i/i-1,x})$$

for the element of $\mathrm{VCN}(\mathcal{G}_{i/i-1,x})$ on which $x_{\{1,\cdots,i\}}$ lies, that is to say:

> *If $x_{\{1,\cdots,i\}} \in X_i(k)$ [cf. the notation given in the discussion preceding Definition 3.1] is a cusp or node of the geometric fiber of the projection $p^{\log}_{i/i-1}\colon X^{\log}_i \to X^{\log}_{i-1}$ over $x^{\log}_{\{1,\cdots,i-1\}}$ corresponding to an edge $e \in \mathrm{Edge}(\mathcal{G}_{i/i-1,x})$, then $z_{i/i-1,x} \stackrel{\mathrm{def}}{=} e$; if $x_{\{1,\cdots,i\}} \in X_i(k)$ is neither a cusp nor node of the geometric fiber of the projection $p^{\log}_{i/i-1}\colon X^{\log}_i \to X^{\log}_{i-1}$ over $x^{\log}_{\{1,\cdots,i-1\}}$ but lies on the irreducible component of the geometric fiber corresponding to a vertex $v \in \mathrm{Edge}(\mathcal{G}_{i/i-1,x})$, then $z_{i/i-1,x} \stackrel{\mathrm{def}}{=} v$.*

Let

$$\alpha \in \mathrm{Out}^{\mathrm{FC}}(\Pi_n)^{\mathrm{brch}}$$

[cf. Definition 4.6, (i)]. Suppose that the element of

$$\mathrm{Aut}^{|\mathrm{Brch}(\mathcal{G})|}(\mathcal{G}) \subseteq (\mathrm{Aut}(\mathcal{G}) \subseteq) \ \mathrm{Out}(\Pi_{\mathcal{G}}) \stackrel{\sim}{\leftarrow} \mathrm{Out}(\Pi_1)$$

[cf. Definition 4.1, (i)] determined by $\alpha \in \mathrm{Out}^{\mathrm{FC}}(\Pi_n)^{\mathrm{brch}}$ [cf. Definition 4.6, (i)] is **contained** *in*

$$\mathrm{Aut}^{|C_x|}(\mathcal{G}) \subseteq \mathrm{Aut}(\mathcal{G})$$

[cf. [CbTpI], Definition 2.6, (i)]. Then there exist

- *a lifting* $\widetilde{\alpha} \in \mathrm{Aut}(\Pi_n)$ *of* α, *and,*
- *for each* $i \in \{1, \cdots, n\}$, *a VCN-subgroup* $\Pi_{z_{i/i-1,x}} \subseteq \Pi_{i/i-1} \xrightarrow{\sim} \Pi_{\mathcal{G}_{i/i-1,x}}$ *[cf. Definition 3.1, (iii)] associated to the element* $z_{i/i-1,x} \in \mathrm{VCN}(\mathcal{G}_{i/i-1,x})$

such that the following properties hold:

(a) For each $i \in \{1, \cdots, n\}$, *the automorphism of* $\Pi_{i/i-1} \xrightarrow{\sim} \Pi_{\mathcal{G}_{i/i-1,x}}$ *determined by* $\widetilde{\alpha}$ ***fixes*** *the VCN-subgroup* $\Pi_{z_{i/i-1,x}} \subseteq \Pi_{i/i-1} \xrightarrow{\sim} \Pi_{\mathcal{G}_{i/i-1,x}}$.

(b) For each $i \in \{1, \cdots, n\}$, *the outomorphism of* $\Pi_{i/i-1} \xrightarrow{\sim} \Pi_{\mathcal{G}_{i/i-1,x}}$ *induced by* $\widetilde{\alpha}$ *is* ***contained*** *in*

$$\mathrm{Aut}^{|\mathrm{Brch}(\mathcal{G}_{i/i-1,x})|}(\mathcal{G}_{i/i-1,x}) \subseteq \mathrm{Out}(\Pi_{\mathcal{G}_{i/i-1,x}}) \xleftarrow{\sim} \mathrm{Out}(\Pi_{i/i-1}).$$

Proof We verify Theorem 4.7 by *induction on n*. If $n = 1$, then Theorem 4.7 follows immediately from the various definitions involved. Now suppose that $n \geq 2$, and that the *induction hypothesis* is in force. In particular, [since the homomorphism $p^{\Pi}_{n/n-1} \colon \Pi_n \twoheadrightarrow \Pi_{n-1}$ is *surjective*] we have a lifting $\widetilde{\alpha} \in \mathrm{Aut}(\Pi_n)$ of α and, for each $i \in \{1, \cdots, n-1\}$, a VCN-subgroup $\Pi_{z_{i/i-1,x}} \subseteq \Pi_{i/i-1} \xrightarrow{\sim} \Pi_{\mathcal{G}_{i/i-1,x}}$ associated to the element $z_{i/i-1,x} \in \mathrm{VCN}(\mathcal{G}_{i/i-1,x})$ such that, for each $i \in \{1, \cdots, n-1\}$, the automorphism of Π_i determined by $\widetilde{\alpha}$ *fixes* $\Pi_{z_{i/i-1,x}} \subseteq \Pi_{i/i-1} \subseteq \Pi_i$, and, moreover, the automorphism of Π_{n-1} determined by $\widetilde{\alpha}$ satisfies the property (b) in the statement of Theorem 4.7. Now we claim that the following assertion holds:

Claim 4.7.A: The outomorphism of $\Pi_{n/n-1} \xrightarrow{\sim} \Pi_{\mathcal{G}_{n/n-1,x}}$ induced by the lifting $\widetilde{\alpha}$ is *contained* in

$$\mathrm{Aut}^{|\mathrm{Brch}(\mathcal{G}_{n/n-1,x})|}(\mathcal{G}_{n/n-1,x}) \subseteq \mathrm{Out}(\Pi_{\mathcal{G}_{n/n-1,x}}) \xleftarrow{\sim} \mathrm{Out}(\Pi_{n/n-1}).$$

To this end, let us first observe that it follows immediately—by replacing $X^{\log}_n$ by the base-change of $p^{\log}_{n/n-2} \colon X^{\log}_n \to X^{\log}_{n-2}$ via a suitable morphism of log schemes $(\mathrm{Spec}\, k)^{\log} \to X^{\log}_{n-2}$ whose image lies on $x_{\{1,\cdots,n-2\}} \in X_{n-2}(k)$—from Lemma 3.2, (iv), that, to verify Claim 4.7.A, we may assume without loss of generality that $n = 2$. Also, one verifies easily, by applying Lemma 3.14, (i) [cf. also [CbTpI], Proposition 2.9, (i)], and possibly replacing, when $z_{1/0,x} \in \mathrm{Vert}(\mathcal{G}_{1/0,x})$,

- $\widetilde{\alpha}$ by the composite of $\widetilde{\alpha}$ with an inner automorphism of $\Pi_n = \Pi_2$ determined by conjugation by a suitable element of $\Pi_n = \Pi_2$ whose image in $\Pi_1 \xrightarrow{\sim} \Pi_{\mathcal{G}_{1/0,x}}$ is *contained* in the closed subgroup $\Pi_{z_{1/0,x}} \subseteq \Pi_{\mathcal{G}_{1/0,x}} \xleftarrow{\sim} \Pi_1$ and
- x by a suitable "x" whose associated "$z_{1/0,x}$" is a node of $\mathcal{G}_{1/0,x}$ that *abuts* to the original $z_{1/0,x} \in \mathrm{Vert}(\mathcal{G}_{1/0,x})$,

that we may assume without loss of generality that $z_{1/0,x} \in \mathrm{Edge}(\mathcal{G}_{1/0,x})$.

Next, let us recall that the automorphism of $\Pi_1 \overset{\sim}{\to} \Pi_{\mathcal{G}_{1/0,x}}$ determined by $\widetilde{\alpha}$ *fixes* the edge-like subgroup $\Pi_{z_{1/0,x}} \subseteq \Pi_1 \overset{\sim}{\to} \Pi_{\mathcal{G}_{1/0,x}}$ associated to the edge $z_{1/0,x}$ of $\mathcal{G}_{1/0,x}$ [cf. the discussion preceding Claim 4.7.A]. Thus, since [we have assumed that] $\alpha \in \mathrm{Out}^{\mathrm{FC}}(\Pi_2)^{\mathrm{brch}}$ [which implies that the outomorphism of $\Pi_1 \overset{\sim}{\to} \Pi_{\mathcal{G}_{1/0,x}}$ determined by α *preserves* the Π_1-conjugacy class of each verticial subgroup of $\Pi_1 \overset{\sim}{\to} \Pi_{\mathcal{G}_{1/0,x}}$], it follows immediately from Lemma 3.13, (i), (ii), that the outomorphism of $\Pi_{\mathcal{G}_{2/1,x}} \overset{\sim}{\leftarrow} \Pi_{2/1}$ induced by $\widetilde{\alpha}$ is *group-theoretically verticial*, hence [cf. [NodNon], Proposition 1.13; [CmbGC], Proposition 1.5, (ii); the fact that α is *C-admissible*] *graphic*, i.e., $\in \mathrm{Aut}(\mathcal{G}_{2/1,x})$. Moreover, since the outomorphism of $\Pi_{\mathcal{G}_{2\in\{2\},x}} \overset{\sim}{\leftarrow} \Pi_1$ induced by $\widetilde{\alpha}$ is, by assumption, *contained* in $\mathrm{Aut}^{|\mathrm{Brch}(\mathcal{G})|}(\mathcal{G})$ [cf. [CmbCsp], Proposition 1.2, (iii)], one verifies easily, by considering the map on vertices/nodes/branches induced by the projection

$$p^{\Pi}_{\{1,2\}/\{2\}}|_{\Pi_{2/1}} : \Pi_{2/1} \twoheadrightarrow \Pi_{\{2\}}$$

[cf. Lemma 3.6, (i), (iv)], that the outomorphism of $\Pi_{\mathcal{G}_{2/1,x}} \overset{\sim}{\leftarrow} \Pi_{2/1}$ induced by $\widetilde{\alpha}$ is *contained* in the subgroup $\mathrm{Aut}^{|\mathrm{Brch}(\mathcal{G}_{2/1,x})|}(\mathcal{G}_{2/1,x})$. This completes the proof of Claim 4.7.A.

On the other hand, one verifies easily from Claim 4.7.A, together with the various definitions involved, that there exist a $\Pi_{n/n-1}$-conjugate $\widetilde{\beta}$ of $\widetilde{\alpha}$ and a VCN-subgroup $\Pi_{z_{n/n-1,x}} \subseteq \Pi_{n/n-1} \overset{\sim}{\to} \Pi_{\mathcal{G}_{n/n-1,x}}$ associated to $z_{n/n-1,x} \in \mathrm{VCN}(\mathcal{G}_{n/n-1,x})$ such that $\widetilde{\beta}$ *fixes* $\Pi_{z_{n/n-1,x}}$. In particular, the lifting $\widetilde{\beta}$ of α and the VCN-subgroups $\Pi_{z_{i/i-1,x}}$ [where $i \in \{1, \cdots, n\}$] satisfy the properties (a), (b) in the statement of Theorem 4.7. This completes the proof of Theorem 4.7. □

Lemma 4.8 (Preservation of Configuration Space Subgroups) *The following hold:*

(i) Let $\alpha \in \mathrm{Out}^{\mathrm{FC}}(\Pi_n)^{\mathrm{brch}}$ *[cf. Definition 4.6, (i)]. Then* α **preserves** *the* Π_n*-conjugacy class of each configuration space subgroup [cf. Definition 4.3] of* Π_n*. Thus, by applying the portion of Lemma 4.4 concerning* **commensurable terminality***, together with Lemma 3.10, (i), we obtain a natural homomorphism*

$$\mathrm{Out}^{\mathrm{FC}}(\Pi_n)^{\mathrm{brch}} \longrightarrow \prod_{v\in\mathrm{Vert}(\mathcal{G})} \mathrm{Out}((\Pi_v)_n).$$

(ii) The displayed homomorphism of (i) **factors** *through*

$$\prod_{v\in\mathrm{Vert}(\mathcal{G})} \mathrm{Out}^{\mathrm{FC}}((\Pi_v)_n)^{\mathcal{G}\text{-node}} \subseteq \prod_{v\in\mathrm{Vert}(\mathcal{G})} \mathrm{Out}((\Pi_v)_n)$$

[cf. Definition 4.6, (ii)].

Proof First, we verify assertion (i). We begin by observing that, in light of the *observation* of Remark 4.3.1 [cf. also [CbTpI], Proposition 2.9, (ii)], to complete the verification of assertion (i), it suffices to verify the following assertion:

Claim 4.8.A: For each $v \in \mathrm{Vert}(\mathcal{G})$, α *preserves* the Π_n-conjugacy class of configuration space subgroups $(\Pi_v)_n \subseteq \Pi_n$ of Π_n associated to v.

To verify Claim 4.8.A, let us observe that, by applying Theorem 4.7 in the case where we take the "x" in the statement of Theorem 4.7 to be such that, for each $i \in \{1, \cdots, n\}$, the element $z_{i/i-1,x} \in \mathrm{Vert}(\mathcal{G}_{i/i-1,x})$ is the vertex of $\mathcal{G}_{i/i-1,x}$ that corresponds [via the various bijections of Lemma 3.6, (iii)] to the vertex v of Claim 4.8.A, we obtain, for each $i \in \{1, \cdots, n\}$, a VCN-subgroup $\Pi_{z_{i/i-1,x}} \subseteq \Pi_{i/i-1} \overset{\sim}{\to} \Pi_{\mathcal{G}_{i/i-1,x}}$ associated to $z_{i/i-1,x} \in \mathrm{VCN}(\mathcal{G}_{i/i-1,x})$ as in the statement of Theorem 4.7, (a). Next, let us observe that one verifies immediately from the *commensurable terminality* [cf. [CmbGC], Proposition 1.2, (ii)] of each of the VCN-subgroups $\Pi_{z_{i/i-1,x}} \subseteq \Pi_{i/i-1}$, where $i \in \{1, \cdots, n\}$, that the Π_n-conjugacy class of the configuration space subgroup $(\Pi_v)_n \subseteq \Pi_n$ *coincides* with the Π_n-conjugacy class of the closed subgroup of Π_n consisting of $\gamma \in \Pi_n$ such that, for each $i \in \{1, \cdots, n\}$, conjugation by γ *preserves* the closed subgroup $\Pi_{z_{i/i-1,x}} \subseteq (\Pi_{i/i-1} \subseteq) \Pi_i$ [so $\Pi_{z_{i/i-1,x}} = (\Pi_v)_{i/i-1}$]. Thus, it follows from Theorem 4.7, (a), that α *preserves* the Π_n-conjugacy class of $(\Pi_v)_n \subseteq \Pi_n$, as desired. This completes the proof of Claim 4.8.A.

Next, we verify assertion (ii). Let $\alpha \in \mathrm{Out}^{\mathrm{FC}}(\Pi_n)^{\mathrm{brch}}$, $v \in \mathrm{Vert}(\mathcal{G})$. Write α_v for the outomorphism of $(\Pi_v)_n$ induced by α [cf. (i)]. Then the *F-admissibility* of α_v follows immediately from the F-admissibility of α [cf. the discussion of Definition 4.3]. The *C-admissibility* of α_v follows immediately from Theorem 4.7 [applied as in the proof of Claim 4.8.A]; [NodNon], Lemma 1.7, together with the definition of C-admissibility. Finally, the fact that $\alpha_v \in \mathrm{Out}^{\mathrm{FC}}((\Pi_v)_n)^{\mathcal{G}\text{-node}}$ follows immediately from the fact that $\alpha \in \mathrm{Out}^{\mathrm{FC}}(\Pi_n)^{\mathrm{brch}}$. This completes the proof of assertion (ii). □

Definition 4.9 We shall write

$$\mathrm{Glu}(\Pi_n) \subseteq \prod_{v \in \mathrm{Vert}(\mathcal{G})} \mathrm{Out}^{\mathrm{FC}}((\Pi_v)_n)^{\mathcal{G}\text{-node}}$$

for the [*closed*] subgroup of $\prod_{v \in \mathrm{Vert}(\mathcal{G})} \mathrm{Out}^{\mathrm{FC}}((\Pi_v)_n)^{\mathcal{G}\text{-node}}$ consisting of "*glueable*" collections of outomorphisms of the various $(\Pi_v)_n$, i.e., the subgroup defined as follows:

(i) Suppose that $n = 1$. Then $\mathrm{Glu}(\Pi_n)$ consists of those collections $(\alpha_v)_{v \in \mathrm{Vert}(\mathcal{G})}$ such that, for every v, $w \in \mathrm{Vert}(\mathcal{G})$, it holds that $\chi_v(\alpha_v) = \chi_w(\alpha_w)$ [cf. [CbTpI], Definition 3.8, (ii)]—where we note that one verifies easily that α_v may be regarded as an element of $\mathrm{Aut}(\mathcal{G}|_v)$.

(ii) Suppose that $n = 2$. Then $\mathrm{Glu}(\Pi_n)$ consists of those collections $(\alpha_v)_{v \in \mathrm{Vert}(\mathcal{G})}$ that satisfy the following condition: Let v, $w \in \mathrm{Vert}(\mathcal{G})$; $e \in \mathcal{N}(v) \cap \mathcal{N}(w)$;

$T \subseteq \Pi_{2/1} \subseteq \Pi_2 = \Pi_n$ a $\{1,2\}$-tripod of Π_n arising from $e \in \mathcal{N}(v) \cap \mathcal{N}(w)$ [cf. Definitions 3.3, (i); 3.7, (i)]. Then one verifies easily from the various definitions involved that there exist Π_n-conjugates T_v, T_w of T such that T_v, T_w are contained in $(\Pi_v)_n$, $(\Pi_w)_n$, respectively, and, moreover,

$$T_v \subseteq (\Pi_v)_{2/1} \subseteq (\Pi_v)_2 = (\Pi_v)_n,$$

$$T_w \subseteq (\Pi_w)_{2/1} \subseteq (\Pi_w)_2 = (\Pi_w)_n$$

are tripods of $(\Pi_v)_n$, $(\Pi_w)_n$ arising from [the cusps of $\mathcal{G}|_v$, $\mathcal{G}|_w$ corresponding to] the node e, respectively. Moreover, since $\alpha_v \in \mathrm{Out}^{\mathrm{FC}}((\Pi_v)_n)^{\mathcal{G}\text{-node}}$, $\alpha_w \in \mathrm{Out}^{\mathrm{FC}}((\Pi_w)_n)^{\mathcal{G}\text{-node}}$, it follows from Theorem 3.16, (iv), that $\alpha_v \in \mathrm{Out}^{\mathrm{FC}}((\Pi_v)_n)[T_v]$, $\alpha_w \in \mathrm{Out}^{\mathrm{FC}}((\Pi_w)_n)[T_w]$; thus, we obtain that $\mathfrak{T}_{T_v}(\alpha_v) \in \mathrm{Out}(T_v) \xrightarrow{\sim} \mathrm{Out}(T)$; $\mathfrak{T}_{T_w}(\alpha_w) \in \mathrm{Out}(T_w) \xrightarrow{\sim} \mathrm{Out}(T)$ [cf. Theorem 3.16, (i)]. Then we *require* that $\mathfrak{T}_{T_v}(\alpha_v) = \mathfrak{T}_{T_w}(\alpha_w)$.

(iii) Suppose that $n \geq 3$. Then $\mathrm{Glu}(\Pi_n)$ consists of those collections $(\alpha_v)_{v \in \mathrm{Vert}(\mathcal{G})}$ that satisfy the following condition: Let $\Pi^{\mathrm{tpd}} \subseteq \Pi_3$ be a 3-central $\{1,2,3\}$-tripod of Π_n [cf. Definitions 3.3, (i); 3.7, (ii)]. Then one verifies easily from the various definitions involved that, for every $v \in \mathrm{Vert}(\mathcal{G})$, there exists a Π_3-conjugate Π_v^{tpd} of Π^{tpd} such that Π_v^{tpd} is contained in $(\Pi_v)_3$, and, moreover, $\Pi_v^{\mathrm{tpd}} \subseteq (\Pi_v)_3$ is a 3-central tripod of $(\Pi_v)_3$. Thus, since $\alpha_v \in \mathrm{Out}^{\mathrm{FC}}((\Pi_v)_n)^{\mathcal{G}\text{-node}}$, we obtain $\mathfrak{T}_{\Pi_v^{\mathrm{tpd}}}(\alpha_v) \in \mathrm{Out}(\Pi_v^{\mathrm{tpd}}) \xrightarrow{\sim} \mathrm{Out}(\Pi^{\mathrm{tpd}})$ [cf. Theorem 3.16, (i), (v)]. Then, for every $v, w \in \mathrm{Vert}(\mathcal{G})$, we *require* that $\mathfrak{T}_{\Pi_v^{\mathrm{tpd}}}(\alpha_v) = \mathfrak{T}_{\Pi_w^{\mathrm{tpd}}}(\alpha_w)$.

Remark 4.9.1 In the notation of Definition 4.9, one verifies easily from the various definitions involved that the natural outer isomorphism $\Pi_1 \xrightarrow{\sim} \Pi_{\mathcal{G}}$ determines a natural isomorphism $\mathrm{Glu}(\Pi_1) \xrightarrow{\sim} \mathrm{Glu}^{\mathrm{brch}}(\mathcal{G})$ [cf. Definition 4.1, (iii)].

Lemma 4.10 (Basic Properties Concerning Groups of Glueable Collections) *For $n \geq 1$, the following hold:*

(i) The natural injections

$$\mathrm{Out}^{\mathrm{FC}}((\Pi_v)_{n+1}) \hookrightarrow \mathrm{Out}^{\mathrm{FC}}((\Pi_v)_n)$$

of [NodNon], Theorem B—where v ranges over the vertices of $\mathcal{G}$—determine an **injection**

$$\mathrm{Glu}(\Pi_{n+1}) \hookrightarrow \mathrm{Glu}(\Pi_n).$$

(ii) The displayed homomorphism of Lemma 4.8, (i),

$$\mathrm{Out}^{\mathrm{FC}}(\Pi_n)^{\mathrm{brch}} \longrightarrow \prod_{v \in \mathrm{Vert}(\mathcal{G})} \mathrm{Out}((\Pi_v)_n)$$

factors *through*

$$\mathrm{Glu}(\Pi_n) \subseteq \prod_{v \in \mathrm{Vert}(\mathcal{G})} \mathrm{Out}((\Pi_v)_n).$$

Proof First, we verify assertion (i). The fact that the image of the composite

$$\mathrm{Glu}(\Pi_{n+1}) \hookrightarrow \prod_{v \in \mathrm{Vert}(\mathcal{G})} \mathrm{Out}^{\mathrm{FC}}((\Pi_v)_{n+1}) \hookrightarrow \prod_{v \in \mathrm{Vert}(\mathcal{G})} \mathrm{Out}^{\mathrm{FC}}((\Pi_v)_n)$$

is *contained* in

$$\prod_{v \in \mathrm{Vert}(\mathcal{G})} \mathrm{Out}^{\mathrm{FC}}((\Pi_v)_n)^{\mathcal{G}\text{-node}} \subseteq \prod_{v \in \mathrm{Vert}(\mathcal{G})} \mathrm{Out}^{\mathrm{FC}}((\Pi_v)_n)$$

follows immediately from the various definitions involved. The fact that the image of the composite

$$\mathrm{Glu}(\Pi_{n+1}) \hookrightarrow \prod_{v \in \mathrm{Vert}(\mathcal{G})} \mathrm{Out}^{\mathrm{FC}}((\Pi_v)_{n+1}) \hookrightarrow \prod_{v \in \mathrm{Vert}(\mathcal{G})} \mathrm{Out}^{\mathrm{FC}}((\Pi_v)_n)$$

is *contained* in

$$\mathrm{Glu}(\Pi_n) \subseteq \prod_{v \in \mathrm{Vert}(\mathcal{G})} \mathrm{Out}^{\mathrm{FC}}((\Pi_v)_n)^{\mathcal{G}\text{-node}}$$

follows immediately from the various definitions involved when $n \geq 3$ and from Theorems 3.16, (iv), (v); 3.18, (ii) [applied to each $(\Pi_v)_{n+1}$!], when $n = 2$. Thus, it remains to verify assertion (i) in the case where $n = 1$. Suppose that $n = 1$. Let $(\alpha_v)_{v \in \mathrm{Vert}(\mathcal{G})} \in \mathrm{Glu}(\Pi_2)$. Write $((\alpha_v)_1)_{v \in \mathrm{Vert}(\mathcal{G})} \in \prod_{v \in \mathrm{Vert}(\mathcal{G})} \mathrm{Out}^{\mathrm{FC}}((\Pi_v)_1)^{\mathcal{G}\text{-node}}$ for the image of $(\alpha_v)_{v \in \mathrm{Vert}(\mathcal{G})}$. Since $\mathbb{G}$ is *connected*, to verify assertion (i) in the case where $n = 1$, it suffices to verify that, for any two vertices v, w of $\mathcal{G}$ such that $\mathcal{N}(v) \cap \mathcal{N}(w) \neq \emptyset$, it holds that $\chi_v((\alpha_v)_1) = \chi_w((\alpha_w)_1)$. Let $x \in X_2(k)$ be a k-valued geometric point of X_2 such that $x_{\{1\}} \in X(k)$ [cf. Definition 3.1, (i)] is a node of $X^{\log}$ corresponding to an element of $\mathcal{N}(v) \cap \mathcal{N}(w) \neq \emptyset$. Then by applying Theorem 4.7 to a suitable lifting $\widetilde{\alpha}_v$ $(\in \mathrm{Aut}^{\mathrm{FC}}((\Pi_v)_2))$ of the outomorphism α_v of $(\Pi_v)_2$ [where we take the "Π_n" in the statement of Theorem 4.7 to be $(\Pi_v)_2$], we conclude that the outomorphism $(\alpha_v)_{2/1}$ of $\Pi_{(\mathcal{G}|_v)_{2\in\{1,2\},x}} \xleftarrow{\sim} (\Pi_v)_{2/1}$ [cf. Definition 3.1, (iii)] determined by $\widetilde{\alpha}_v$ is *graphic* and *fixes* each of the vertices of $(\mathcal{G}|_v)_{2\in\{1,2\},x}$. Thus, if we write $(\alpha_v)_{\{2\}}$ for the outomorphism of the "$\Pi_{\{2\}}$" that occurs in the case where we take "Π_2" to be $(\Pi_v)_2$, then it follows from [CmbCsp], Proposition 1.2, (iii), together with the *C-admissibility* of $(\alpha_v)_1$, that $(\alpha_v)_{\{2\}}$ is *C*-

admissible, i.e., $\in \mathrm{Aut}(\mathcal{G}|_v)$. Now, for a [$\{1,2\}$-]tripod $T_v \subseteq (\Pi_v)_2$ arising from the *cusp* $x_{\{1\}}$ of $\mathcal{G}|_v$ [cf. Definitions 3.3, (i); 3.7, (i)], we compute:

$$\begin{aligned}
\chi_{\mathcal{G}|_v}((\alpha_v)_1) &= \chi_{\mathcal{G}|_v}((\alpha_v)_{\{2\}}) && \text{[cf. [CmbCsp], Proposition 1.2, (iii)]}\\
&= \chi_{(\mathcal{G}|_v)_{2\in\{1,2\},x}}((\alpha_v)_{2/1}) && \text{[cf. [CbTpI], Corollary 3.9, (iv)]}\\
&= \chi_{T_v}((\alpha_v)_{2/1}|_{T_v}) && \text{[cf. [CbTpI], Corollary 3.9, (iv)]}
\end{aligned}$$

[where we refer to Lemma 3.12, (i), concerning "$(\alpha_v)_{2/1}|_{T_v}$", and we write χ_{T_v} for the "χ" associated to the vertex of $(\mathcal{G}|_v)_{2\in\{1,2\},x}$ corresponding to T_v]. Moreover, by applying a similar argument to the above argument, we conclude that there exists a lifting $\widetilde{\alpha}_w$ of α_w such that the outomorphism $(\alpha_w)_{2/1}$ of $\Pi_{(\mathcal{G}|_w)_{2\in\{1,2\},x}} \overset{\sim}{\leftarrow} (\Pi_w)_{2/1}$ determined by $\widetilde{\alpha}_w$ is *graphic* [and *fixes* each of the vertices of $(\mathcal{G}|_w)_{2\in\{1,2\},x}$], and, moreover, for a [$\{1,2\}$-]tripod $T_w \subseteq (\Pi_w)_2$ arising from the *cusp* $x_{\{1\}}$ of $\mathcal{G}|_w$, it holds that $\chi_{\mathcal{G}|_w}((\alpha_w)_1) = \chi_{T_w}((\alpha_w)_{2/1}|_{T_w})$. On the other hand, since $(\alpha_v)_{v\in\mathrm{Vert}(\mathcal{G})} \in \mathrm{Glu}(\Pi_2)$, it holds that $\chi_{T_v}((\alpha_v)_{2/1}|_{T_v}) = \chi_{T_w}((\alpha_w)_{2/1}|_{T_w})$. In particular, we obtain that $\chi_{\mathcal{G}|_v}((\alpha_v)_1) = \chi_{\mathcal{G}|_w}((\alpha_w)_1)$. This completes the proof of assertion (i).

Next, we verify assertion (ii). If $n = 1$, then assertion (ii) amounts to Theorem 4.2, (ii) [cf. also Remark 4.9.1]. If $n \geq 2$, then assertion (ii) follows immediately from Lemma 4.8, (ii), together with the fact that the homomorphism "$\mathfrak{T}_T$" of Theorem 3.16, (i), does *not depend* on the choice of "T" among its conjugates. This completes the proof of assertion (ii). □

Definition 4.11 We shall write ρ_n^{brch} for the homomorphism

$$\mathrm{Out}^{\mathrm{FC}}(\Pi_n)^{\mathrm{brch}} \longrightarrow \mathrm{Glu}(\Pi_n)$$

determined by the factorization of Lemma 4.10, (ii).

Lemma 4.12 (Glueable Collections in the Case of Precisely One Node) *Suppose that* $n = 2$, *and that* $\#\mathrm{Node}(\mathcal{G}) = 1$. *Let* $\widetilde{v}, \widetilde{w} \in \mathrm{Vert}(\widetilde{\mathcal{G}})$ *be distinct elements such that* $\mathcal{N}(\widetilde{v}) \cap \mathcal{N}(\widetilde{w}) \neq \emptyset$. *Write* $\widetilde{e} \in \mathrm{Node}(\widetilde{\mathcal{G}})$ *for the unique element of* $\mathcal{N}(\widetilde{v}) \cap \mathcal{N}(\widetilde{w})$ *[cf. [NodNon], Lemma 1.8];* $\Pi_{\widetilde{v}}, \Pi_{\widetilde{w}}, \Pi_{\widetilde{e}} \subseteq \Pi_{\mathcal{G}} \overset{\sim}{\leftarrow} \Pi_1$ *for the VCN-subgroups of* $\Pi_{\mathcal{G}} \overset{\sim}{\leftarrow} \Pi_1$ *associated to* $\widetilde{v}, \widetilde{w}, \widetilde{e} \in \mathrm{VCN}(\widetilde{\mathcal{G}})$, *respectively;* $v \overset{\mathrm{def}}{=} \widetilde{v}(\mathcal{G})$; $w \overset{\mathrm{def}}{=} \widetilde{w}(\mathcal{G})$; $e \overset{\mathrm{def}}{=} \widetilde{e}(\mathcal{G})$. *[Thus, one verifies easily that* $\Pi_{\widetilde{e}} = \Pi_{\widetilde{v}} \cap \Pi_{\widetilde{w}}$ *[cf. [NodNon], Lemma 1.9, (i)], that* $\mathrm{Vert}(\mathcal{G}) = \{v, w\}$, *and that if* $\mathcal{G}$ *is* **noncyclically primitive** *(respectively,* **cyclically primitive***) [cf. [CbTpI], Definition 4.1], then* $v \neq w$ *(respectively,* $v = w$*).] Let* $x \in X_2(k)$ *be a* k*-valued geometric point of* X_2 *such that* $x_{\{1\}} \in X(k)$ *[cf. Definition 3.1, (i)] lies on the unique node of* $X^{\log}$ *[i.e., which corresponds to* e*]. Write* $\mathcal{G}_{2/1} \overset{\mathrm{def}}{=} \mathcal{G}_{2\in\{1,2\},x}$ *[cf. Definition 3.1, (iii)];* $\widetilde{\mathcal{G}}_{2/1} \to \mathcal{G}_{2/1}$ *for the profinite étale covering corresponding to* $\Pi_{\mathcal{G}_{2/1}} \overset{\sim}{\leftarrow} \Pi_{2/1}$; v^{new} *for the "*$v^{\mathrm{new}}_{2,1,x}$*" of Lemma 3.6, (iv). For each* $z \in \mathrm{Vert}(\mathcal{G})$, *write* $z^{\circ} \in \mathrm{Vert}(\mathcal{G}_{2/1})$ *for the vertex of* $\mathcal{G}_{2/1}$ *that corresponds to* z *via the bijections of Lemma 3.6, (i), (iv). [Thus, it follows from*

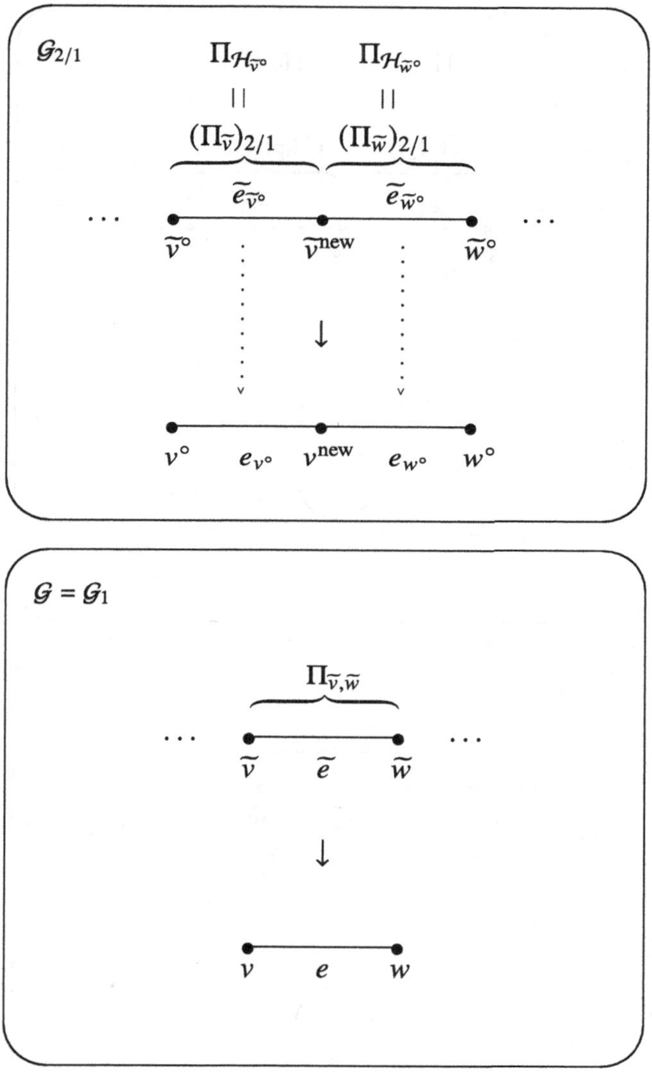

Fig. 4.1 The noncyclically primitive case

Lemma 3.6, (iv), that $\mathrm{Vert}(\mathcal{G}_{2/1}) = \{v^{\text{new}}, v^\circ, w^\circ\}$*.] Then the following hold [cf. also Figs. 4.1, and 4.2]:*

(i) *Let* $(\Pi_{\widetilde{v}})_2 \subseteq \Pi_2$ *be a configuration space subgroup of* Π_2 *associated to* v *[cf. Definition 4.3] such that the image of the composite* $(\Pi_{\widetilde{v}})_2 \hookrightarrow \Pi_2 \overset{p^{\Pi}_{2/1}}{\twoheadrightarrow} \Pi_1$ **coincides** *with* $\Pi_{\widetilde{v}} \subseteq \Pi_{\mathcal{G}} \overset{\sim}{\leftarrow} \Pi_1$*. Also, let us fix a verticial subgroup*

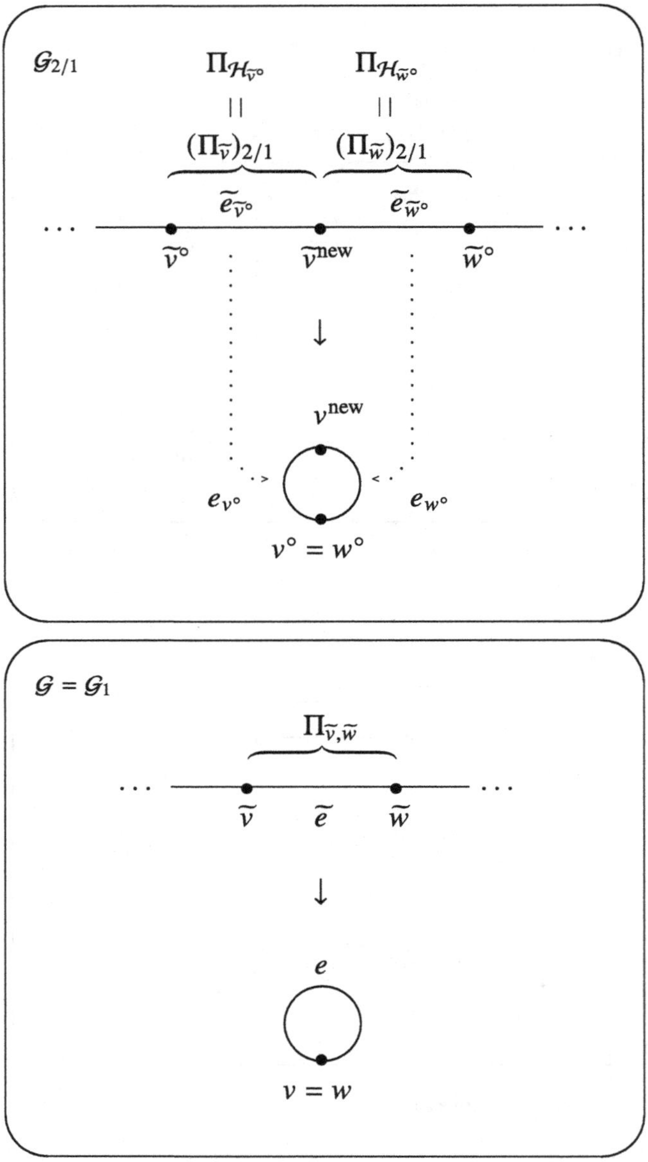

Fig. 4.2 The cyclically primitive case

$\Pi_{\widetilde{v}^{\mathrm{new}}} \subseteq \Pi_{\mathcal{G}_{2/1}} \overset{\sim}{\leftarrow} \Pi_{2/1}$ *of* $\Pi_{\mathcal{G}_{2/1}} \overset{\sim}{\leftarrow} \Pi_{2/1}$ *associated to a* $\widetilde{v}^{\mathrm{new}} \in \mathrm{Vert}(\widetilde{\mathcal{G}}_{2/1})$ *that lies over* $v^{\mathrm{new}} \in \mathrm{Vert}(\mathcal{G}_{2/1})$ *and is* **contained** *in* $(\Pi_{\widetilde{v}})_2$*. Then there exists a* **unique** *configuration space subgroup* $(\Pi_{\widetilde{w}})_2 \subseteq \Pi_2$ *of* Π_2 *associated to* w

[cf. Definition 4.3] such that $\Pi_{\widetilde{v}^{\mathrm{new}}} = (\Pi_{\widetilde{v}})_{2/1} \cap (\Pi_{\widetilde{w}})_{2/1}$*—where we write* $(\Pi_{\widetilde{v}})_{2/1} \stackrel{\mathrm{def}}{=} \Pi_{2/1} \cap (\Pi_{\widetilde{v}})_2$*;* $(\Pi_{\widetilde{w}})_{2/1} \stackrel{\mathrm{def}}{=} \Pi_{2/1} \cap (\Pi_{\widetilde{w}})_2$*—and, moreover, the image of the composite* $(\Pi_{\widetilde{w}})_2 \hookrightarrow \Pi_2 \stackrel{p^{\Pi}_{2/1}}{\twoheadrightarrow} \Pi_1$ **coincides** *with* $\Pi_{\widetilde{w}} \subseteq \Pi_1$.

(ii) In the situation of (i), the natural homomorphism

$$\varinjlim(\Pi_{\widetilde{v}} \hookleftarrow \Pi_{\widetilde{e}} \hookrightarrow \Pi_{\widetilde{w}}) \longrightarrow \Pi_1$$

—where the inductive limit is taken in the category of pro-Σ groups—is **injective**, *and its image is* **commensurably terminal** *in* Π_1. *Write* $\Pi_{\widetilde{v},\widetilde{w}} \subseteq \Pi_1$ *for the image of the above homomorphism;* $\Pi_2|_{\Pi_{\widetilde{v},\widetilde{w}}}$ $(\subseteq \Pi_2)$ *for the fiber product of* $\Pi_2 \stackrel{p^{\Pi}_{2/1}}{\twoheadrightarrow} \Pi_1$ *and* $\Pi_{\widetilde{v},\widetilde{w}} \hookrightarrow \Pi_1$. *Thus, we have an* **exact** *sequence of profinite groups*

$$1 \longrightarrow \Pi_{2/1} \longrightarrow \Pi_2|_{\Pi_{\widetilde{v},\widetilde{w}}} \longrightarrow \Pi_{\widetilde{v},\widetilde{w}} \longrightarrow 1.$$

Finally, if $\mathcal{G}$ is **noncyclically primitive**, *then* $\Pi_{\widetilde{v},\widetilde{w}} = \Pi_1$, $\Pi_2|_{\Pi_{\widetilde{v},\widetilde{w}}} = \Pi_2$.

(iii) In the situation of (ii), for each $\widetilde{z} \in \{\widetilde{v}, \widetilde{w}\}$, *let* $\Pi_{\widetilde{z}^\circ} \subseteq \Pi_{\mathcal{G}_{2/1}} \stackrel{\sim}{\leftarrow} \Pi_{2/1}$ *be a verticial subgroup of* $\Pi_{\mathcal{G}_{2/1}} \stackrel{\sim}{\leftarrow} \Pi_{2/1}$ *associated to a* $\widetilde{z}^\circ \in \mathrm{Vert}(\widetilde{\mathcal{G}}_{2/1})$ *that lies over* $z^\circ \in \mathrm{Vert}(\mathcal{G}_{2/1})$ *such that* $\Pi_{\widetilde{z}^\circ} \subseteq (\Pi_{\widetilde{z}})_{2/1}$ *[cf. (i)], and, moreover,* $\Pi_{\widetilde{z}^\circ} \cap \Pi_{\widetilde{v}^{\mathrm{new}}} \neq \{1\}$. *Thus,* $\Pi_{\widetilde{e}_{\widetilde{z}^\circ}} \stackrel{\mathrm{def}}{=} \Pi_{\widetilde{z}^\circ} \cap \Pi_{\widetilde{v}^{\mathrm{new}}}$ *is the nodal subgroup of* $\Pi_{\mathcal{G}_{2/1}} \stackrel{\sim}{\leftarrow} \Pi_{2/1}$ *associated to the unique element* $\widetilde{e}_{\widetilde{z}^\circ}$ *of* $\mathcal{N}(\widetilde{z}^\circ) \cap \mathcal{N}(\widetilde{v}^{\mathrm{new}})$ *[cf. [NodNon], Lemma 1.9, (i)]. Write* $e_{z^\circ} \stackrel{\mathrm{def}}{=} \widetilde{e}_{\widetilde{z}^\circ}(\mathcal{G}_{2/1})$. *Then the natural homomorphism*

$$\varinjlim(\Pi_{\widetilde{z}^\circ} \hookleftarrow \Pi_{\widetilde{e}_{\widetilde{z}^\circ}} \hookrightarrow \Pi_{\widetilde{v}^{\mathrm{new}}}) \longrightarrow (\Pi_{\widetilde{z}})_{2/1}$$

—where the inductive limit is taken in the category of pro-Σ groups—is an **isomorphism**. *Write* $\mathbb{G}^{\dagger}_{z^\circ}$ *for the sub-semi-graph of* **PSC-type** *[cf. [CbTpI], Definition 2.2, (i)] of the underlying semi-graph of* $\mathcal{G}_{2/1}$ *whose set of vertices* $= \{\widetilde{z}(\mathcal{G})^\circ, v^{\mathrm{new}}\}$*;* $T_{z^\circ} \stackrel{\mathrm{def}}{=} (\mathrm{Node}(\mathcal{G}_{2/1}) \setminus \{e_{z^\circ}\}) \cap \mathrm{Node}(\mathcal{G}_{2/1}|_{\mathbb{G}^{\dagger}_{z^\circ}}) \subseteq \mathrm{Node}(\mathcal{G}_{2/1})$ *[cf. [CbTpI], Definition 2.2, (ii)]. Then the natural homomorphism of the above display allows one to identify* $(\Pi_{\widetilde{z}})_{2/1}$ *with the [pro-Σ] fundamental group* $\Pi_{\mathcal{H}_{\widetilde{z}^\circ}}$ *of*

$$\mathcal{H}_{z^\circ} \stackrel{\mathrm{def}}{=} (\mathcal{G}_{2/1}|_{\mathbb{G}^{\dagger}_{z^\circ}})_{\succ T_{z^\circ}}$$

[cf. [CbTpI], Definition 2.5, (ii)].

(iv) In the situation of (iii), let $(\alpha_z)_{z \in \mathrm{Vert}(\mathcal{G})} \in \mathrm{Glu}(\Pi_2)$. *Write* $((\alpha_z)_1)_{z \in \mathrm{Vert}(\mathcal{G})} \in \mathrm{Glu}(\Pi_1)$ *for the image of* $(\alpha_z)_{z \in \mathrm{Vert}(\mathcal{G})} \in \mathrm{Glu}(\Pi_2)$ *via the injection of*

Lemma 4.10, (i). Let $\alpha_1 \in \mathrm{Aut}^{|\mathrm{Brch}(\mathcal{G})|}(\mathcal{G})$ *be such that* $\rho_1^{\mathrm{brch}}(\alpha_1) = ((\alpha_z)_1)_{z\in\mathrm{Vert}(\mathcal{G})} \in \mathrm{Glu}(\Pi_1)$ *[cf. Theorem 4.2, (iii); Definition 4.11]. Then the outomorphism* α_1 *of* Π_1 **preserves** *the* Π_1*-conjugacy class of* $\Pi_{\tilde{v},\tilde{w}} \subseteq \Pi_1$*. Thus, by applying the portion of (ii) concerning commensurable terminality, we obtain [cf. Lemma 3.10, (i)] a restricted outomorphism* $\alpha_1|_{\Pi_{\tilde{v},\tilde{w}}} \in \mathrm{Out}(\Pi_{\tilde{v},\tilde{w}})$*.*

(v) *In the situation of (iv), there exists an outomorphism* $\beta_{\tilde{v},\tilde{w}}[\alpha_1]$ *of* $\Pi_2|_{\Pi_{\tilde{v},\tilde{w}}}$ *that satisfies the following conditions:*

(1) $\beta_{\tilde{v},\tilde{w}}[\alpha_1]$ **preserves** $\Pi_{2/1} \subseteq \Pi_2|_{\Pi_{\tilde{v},\tilde{w}}}$ *and the* $\Pi_2|_{\Pi_{\tilde{v},\tilde{w}}}$*-conjugacy classes of* $(\Pi_{\tilde{v}})_2, (\Pi_{\tilde{w}})_2 \subseteq \Pi_2|_{\Pi_{\tilde{v},\tilde{w}}}$*.*

(2) *There exists an automorphism* $\widetilde{\beta}_{\tilde{v},\tilde{w}}[\alpha_1]$ *of* $\Pi_2|_{\Pi_{\tilde{v},\tilde{w}}}$ *that* **lifts** *the outomorphism* $\beta_{\tilde{v},\tilde{w}}[\alpha_1]$ *such that the outomorphism of* $\Pi_{\mathcal{G}_{2/1}} \overset{\sim}{\leftarrow} \Pi_{2/1}$ *determined by* $\widetilde{\beta}_{\tilde{v},\tilde{w}}[\alpha_1]$ *[cf. (1)] is* **contained** *in* $\mathrm{Aut}^{|\mathrm{Brch}(\mathcal{G}_{2/1})|}(\mathcal{G}_{2/1}) \subseteq \mathrm{Out}(\Pi_{\mathcal{G}_{2/1}}) \overset{\sim}{\leftarrow} \mathrm{Out}(\Pi_{2/1})$*.*

(3) *For each* $\tilde{z} \in \{\tilde{v}, \tilde{w}\}$*, the outomorphism* $\beta_{\tilde{v},\tilde{w}}[\alpha_1]|_{(\Pi_{\tilde{z}})_2}$ *of* $(\Pi_{\tilde{z}})_2$ *determined by* $\beta_{\tilde{v},\tilde{w}}[\alpha_1]$ *[i.e., obtained by applying (1) and Lemma 3.10, (i)—where we note that* $(\Pi_{\tilde{z}})_2$ *is* **commensurably terminal** *in* Π_2 *[cf. Lemma 4.4], hence also in* $\Pi_2|_{\Pi_{\tilde{v},\tilde{w}}}$*]* **coincides** *with* $\alpha_{\tilde{z}(\mathcal{G})}$ *[cf. the notation of (iv)].*

(4) *The outomorphism of* $\Pi_{\tilde{v},\tilde{w}}$ *induced by* $\beta_{\tilde{v},\tilde{w}}[\alpha_1]$ *[cf. (1)]* **coincides** *with* $\alpha_1|_{\Pi_{\tilde{v},\tilde{w}}}$ *[cf. (iv)].*

Here, we observe, in the context of (2), that the outer isomorphism $\Pi_{2/1} \overset{\sim}{\to} \Pi_{\mathcal{G}_{2/1}}$ *[i.e., which gives rise to "the" closed subgroup* $\mathrm{Aut}^{|\mathrm{Brch}(\mathcal{G}_{2/1})|}(\mathcal{G}_{2/1}) \subseteq \mathrm{Out}(\Pi_{\mathcal{G}_{2/1}}) \overset{\sim}{\leftarrow} \mathrm{Out}(\Pi_{2/1})$*] may be characterized, up to composition with elements of the subgroup* $\mathrm{Aut}^{|\mathrm{Brch}(\mathcal{G}_{2/1})|}(\mathcal{G}_{2/1}) \subseteq \mathrm{Out}(\Pi_{\mathcal{G}_{2/1}}) \overset{\sim}{\leftarrow} \mathrm{Out}(\Pi_{2/1})$*, as the group-theoretically cuspidal [cf. [CmbGC], Definition 1.4, (iv)] outer isomorphism such that the semi-graph of anabelioids structure on* $\mathcal{G}_{2/1}$ *is the semi-graph of anabelioids structure determined [cf. [NodNon], Theorem A] by the resulting composite*

$$\Pi_{\tilde{e}} \hookrightarrow \Pi_{\mathcal{G}} \overset{\sim}{\leftarrow} \Pi_1 \to \mathrm{Out}(\Pi_{2/1}) \overset{\sim}{\to} \mathrm{Out}(\Pi_{\mathcal{G}_{2/1}})$$

—where the third arrow is the outer action determined by the exact sequence $1 \to \Pi_{2/1} \to \Pi_2 \overset{p^{\Pi}_{2/1}}{\to} \Pi_1 \to 1$*—in a fashion compatible with the projection* $p^{\Pi}_{\{1,2\}/\{2\}}|_{\Pi_{2/1}} : \Pi_{2/1} \twoheadrightarrow \Pi_{\{2\}}$ *and the given outer isomorphisms* $\Pi_{\{2\}} \overset{\sim}{\to} \Pi_1 \overset{\sim}{\to} \Pi_{\mathcal{G}}$*.*

Proof First, we verify assertion (i). The existence of such a $(\Pi_{\tilde{w}})_2 \subseteq \Pi_2$ follows immediately from the various definitions involved. Thus, it remains to verify the *uniqueness* of such a $(\Pi_{\tilde{w}})_2$. Let $(\Pi_{\tilde{w}})_2 \subseteq \Pi_2$ be as in assertion (i) and $\gamma \in \Pi_2$ an element such that the conjugate $(\Pi_{\tilde{w}})_2^{\gamma}$ of $(\Pi_{\tilde{w}})_2$ by γ satisfies the condition on "$(\Pi_{\tilde{w}})_2$" stated in assertion (i). Then since $\Pi_{\tilde{w}}$ is *commensurably terminal* in

Π_1 [cf. [CmbGC], Proposition 1.2, (ii)], it holds that the image of γ via $p^{\Pi}_{2/1}$ is *contained* in $\Pi_{\widetilde{w}}$. Thus—by multiplying γ by a suitable element of $(\Pi_{\widetilde{w}})_2$—we may assume without loss of generality that $\gamma \in \Pi_{2/1}$. In particular, since $\Pi_{\widetilde{v}^{\mathrm{new}}} \subseteq (\Pi_{\widetilde{w}})_{2/1} \cap (\Pi_{\widetilde{w}})^{\gamma}_{2/1}$—where we write $(\Pi_{\widetilde{w}})^{\gamma}_{2/1} \stackrel{\mathrm{def}}{=} \Pi_{2/1} \cap (\Pi_{\widetilde{w}})^{\gamma}_{2}$— is *not abelian* [cf. [CmbGC], Remark 1.1.3], it follows immediately from [NodNon], Lemma 1.9, (i), that $(\Pi_{\widetilde{w}})_{2/1} = (\Pi_{\widetilde{w}})^{\gamma}_{2/1}$. Thus, since $(\Pi_{\widetilde{w}})_{2/1}$ is *commensurably terminal* in $\Pi_{2/1}$ [cf. [CmbGC], Proposition 1.2, (ii)], it holds that $\gamma \in (\Pi_{\widetilde{w}})_{2/1}$. This completes the proof of assertion (i).

Assertions (ii), (iii), (iv) follow immediately from the various definitions involved [cf. also [CmbGC], Propositions 1.2, (ii), and 1.5, (i), as well as the proofs of [CmbCsp], Proposition 1.5, (iii); [CbTpI], Proposition 2.11].

Finally, we verify assertion (v). It follows immediately from the definition of "$\mathrm{Out}^{\mathrm{FC}}((\Pi_{(-)})_2)^{\mathcal{G}\text{-node}}$" [cf. Definitions 4.6, (ii); 4.9] that, for each $\widetilde{z} \in \{\widetilde{v}, \widetilde{w}\}$, there exists a lifting $\widetilde{\alpha}_{\widetilde{z}} \in \mathrm{Aut}((\Pi_{\widetilde{z}})_2)$ of $\alpha_{\widetilde{z}(\mathcal{G})}$ such that if we write $(\widetilde{\alpha}_{\widetilde{z}})_1$ for the automorphism of $\Pi_{\widetilde{z}}$ determined by $\widetilde{\alpha}_{\widetilde{z}}$, then $(\widetilde{\alpha}_{\widetilde{z}})_1(\Pi_{\widetilde{e}}) = \Pi_{\widetilde{e}}$. Next, let us observe that it follows immediately from assertion (ii) that the automorphisms $(\widetilde{\alpha}_{\widetilde{v}})_1$, $(\widetilde{\alpha}_{\widetilde{w}})_1$ [i.e., determined by the liftings $\widetilde{\alpha}_{\widetilde{v}}$, $\widetilde{\alpha}_{\widetilde{w}}$] determine an automorphism $\widetilde{\alpha}_1|_{\Pi_{\widetilde{v},\widetilde{w}}}$ of $\Pi_{\widetilde{v},\widetilde{w}}$. Moreover, let us also observe that it follows immediately from Theorem 4.2, (iii) [cf. also the definition of *profinite Dehn multi-twists* given in [CbTpI], Definition 4.4], that the assignment "$\alpha_1 \mapsto \alpha_1|_{\Pi_{\widetilde{v},\widetilde{w}}}$" implicit in assertion (iv) is *injective*. Thus, one verifies immediately from the definition of profinite Dehn multi-twists that one may choose the respective liftings $\widetilde{\alpha}_{\widetilde{v}}$, $\widetilde{\alpha}_{\widetilde{w}}$ of α_v, α_w so that $(\widetilde{\alpha}_{\widetilde{v}})_1(\Pi_{\widetilde{e}}) = (\widetilde{\alpha}_{\widetilde{w}})_1(\Pi_{\widetilde{e}}) = \Pi_{\widetilde{e}}$, and, moreover, the outomorphism of $\Pi_{\widetilde{v},\widetilde{w}}$ determined by the resulting automorphism $\widetilde{\alpha}_1|_{\Pi_{\widetilde{v},\widetilde{w}}}$ *coincides* with the outomorphism $\alpha_1|_{\Pi_{\widetilde{v},\widetilde{w}}}$ of assertion (iv).

Now we claim that the following assertion holds:

Claim 4.12.A: Write $(\widetilde{\alpha}_{\widetilde{z}})_{2/1}$ for the automorphism of $(\Pi_{\widetilde{z}})_{2/1}$ determined by $\widetilde{\alpha}_{\widetilde{z}}$ and $(\alpha_{\widetilde{z}})_{2/1}$ for the outomorphism of $(\Pi_{\widetilde{z}})_{2/1}$ determined by $(\widetilde{\alpha}_{\widetilde{z}})_{2/1}$. Then—relative to the natural identification $\Pi_{\mathcal{H}_{\widetilde{z}^\circ}} \xrightarrow{\sim} (\Pi_{\widetilde{z}})_{2/1}$ of assertion (iii)—it holds that

$$(\alpha_{\widetilde{z}})_{2/1} \in \mathrm{Aut}^{|\mathrm{Brch}(\mathcal{H}_{z^\circ})|}(\mathcal{H}_{z^\circ}) \quad (\subseteq \mathrm{Out}(\Pi_{\mathcal{H}_{\widetilde{z}^\circ}}) \xrightarrow{\sim} \mathrm{Out}((\Pi_{\widetilde{z}})_{2/1})).$$

Indeed, careful inspection of the various definitions involved reveals that Claim 4.12.A follows immediately from Theorem 4.7 [together with the *commensurable terminality* of the subgroup $\Pi_{\widetilde{e}} \subseteq \Pi_{\widetilde{z}}$—cf. [CmbGC], Proposition 1.2, (ii)]. Thus—by replacing $\widetilde{\alpha}_{\widetilde{z}}$ by the composite of $\widetilde{\alpha}_{\widetilde{z}}$ with an inner automorphism determined by conjugation by a suitable element of $(\Pi_{\widetilde{z}})_{2/1}$—we may assume without loss of generality that $\widetilde{\alpha}_{\widetilde{z}}(\Pi_{\widetilde{e}_{\widetilde{z}^\circ}}) = \Pi_{\widetilde{e}_{\widetilde{z}^\circ}}$. Moreover, since [cf. Claim 4.12.A] $\widetilde{\alpha}_{\widetilde{z}}$ *preserves* the $(\Pi_{\widetilde{z}})_{2/1}$-conjugacy classes of $\Pi_{\widetilde{z}^\circ}$ and $\Pi_{\widetilde{v}^{\mathrm{new}}}$, and the verticial subgroups $\Pi_{\widetilde{z}^\circ}$, $\Pi_{\widetilde{v}^{\mathrm{new}}} \subseteq \Pi_{\mathcal{G}_{2/1}} \xleftarrow{\sim} \Pi_{2/1}$ are the *unique* verticial subgroups of $\Pi_{\mathcal{G}_{2/1}} \xleftarrow{\sim} \Pi_{2/1}$ associated to $\widetilde{z}(\mathcal{G})^\circ$, $v^{\mathrm{new}} \in \mathrm{Vert}(\mathcal{G}_{2/1})$, respectively, such that $\Pi_{\widetilde{e}_{\widetilde{z}^\circ}} \subseteq \Pi_{\widetilde{z}^\circ}$, $\Pi_{\widetilde{e}_{\widetilde{z}^\circ}} \subseteq \Pi_{\widetilde{v}^{\mathrm{new}}}$ [cf. [CmbGC], Proposition 1.5, (i)], we thus conclude that $\widetilde{\alpha}_{\widetilde{z}}(\Pi_{\widetilde{z}^\circ}) = \Pi_{\widetilde{z}^\circ}$, $\widetilde{\alpha}_{\widetilde{z}}(\Pi_{\widetilde{v}^{\mathrm{new}}}) = \Pi_{\widetilde{v}^{\mathrm{new}}}$.

Next, write $(\alpha_{\widetilde{z}})_{\widetilde{z}^\circ}$, $(\alpha_{\widetilde{z}})_{\widetilde{v}^{\mathrm{new}}}$ for the respective outomorphisms of $\Pi_{\widetilde{z}^\circ}$, $\Pi_{\widetilde{v}^{\mathrm{new}}}$ determined by $\widetilde{\alpha}_{\widetilde{z}}$. Now we claim that the following assertion holds:

Claim 4.12.B: It holds that

$$(\alpha_{\widetilde{v}})_{\widetilde{v}^{\mathrm{new}}} = (\alpha_{\widetilde{w}})_{\widetilde{v}^{\mathrm{new}}}.$$

Moreover, if $v = w$, i.e., $\mathcal{G}$ is *cyclically primitive*, then—relative to the natural outer isomorphism $\Pi_{\widetilde{v}^\circ} \xrightarrow{\sim} \Pi_{\widetilde{w}^\circ}$ [where we note that if $v = w$, then $\Pi_{\widetilde{v}^\circ}$ is a $\Pi_{2/1}$-*conjugate* of $\Pi_{\widetilde{w}^\circ}$]—it holds that

$$(\alpha_{\widetilde{v}})_{\widetilde{v}^\circ} = (\alpha_{\widetilde{w}})_{\widetilde{w}^\circ}.$$

Indeed, the equality $(\alpha_{\widetilde{v}})_{\widetilde{v}^{\mathrm{new}}} = (\alpha_{\widetilde{w}})_{\widetilde{v}^{\mathrm{new}}}$ follows from the definition of $\mathrm{Glu}(\Pi_2)$. Next, suppose that $\mathcal{G}$ is *cyclically primitive*. To verify the equality $(\alpha_{\widetilde{v}})_{\widetilde{v}^\circ} = (\alpha_{\widetilde{w}})_{\widetilde{w}^\circ}$, let us observe that, for each $\widetilde{z} \in \{\widetilde{v}, \widetilde{w}\}$, the composite $\Pi_{\widetilde{z}^\circ} \hookrightarrow \Pi_2 \overset{p^{\Pi}_{\{1,2\}/\{2\}}}{\twoheadrightarrow} \Pi_{\{2\}} \xrightarrow{\sim} \Pi_{\mathcal{G}}$ is *injective* [and its image is a verticial subgroup of $\Pi_{\mathcal{G}}$ associated to $\widetilde{z}(\mathcal{G}) \in \mathrm{Vert}(\mathcal{G})$]. Thus, to verify the equality $(\alpha_{\widetilde{v}})_{\widetilde{v}^\circ} = (\alpha_{\widetilde{w}})_{\widetilde{w}^\circ}$, it suffices to verify that the outomorphism of the image of $\Pi_{\widetilde{v}^\circ}$ in $\Pi_{\{2\}}$ induced by $(\alpha_{\widetilde{v}})_{\widetilde{v}^\circ}$ *coincides* with the outomorphism of the image of $\Pi_{\widetilde{w}^\circ}$ in $\Pi_{\{2\}}$ induced by $(\alpha_{\widetilde{w}})_{\widetilde{w}^\circ}$. On the other hand, this follows immediately from the fact that both $\widetilde{\alpha}_{\widetilde{v}}$ and $\widetilde{\alpha}_{\widetilde{w}}$ are liftings of the *same* outomorphism $\alpha_v = \alpha_w$ of "$(\Pi_v)_2$"="$(\Pi_w)_2$" [cf. [CmbCsp], Proposition 1.2, (iii)]. This completes the proof of Claim 4.12.B.

Next, let us observe that it follows immediately from the various definitions involved that if $\mathcal{G}$ is *noncyclically primitive* (respectively, *cyclically primitive*), then $\#\mathrm{Vert}((\mathcal{G}_{2/1})_{\rightsquigarrow\{e_{v^\circ}\}}) = 2$ (respectively, $= 1$), and that, relative to the correspondence discussed in [CbTpI], Proposition 2.9, (i), (3), $\mathcal{H}_{v^\circ}$ and $\mathcal{G}_{2/1}|_{w^\circ(\mathcal{G})}$ (respectively, $\mathcal{H}_{v^\circ}$) correspond(s) to the two vertices (respectively, the unique vertex) of $(\mathcal{G}_{2/1})_{\rightsquigarrow\{e_{v^\circ}\}}$.

Next, let us observe the following equalities [cf. the notation of [CbTpI], Definition 3.8, (ii)]:

$$\begin{aligned}
\chi_{\mathcal{H}_{v^\circ}}((\alpha_{\widetilde{v}})_{2/1}) &= \chi_{\mathcal{H}_{z^\circ}|_{v^{\mathrm{new}}}}((\alpha_{\widetilde{v}})_{\widetilde{v}^{\mathrm{new}}}) && \text{[cf. [CbTpI], Corollary 3.9, (iv)]}\\
&= \chi_{\mathcal{H}_{v^\circ}|_{v^{\mathrm{new}}}}((\alpha_{\widetilde{w}})_{\widetilde{v}^{\mathrm{new}}}) && \text{[cf. Claim 4.12.B]}\\
&= \chi_{\mathcal{H}_{w^\circ}}((\alpha_{\widetilde{w}})_{2/1}) && \text{[cf. [CbTpI], Corollary 3.9, (iv)]}\\
&= \chi_{\mathcal{G}_{2/1}|_{w^\circ(\mathcal{G})}}((\alpha_{\widetilde{w}})_{\widetilde{w}^\circ}) && \text{[cf. [CbTpI], Corollary 3.9, (iv)].}
\end{aligned}$$

Now it follows immediately from these equalities, together with Claim 4.12.A, that the data

$$((\alpha_{\widetilde{v}})_{2/1}, (\alpha_{\widetilde{w}})_{\widetilde{w}^\circ}) \in \mathrm{Aut}(\mathcal{H}_{v^\circ}) \times \mathrm{Aut}(\mathcal{G}_{2/1}|_{w^\circ(\mathcal{G})})$$

$$(\text{respectively,} \quad (\alpha_{\widetilde{v}})_{2/1} \in \mathrm{Aut}(\mathcal{H}_{v^\circ}))$$

may be regarded as an element of $\mathrm{Glu}^{\mathrm{brch}}((\mathcal{G}_{2/1})_{\rightsquigarrow\{e_{v^\circ}\}})$ [cf. Definition 4.1, (iii)]. Thus, by applying the exact sequence of Theorem 4.2, (iii) [cf. also Remark 4.9.1], we conclude that there exists an element

$$\alpha_{2/1}[\widetilde{v}] \in \mathrm{Aut}^{|\mathrm{Brch}((\mathcal{G}_{2/1})_{\rightsquigarrow\{e_{v^\circ}\}})|}((\mathcal{G}_{2/1})_{\rightsquigarrow\{e_{v^\circ}\}})$$

of a collection of outomorphisms of

$$\Pi_{(\mathcal{G}_{2/1})_{\rightsquigarrow\{e_{v^\circ}\}}} \overset{\Phi_{(\mathcal{G}_{2/1})_{\rightsquigarrow\{e_{v^\circ}\}}}}{\underset{\sim}{\longrightarrow}} \Pi_{\mathcal{G}_{2/1}} \overset{\sim}{\longrightarrow} \Pi_{2/1}$$

[i.e., contained in the image of $\mathrm{Aut}((\mathcal{G}_{2/1})_{\rightsquigarrow\{e_{v^\circ}\}}) \hookrightarrow \mathrm{Out}(\Pi_{2/1})$—cf. [CbTpI], Definition 2.10] that admits a natural structure of *torsor* over

$$\mathrm{Dehn}((\mathcal{G}_{2/1})_{\rightsquigarrow\{e_{v^\circ}\}}) \ (\subseteq \mathrm{Aut}^{|\mathrm{Brch}((\mathcal{G}_{2/1})_{\rightsquigarrow\{e_{v^\circ}\}})|}((\mathcal{G}_{2/1})_{\rightsquigarrow\{e_{v^\circ}\}})).$$

A similar argument yields the existence of an element

$$\alpha_{2/1}[\widetilde{w}] \in \mathrm{Aut}^{|\mathrm{Brch}((\mathcal{G}_{2/1})_{\rightsquigarrow\{e_{w^\circ}\}})|}((\mathcal{G}_{2/1})_{\rightsquigarrow\{e_{w^\circ}\}})$$

of a collection of outomorphisms of

$$\Pi_{(\mathcal{G}_{2/1})_{\rightsquigarrow\{e_{w^\circ}\}}} \overset{\Phi_{(\mathcal{G}_{2/1})_{\rightsquigarrow\{e_{w^\circ}\}}}}{\underset{\sim}{\longrightarrow}} \Pi_{\mathcal{G}_{2/1}} \overset{\sim}{\longrightarrow} \Pi_{2/1}$$

[i.e., contained in the image of $\mathrm{Aut}((\mathcal{G}_{2/1})_{\rightsquigarrow\{e_{w^\circ}\}}) \hookrightarrow \mathrm{Out}(\Pi_{2/1})$] that admits a natural structure of *torsor* over

$$\mathrm{Dehn}((\mathcal{G}_{2/1})_{\rightsquigarrow\{e_{w^\circ}\}}) \ (\subseteq \mathrm{Aut}^{|\mathrm{Brch}((\mathcal{G}_{2/1})_{\rightsquigarrow\{e_{w^\circ}\}})|}((\mathcal{G}_{2/1})_{\rightsquigarrow\{e_{w^\circ}\}})).$$

Now we claim that the following assertion holds:

Claim 4.12.C: For each $\widetilde{z} \in \{\widetilde{v}, \widetilde{w}\}$, the automorphism $(\widetilde{\alpha}_{\widetilde{z}})_1$ of $\Pi_{\widetilde{z}}$ is *compatible* with the outomorphism $\alpha_{2/1}[\widetilde{z}]$ of $\Pi_{2/1}$ relative to the homomorphism $\Pi_{\widetilde{z}} \hookrightarrow \Pi_1 \to \mathrm{Out}(\Pi_{2/1})$—where the second arrow is the natural outer action determined by the exact sequence

$$1 \longrightarrow \Pi_{2/1} \longrightarrow \Pi_2 \overset{p^{\Pi}_{2/1}}{\longrightarrow} \Pi_1 \longrightarrow 1.$$

Indeed, to verify the *compatibility* of $(\widetilde{\alpha}_{\widetilde{v}})_1$ and $\alpha_{2/1}[\widetilde{v}]$, it follows immediately from the various definitions involved that it suffices to verify that, for each $\sigma \in \Pi_{\widetilde{v}}$, if we write $\tau \overset{\mathrm{def}}{=} (\widetilde{\alpha}_{\widetilde{v}})_1(\sigma) \in \Pi_{\widetilde{v}}$, then there exist liftings $\widetilde{\sigma}, \widetilde{\tau} \in \Pi_2$ of $\sigma, \tau \in \Pi_{\widetilde{v}}$, respectively, such that the equality [which is in fact independent of the choice of liftings]

$$\alpha_{2/1}[\tilde{v}] \circ [\mathrm{Inn}(\tilde{\sigma})] \circ \alpha_{2/1}[\tilde{v}]^{-1} = [\mathrm{Inn}(\tilde{\tau})] \quad \in \mathrm{Out}(\Pi_{2/1})$$

—where we write "Inn(−)" for the automorphism of $\Pi_{2/1}$ determined by conjugation by "(−)" and "[Inn(−)]" for the outomorphism of $\Pi_{2/1}$ determined by this automorphism—holds. To this end, let $\tilde{\sigma} \in (\Pi_{\tilde{v}})_2$ be a lifting of $\sigma \in \Pi_{\tilde{v}}$. Then since $(\Pi_{\tilde{v}})_{2/1} \subseteq (\Pi_{\tilde{v}})_2$ is *normal*, $\mathrm{Inn}(\tilde{\sigma})$ *preserves* $(\Pi_{\tilde{v}})_{2/1}$.

Next, let us *observe* that the semi-graph of anabelioids structure of $(\mathcal{G}_{2/1})_{\leadsto\{e_{v^\circ}\}}$ [with respect to which w° is a vertex if $\mathcal{G}$ is *noncyclically primitive* and, moreover, with respect to which e_{w° is a node in both the *cyclically primitive* and *noncyclically primitive* cases] may be thought of as the semi-graph of anabelioids structure on the fiber subgroup $\Pi_{2/1}$ [cf. Definition 3.1, (iii)] arising from a point of $X^{\log}$ that lies in the 1-interior of the irreducible component of $X^{\log}$ corresponding to v. Now it follows immediately from this *observation* that $\mathrm{Inn}(\tilde{\sigma})$ *preserves* the $\Pi_{2/1}$-conjugacy class of $\Pi_{\tilde{w}^\circ}$, as well as the $\Pi_{2/1}$-conjugacy class of $\Pi_{\tilde{e}_{\tilde{w}^\circ}} = (\Pi_{\tilde{v}})_{2/1} \cap \Pi_{\tilde{w}^\circ}$ if $\mathcal{G}$ is *noncyclically primitive* (respectively, *preserves* the $\Pi_{2/1}$-conjugacy class of $\Pi_{\tilde{e}_{\tilde{w}^\circ}}$ if $\mathcal{G}$ is *cyclically primitive*). By considering the various $\Pi_{2/1}$-conjugates of $\Pi_{\tilde{e}_{\tilde{w}^\circ}}$ and $\Pi_{\tilde{w}^\circ}$ and applying [CmbGC], Propositions 1.2, (ii); 1.5, (i), we thus conclude that $\mathrm{Inn}(\tilde{\sigma})$ *preserves* the $(\Pi_{\tilde{v}})_{2/1}$-conjugacy classes of $\Pi_{\tilde{e}_{\tilde{w}^\circ}}$, $\Pi_{\tilde{w}^\circ}$ if $\mathcal{G}$ is *noncyclically primitive* (respectively, *preserves* the $(\Pi_{\tilde{v}})_{2/1}$-conjugacy class of $\Pi_{\tilde{e}_{\tilde{w}^\circ}}$ if $\mathcal{G}$ is *cyclically primitive*). In particular—by multiplying $\tilde{\sigma}$ by a suitable element of $(\Pi_{\tilde{v}})_{2/1}$—we may assume without loss of generality that $\mathrm{Inn}(\tilde{\sigma})$ *preserves* $(\Pi_{\tilde{v}})_{2/1}$, $\Pi_{\tilde{w}^\circ}$, and $\Pi_{\tilde{e}_{\tilde{w}^\circ}}$ in the *noncyclically primitive* case (respectively, *preserves* $(\Pi_{\tilde{v}})_{2/1}$ and $\Pi_{\tilde{e}_{\tilde{w}^\circ}}$ in the *cyclically primitive* case).

Next, let us observe that one verifies easily [cf. Lemma 3.6, (iv)] that the composite $\Pi_{\tilde{e}_{\tilde{w}^\circ}} \hookrightarrow \Pi_2 \overset{p^{\Pi}_{\{1,2\}/\{2\}}}{\twoheadrightarrow} \Pi_{\{2\}}$ *surjects* onto a nodal subgroup of $\Pi_{\mathcal{G}} \overset{\sim}{\leftarrow} \Pi_{\{2\}}$ associated to $e \in \mathrm{Node}(\mathcal{G})$. Thus, since $\mathrm{Inn}(\tilde{\sigma})$ *preserves* $\Pi_{\tilde{e}_{\tilde{w}^\circ}}$, it follows [cf. [CmbGC], Proposition 1.2, (ii)] that the image of $\tilde{\sigma} \in \Pi_2$ via $\Pi_2 \overset{p^{\Pi}_{\{1,2\}/\{2\}}}{\twoheadrightarrow} \Pi_{\{2\}}$ is *contained* in the image of the composite $\Pi_{\tilde{e}_{\tilde{w}^\circ}} \hookrightarrow \Pi_2 \overset{p^{\Pi}_{\{1,2\}/\{2\}}}{\twoheadrightarrow} \Pi_{\{2\}}$. In particular—by multiplying $\tilde{\sigma}$ by a suitable element of $\Pi_{\tilde{e}_{\tilde{w}^\circ}}$ ($\subseteq (\Pi_{\tilde{v}})_{2/1}$)—we may assume without loss of generality that $\tilde{\sigma} \in \mathrm{Ker}(p^{\Pi}_{\{1,2\}/\{2\}})$. A similar argument implies that there exists a lifting $\tilde{\tau} \in (\Pi_{\tilde{v}})_2$ of $\tau = (\tilde{\alpha}_{\tilde{v}})_1(\sigma) \in \Pi_{\tilde{v}}$ such that $\mathrm{Inn}(\tilde{\tau})$ *preserves* $(\Pi_{\tilde{v}})_{2/1}$, $\Pi_{\tilde{w}^\circ}$, $\Pi_{\tilde{e}_{\tilde{w}^\circ}}$ if $\mathcal{G}$ is *noncyclically primitive* (respectively, *preserves* $(\Pi_{\tilde{v}})_{2/1}$ and $\Pi_{\tilde{e}_{\tilde{w}^\circ}}$ if $\mathcal{G}$ is *cyclically primitive*), and, moreover, $\tilde{\tau} \in \mathrm{Ker}(p^{\Pi}_{\{1,2\}/\{2\}})$.

Now since the automorphisms $(\tilde{\alpha}_{\tilde{v}})_{2/1}$, $(\tilde{\alpha}_{\tilde{v}})_1$ of $(\Pi_{\tilde{v}})_{2/1}$, $\Pi_{\tilde{v}}$, respectively, arise from the automorphism $\tilde{\alpha}_{\tilde{v}}$ of $(\Pi_{\tilde{v}})_2$, it follows immediately from the construction of $\alpha_{2/1}[\tilde{v}]$ that the equality

$$\alpha_{2/1}[\tilde{v}] \circ [\mathrm{Inn}(\tilde{\sigma})] \circ \alpha_{2/1}[\tilde{v}]^{-1} = [\mathrm{Inn}(\tilde{\tau})]$$

holds upon restriction to [an equality of outomorphisms of] $(\Pi_{\widetilde{v}})_{2/1}$. Moreover, if $\mathcal{G}$ is *noncyclically primitive*, then since the composite $\Pi_{\widetilde{w}^\circ} \hookrightarrow \Pi_2 \overset{p^\Pi_{\{1,2\}/\{2\}}}{\twoheadrightarrow} \Pi_{\{2\}}$ is *injective* [and its image is a verticial subgroup of $\Pi_{\mathcal{G}} \overset{\sim}{\leftarrow} \Pi_{\{2\}}$ associated to $w \in \mathrm{Vert}(\mathcal{G})$—cf. Lemma 3.6, (iv)], to verify the restriction of the equality

$$\alpha_{2/1}[\widetilde{v}] \circ [\mathrm{Inn}(\widetilde{\sigma})] \circ \alpha_{2/1}[\widetilde{v}]^{-1} = [\mathrm{Inn}(\widetilde{\tau})]$$

to [an equality of outomorphisms of] $\Pi_{\widetilde{w}^\circ}$, it suffices to verify that the outomorphism of the image of $\Pi_{\widetilde{w}^\circ}$ in $\Pi_{\{2\}}$ induced by the product

$$\alpha_{2/1}[\widetilde{v}] \circ [\mathrm{Inn}(\widetilde{\sigma})] \circ \alpha_{2/1}[\widetilde{v}]^{-1} \circ [\mathrm{Inn}(\widetilde{\tau})]^{-1}$$

is *trivial*. On the other hand, this follows immediately from the fact that $\widetilde{\sigma}, \widetilde{\tau} \in \mathrm{Ker}(p^\Pi_{\{1,2\}/\{2\}})$.

Thus, in summary, the restriction of the equality in question [i.e., in the discussion immediately following Claim 4.12.C] to [an equality of outomorphisms of] $(\Pi_{\widetilde{v}})_{2/1}$ holds. Moreover, if $\mathcal{G}$ is *noncyclically primitive*, then the restriction of the equality in question to [an equality of outomorphisms of] $\Pi_{\widetilde{w}^\circ}$ holds. In particular, it follows immediately from the displayed exact sequence of Theorem 4.2, (iii) [cf. also Remark 4.9.1], that the product

$$\alpha_{2/1}[\widetilde{v}] \circ [\mathrm{Inn}(\widetilde{\sigma})] \circ \alpha_{2/1}[\widetilde{v}]^{-1} \circ [\mathrm{Inn}(\widetilde{\tau})]^{-1}$$

is *contained* in $\mathrm{Dehn}((\mathcal{G}_{2/1})_{\rightsquigarrow\{e_{v^\circ}\}})$. Thus—by considering the outomorphism of $\Pi_{\{2\}}$ induced by the above product—one verifies easily from [CbTpI], Theorem 4.8, (iv), together with the fact that $\widetilde{\sigma}, \widetilde{\tau} \in \mathrm{Ker}(p^\Pi_{\{1,2\}/\{2\}})$, that the equality in question holds. This completes the proof of the *compatibility* of $(\widetilde{\alpha}_{\widetilde{v}})_1$ and $\alpha_{2/1}[\widetilde{v}]$. The *compatibility* of $(\widetilde{\alpha}_{\widetilde{w}})_1$ and $\alpha_{2/1}[\widetilde{w}]$ follows from a similar argument. This completes the proof of Claim 4.12.C.

Next, we claim that the following assertion holds:

Claim 4.12.D: The difference $\alpha_{2/1}[\widetilde{v}] \circ \alpha_{2/1}[\widetilde{w}]^{-1} \in \mathrm{Out}(\Pi_{2/1})$ is *contained* in $\mathrm{Dehn}(\mathcal{G}_{2/1})$ $(\subseteq \mathrm{Out}(\Pi_{\mathcal{G}_{2/1}}) \overset{\sim}{\leftarrow} \mathrm{Out}(\Pi_{2/1}))$.

Indeed, this follows immediately from the two displayed equalities of Claim 4.12.B, together with the construction of $\alpha_{2/1}[\widetilde{v}]$, $\alpha_{2/1}[\widetilde{w}]$. This completes the proof of Claim 4.12.D.

Thus, it follows immediately from Claim 4.12.D, together with the existence of the natural isomorphism

$$\mathrm{Dehn}((\mathcal{G}_{2/1})_{\rightsquigarrow\{e_{v^\circ}\}}) \oplus \mathrm{Dehn}((\mathcal{G}_{2/1})_{\rightsquigarrow\{e_{w^\circ}\}}) \overset{\sim}{\longrightarrow} \mathrm{Dehn}(\mathcal{G}_{2/1})$$

[cf. [CbTpI], Theorem 4.8, (ii), (iv)], that—by replacing $\alpha_{2/1}[\widetilde{v}]$, $\alpha_{2/1}[\widetilde{w}]$ by the composites of $\alpha_{2/1}[\widetilde{v}]$, $\alpha_{2/1}[\widetilde{w}]$ with suitable elements of $\mathrm{Dehn}((\mathcal{G}_{2/1})_{\rightsquigarrow\{e_{v^\circ}\}})$,

$\mathrm{Dehn}((\mathcal{G}_{2/1})_{\rightsquigarrow\{e_{w^\circ}\}})$, respectively [where we recall that the outomorphisms $\alpha_{2/1}[\widetilde{v}]$, $\alpha_{2/1}[\widetilde{w}]$ belong to *torsors* over $\mathrm{Dehn}((\mathcal{G}_{2/1})_{\rightsquigarrow\{e_{v^\circ}\}})$, $\mathrm{Dehn}((\mathcal{G}_{2/1})_{\rightsquigarrow\{e_{w^\circ}\}})$, respectively]—we may assume without loss of generality that

$$\alpha_{2/1}[\widetilde{v}] = \alpha_{2/1}[\widetilde{w}].$$

Write $\beta_{2/1} \overset{\mathrm{def}}{=} \alpha_{2/1}[\widetilde{v}] = \alpha_{2/1}[\widetilde{w}]$. Then it follows immediately from Claim 4.12.C, together with the fact that $\Pi_{\widetilde{v},\widetilde{w}}$ is *topologically generated* by $\Pi_{\widetilde{v}}$, $\Pi_{\widetilde{w}} \subseteq \Pi_{\widetilde{v},\widetilde{w}}$ [cf. assertion (ii)], that the outomorphism $\beta_{2/1}$ of $\Pi_{2/1}$ is *compatible* with the automorphism $\widetilde{\alpha}_1|_{\Pi_{\widetilde{v},\widetilde{w}}}$ of $\Pi_{\widetilde{v},\widetilde{w}}$ [i.e., the automorphism induced by $(\widetilde{\alpha}_{\widetilde{v}})_1$, $(\widetilde{\alpha}_{\widetilde{w}})_1$—cf. the discussion immediately preceding Claim 4.12.A], relative to the composite $\Pi_{\widetilde{v},\widetilde{w}} \hookrightarrow \Pi_1 \to \mathrm{Out}(\Pi_{2/1})$, where the second arrow is the outer action determined by the displayed exact sequence of Claim 4.12.C. In particular, by considering the natural isomorphism $\Pi_2|_{\Pi_{\widetilde{v},\widetilde{w}}} \xrightarrow{\sim} \Pi_{2/1} \overset{\mathrm{out}}{\rtimes} \Pi_{\widetilde{v},\widetilde{w}}$ [cf. the discussion entitled "*Topological groups*" in [CbTpI], §0], we obtain an outomorphism $\beta_{\widetilde{v},\widetilde{w}}$ of $\Pi_2|_{\Pi_{\widetilde{v},\widetilde{w}}}$ which, by *construction*, satisfies the four conditions listed in assertion (v). This completes the proof of assertion (v). □

Lemma 4.13 (Glueability of Combinatorial Cuspidalizations in the Case of Precisely One Node) *Suppose that $n = 2$, and that $\#\mathrm{Node}(\mathcal{G}) = 1$. Then ρ_2^{brch} [cf. Definition 4.11] is* **surjective**.

Proof If $\mathcal{G}$ is *noncyclically primitive* [cf. [CbTpI], Definition 4.1], then the *surjectivity* of ρ_2^{brch} follows immediately from Lemma 4.12, (v) [cf. also [CmbCsp], Proposition 1.2, (i)], together with the fact that the natural injection $\Pi_{\widetilde{v},\widetilde{w}} \hookrightarrow \Pi_1$ is an *isomorphism* [cf. Lemma 4.12, (ii)]. Thus, it remains to verify the *surjectivity* of ρ_2^{brch} in the case where $\mathcal{G}$ is *cyclically primitive* [cf. [CbTpI], Definition 4.1]. Since we are in the situation of [CbTpI], Lemma 4.3, we shall apply the notational conventions established in [CbTpI], Lemma 4.3. Also, we shall write $\mathrm{Vert}(\mathcal{G}) = \{v\}$, $\mathrm{Node}(\mathcal{G}) = \{e\}$. Let $x \in X_2(k)$ be a k-rational geometric point of X_2 such that $x_{\{1\}} \in X(k)$ [cf. Definition 3.1, (i)] lies on the unique node of $X^{\log}$ [i.e., which corresponds to e].

Recall from [CbTpI], Lemma 4.3, (i), that we have a natural exact sequence

$$1 \longrightarrow \pi_1^{\mathrm{temp}}(\mathcal{G}_\infty) \longrightarrow \pi_1^{\mathrm{temp}}(\mathcal{G}) \longrightarrow \pi_1^{\mathrm{top}}(\mathbb{G}) \longrightarrow 1.$$

Let $\gamma_\infty \in \pi_1^{\mathrm{top}}(\mathbb{G})$ be a generator of $\pi_1^{\mathrm{top}}(\mathbb{G})$ ($\simeq \mathbb{Z}$) and $\widetilde{\gamma}_\infty \in \pi_1^{\mathrm{temp}}(\mathcal{G})$ a lifting of γ_∞. By abuse of notation, write $\widetilde{\gamma}_\infty \in \Pi_{\mathcal{G}} \xleftarrow{\sim} \Pi_1$ for the image of $\widetilde{\gamma}_\infty \in \pi_1^{\mathrm{temp}}(\mathcal{G})$ via the natural injection $\pi_1^{\mathrm{temp}}(\mathcal{G}) \hookrightarrow \Pi_{\mathcal{G}} \xleftarrow{\sim} \Pi_1$ [cf. the evident pro-Σ generalization of [SemiAn], Proposition 3.6, (iii); [RZ], Proposition 3.3.15]. Next, let us fix a verticial subgroup

$$\Pi_{\widetilde{v}(0)}^{\mathrm{temp}} \subseteq (\pi_1^{\mathrm{temp}}(\mathcal{G}_\infty) \subseteq)\ \pi_1^{\mathrm{temp}}(\mathcal{G})$$

of $\pi_1^{\text{temp}}(\mathcal{G})$ that corresponds to a vertex $\widetilde{v}(0) \in \text{Vert}(\widetilde{\mathcal{G}})$ that *lifts* the vertex $V(0) \in \text{Vert}(\mathcal{G}_\infty)$ [cf. [CbTpI], Lemma 4.3, (iii)]. Thus, for each integer $a \in \mathbb{Z}$, by forming the conjugate of $\Pi_{\widetilde{v}(0)}^{\text{temp}}$ by $\widetilde{\gamma}_\infty^a$, we obtain a verticial subgroup

$$\Pi_{\widetilde{v}(a)}^{\text{temp}} \subseteq (\pi_1^{\text{temp}}(\mathcal{G}_\infty) \subseteq)\ \pi_1^{\text{temp}}(\mathcal{G})$$

of $\pi_1^{\text{temp}}(\mathcal{G})$ associated to some vertex $\widetilde{v}(a) \in \text{Vert}(\widetilde{\mathcal{G}})$ that *lifts* the vertex $V(a) \in \text{Vert}(\mathcal{G}_\infty)$ [cf. [CbTpI], Lemma 4.3, (iii), (vi)]. Write

$$\Pi_{\widetilde{v}(a)} \subseteq \Pi_{\mathcal{G}}$$

for the image of $\Pi_{\widetilde{v}(a)}^{\text{temp}} \subseteq \pi_1^{\text{temp}}(\mathcal{G})$ in $\Pi_{\mathcal{G}}$.

Next, let us suppose that $\widetilde{\gamma}_\infty$ was chosen in such a way that, for each $a \in \mathbb{Z}$, the intersection $\mathcal{N}(\widetilde{v}(a)) \cap \mathcal{N}(\widetilde{v}(a+1))$ consists of a *unique* node $\widetilde{n}(a, a+1) \in \text{Node}(\widetilde{\mathcal{G}})$ that *lifts* the node $N(a+1) \in \text{Node}(\mathcal{G}_\infty)$ [cf. [CbTpI], Lemma 4.3, (iii)]. [One verifies easily that such a $\widetilde{\gamma}_\infty$ always exists.] Then let us observe that, for each $a \le b \in \mathbb{Z}$, we have a natural morphism of semi-graphs of anabelioids $\mathcal{G}_{[a,b]} \to \mathcal{G}_\infty$ [cf. [CbTpI], Lemma 4.3, (iv)], which induces *injections* [cf. the evident pro-Σ generalizations of [SemiAn], Example 2.10; [SemiAn], Proposition 2.5, (i); [SemiAn], Proposition 3.6, (iii); [RZ], Proposition 3.3.15]

$$\pi_1^{\text{temp}}(\mathcal{G}_{[a,b]}) \hookrightarrow \pi_1^{\text{temp}}(\mathcal{G}_\infty), \quad \Pi_{\mathcal{G}_{[a,b]}} \hookrightarrow \Pi_{\mathcal{G}}$$

—where we write, respectively, $\pi_1^{\text{temp}}(\mathcal{G}_{[a,b]})$, $\Pi_{\mathcal{G}_{[a,b]}}$ for the tempered, pro-Σ fundamental groups of the semi-graph of anabelioids $\mathcal{G}_{[a,b]}$ of pro-Σ PSC-type—which are well-defined up to composition with *inner automorphisms*. By choosing appropriate basepoints [cf. also our choice of $\widetilde{\gamma}_\infty$], these inner automorphism indeterminacies may be eliminated in such a way that, for each $a \le c \le b$, the resulting injections are *compatible* with one another and, moreover, their images *contain* the subgroups $\Pi_{\widetilde{v}(c)}^{\text{temp}} \subseteq \pi_1^{\text{temp}}(\mathcal{G}_\infty)$, $\Pi_{\widetilde{v}(c)} \subseteq \Pi_{\mathcal{G}} \xleftarrow{\sim} \Pi_1$, respectively. Then, relative to the resulting inclusions, $\Pi_{\widetilde{v}(c)}^{\text{temp}}$, $\Pi_{\widetilde{v}(c)}$ form verticial subgroups of $\pi_1^{\text{temp}}(\mathcal{G}_{[a,b]})$, $\Pi_{\mathcal{G}_{[a,b]}}$ associated to the vertex of $\mathcal{G}_{[a,b]}$ corresponding to $V(c)$ [cf. [CbTpI], Lemma 4.3, (iii)]. In particular, we have a natural isomorphism

$$\Pi_{[a,a+1]} \stackrel{\text{def}}{=} \Pi_{\widetilde{v}(a),\widetilde{v}(a+1)} \xrightarrow{\sim} \Pi_{\mathcal{G}_{[a,a+1]}}$$

[cf. Lemma 4.12, (ii)]. Let us write

$$\Pi_2|_{[a,a+1]} \stackrel{\text{def}}{=} \Pi_2|_{\Pi_{[a,a+1]}} \subseteq \Pi_2$$

[cf. Lemma 4.12, (ii)];

$$\Pi_{[a]} \stackrel{\mathrm{def}}{=} \Pi_{\widetilde{v}(a)};$$

$$\Pi_2|_{[a]} \stackrel{\mathrm{def}}{=} \Pi_2 \times_{\Pi_1} \Pi_{[a]} \subseteq \Pi_2|_{[a-1,a]}, \quad \Pi_2|_{[a,a+1]}.$$

Next, we claim that the following assertion holds:

Claim 4.13.A: The profinite group $\Pi_{\mathcal{G}}$ is *topologically generated* by $\Pi_{[0]} \subseteq \Pi_{\mathcal{G}}$ and $\widetilde{\gamma}_\infty \in \Pi_{\mathcal{G}}$.

Indeed, let us first observe that it follows immediately from a similar argument to the argument applied in the proof of [CmbCsp], Proposition 1.5, (iii) [i.e., in essence, from the "*van Kampen Theorem*" in elementary algebraic topology], that

- the image of the natural homomorphism

$$\varinjlim_{a \geq 0} \pi_1^{\mathrm{temp}}(\mathcal{G}_{[-a,a]}) \longrightarrow \pi_1^{\mathrm{temp}}(\mathcal{G}_\infty)$$

  —where the inductive limit is taken in the category of tempered groups [cf. [SemiAn], Definition 3.1, (i); [SemiAn], Example 2.10; [SemiAn], Proposition 3.6, (i)]—is *dense*;
- for each nonnegative integer a, the tempered group $\pi_1^{\mathrm{temp}}(\mathcal{G}_{[-a,a]})$ [cf. [SemiAn], Example 2.10; [SemiAn], Proposition 3.6, (i)] is *topologically generated* by $\Pi^{\mathrm{temp}}_{\widetilde{v}(-a)}, \ldots, \Pi^{\mathrm{temp}}_{\widetilde{v}(a)} \subseteq \pi_1^{\mathrm{temp}}(\mathcal{G}_{[-a,a]})$.

In particular, it follows immediately from the exact sequence of [CbTpI], Lemma 4.3, (i), that the tempered group $\pi_1^{\mathrm{temp}}(\mathcal{G})$ [cf. [SemiAn], Example 2.10; [SemiAn], Proposition 3.6, (i)] is *topologically generated* by $\Pi^{\mathrm{temp}}_{\widetilde{v}(0)} \subseteq \pi_1^{\mathrm{temp}}(\mathcal{G})$ and $\widetilde{\gamma}_\infty \in \pi_1^{\mathrm{temp}}(\mathcal{G})$. Thus, Claim 4.13.A follows immediately from the fact that the image of the natural injection $\pi_1^{\mathrm{temp}}(\mathcal{G}) \hookrightarrow \Pi_{\mathcal{G}}$ is *dense*. This completes the proof of Claim 4.13.A.

For $a \in \mathbb{Z}$, let us write

$$\mathcal{G}_{2/1}^{[a,a+1]} \stackrel{\mathrm{def}}{=} \mathcal{G}_{2 \in \{1,2\},x}$$

[cf. Definition 3.1, (iii)], where we *fix* isomorphisms

$$\Pi_{2/1} \stackrel{\sim}{\longrightarrow} \Pi_{\mathcal{G}_{2/1}^{[a,a+1]}}, \quad \Pi_{\{2\}} \stackrel{\sim}{\longrightarrow} \Pi_{\mathcal{G}_{2 \in \{2\},x}} = \Pi_{\mathcal{G}}$$

[the latter of which is to be understood as being *independent* of $a \in \mathbb{Z}$] as in [i.e., that belong to the collections of isomorphisms that constitute the outer isomorphisms of the final display of] Definition 3.1, (iii), to be isomorphisms [cf. the discussion of the final portion of Lemma 4.12, (v)] such that the semi-graph of anabelioids structure on $\mathcal{G}_{2/1}^{[a,a+1]}$ is the semi-graph of anabelioids structure determined by the resulting composite

$$\Pi_{\widetilde{n}(a,a+1)} \hookrightarrow \Pi_{\mathcal{G}} \overset{\sim}{\leftarrow} \Pi_1 \to \mathrm{Out}(\Pi_{2/1}) \overset{\sim}{\to} \mathrm{Out}(\Pi_{\mathcal{G}_{2/1}^{[a,a+1]}})$$

—where we write $\Pi_{\widetilde{n}(a,a+1)} \subseteq \Pi_{\mathcal{G}}$ for the nodal subgroup of $\Pi_{\mathcal{G}}$ associated to the *unique* element $\widetilde{n}(a, a+1) \in \mathcal{N}(\widetilde{v}(a)) \cap \mathcal{N}(\widetilde{v}(a+1))$, and the third arrow arises from the outer action determined by the exact sequence $1 \to \Pi_{2/1} \to \Pi_2 \overset{p_{2/1}^{\Pi}}{\to} \Pi_1 \to 1$—in a fashion compatible with the projection $p_{\{1,2\}/\{2\}}^{\Pi}|_{\Pi_{2/1}} : \Pi_{2/1} \twoheadrightarrow \Pi_{\{2\}}$ and the isomorphisms $\Pi_{\{2\}} \overset{\sim}{\to} \Pi_{\mathcal{G}} \overset{\sim}{\leftarrow} \Pi_1$ [cf. Definition 3.1, (ii)]. Here, we note that, for a, $b \in \mathbb{Z}$, there exist natural isomorphisms $\mathcal{G}_{2/1}^{[a,a+1]} \overset{\sim}{\to} \mathcal{G}_{2\in\{1,2\},x} \overset{\sim}{\to} \mathcal{G}_{2/1}^{[b,b+1]}$ of semi-graphs of anabelioids of pro-Σ PSC-type [induced by conjugation by $\widetilde{\gamma}_{\infty}^{b-a}$]. On the other hand, it is not difficult to show [although we shall not use this fact in the present proof!] that the well-known *injectivity* of the homomorphism $\Pi_1 \to \mathrm{Out}(\Pi_{2/1})$ of the above display [cf. [CbTpI], Lemma 5.4, (i), (ii), (iii); [CbTpI], Theorem 4.8, (iv); [Asd], Theorem 1; [Asd], the Remark following the proof of Theorem 1; [CmbGC], Proposition 1.2, (i), (ii)] implies that when $a \neq b$, the composite

$$\Pi_{\mathcal{G}_{2/1}^{[a,a+1]}} \overset{\sim}{\leftarrow} \Pi_{2/1} \overset{\sim}{\to} \Pi_{\mathcal{G}_{2/1}^{[b,b+1]}}$$

in fact *fails to be graphic*!

For each $a \in \mathbb{Z}$, let us write

$$\mathcal{G}_{2/1}^{[a,a+1]\rightsquigarrow[a]} \overset{\text{def}}{=} (\mathcal{G}_{2/1}^{[a,a+1]})_{\rightsquigarrow\{e_{v(a)^\circ}\}}, \quad \mathcal{G}_{2/1}^{[a,a+1]\rightsquigarrow[a+1]} \overset{\text{def}}{=} (\mathcal{G}_{2/1}^{[a,a+1]})_{\rightsquigarrow\{e_{v(a+1)^\circ}\}}$$

—where we write $e_{v(a)^\circ}$, $e_{v(a+1)^\circ}$ for the nodes "e_{z°" of Lemma 4.12, (ii), that occur, respectively, in the cases where the pair "$(\mathcal{G}_{2/1}, \widetilde{z}^\circ)$" is taken to be $(\mathcal{G}_{2/1}^{[a,a+1]}, \widetilde{v}(a)^\circ)$; $(\mathcal{G}_{2/1}^{[a,a+1]}, \widetilde{v}(a+1)^\circ)$. Then one verifies easily [cf. Lemma 4.12, (i), (iii)] that the composite

$$\Pi_{\mathcal{G}_{2/1}^{[a-1,a]\rightsquigarrow[a]}} \overset{\sim}{\to} \Pi_{\mathcal{G}_{2/1}^{[a-1,a]}} \overset{\sim}{\leftarrow} \Pi_{2/1} \overset{\sim}{\to} \Pi_{\mathcal{G}_{2/1}^{[a,a+1]}} \overset{\sim}{\leftarrow} \Pi_{\mathcal{G}_{2/1}^{[a,a+1]\rightsquigarrow[a]}}$$

—where the first and fourth arrows are the *natural specialization outer isomorphisms* [cf. [CbTpI], Definition 2.10], and the second and third arrows are the isomorphisms fixed above—is *graphic*. In light of this observation, it makes sense to write

$$\mathcal{G}_{2/1}^{[a]} \overset{\text{def}}{=} \mathcal{G}_{2/1}^{[a-1,a]\rightsquigarrow[a]} \overset{\sim}{\to} \mathcal{G}_{2/1}^{[a,a+1]\rightsquigarrow[a]}$$

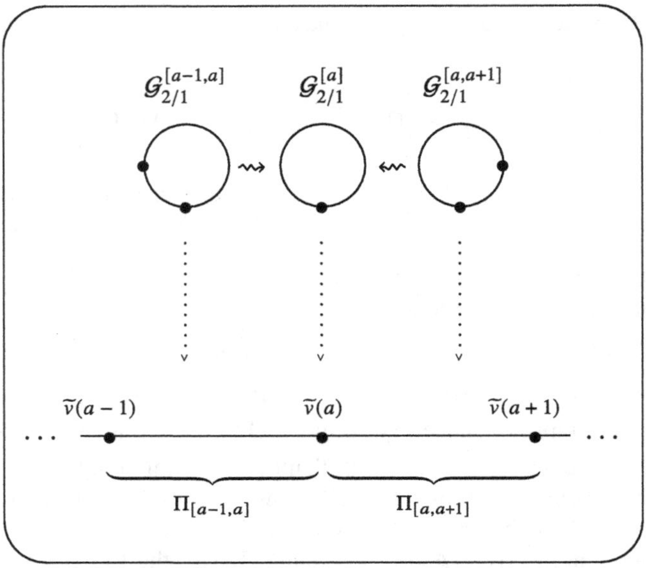

Fig. 4.3 $\mathcal{G}_{2/1}^{[a-1,a]}$, $\mathcal{G}_{2/1}^{[a]}$, and $\mathcal{G}_{2/1}^{[a,a+1]}$

[cf. Fig. 4.3]. This notation allows us to express the *graphicity* observed above in the following way:

The composites

$$\Pi_{[a]} \hookrightarrow \Pi_{\mathcal{G}} \xleftarrow{\sim} \Pi_1 \to \mathrm{Out}(\Pi_{2/1}) \xrightarrow{\sim} \mathrm{Out}(\Pi_{\mathcal{G}_{2/1}^{[a-1,a]}}) \xleftarrow{\sim} \mathrm{Out}(\Pi_{\mathcal{G}_{2/1}^{[a]}}),$$

$$\Pi_{[a]} \hookrightarrow \Pi_{\mathcal{G}} \xleftarrow{\sim} \Pi_1 \to \mathrm{Out}(\Pi_{2/1}) \xrightarrow{\sim} \mathrm{Out}(\Pi_{\mathcal{G}_{2/1}^{[a,a+1]}}) \xleftarrow{\sim} \mathrm{Out}(\Pi_{\mathcal{G}_{2/1}^{[a]}})$$

—where the third arrows in each line of the display arise from the outer action determined by the exact sequence $1 \to \Pi_{2/1} \to \Pi_2 \overset{p_{2/1}^{\Pi}}{\to} \Pi_1 \to 1$, the fourth arrows are the isomorphisms induced by the isomorphisms $\Pi_{2/1} \xrightarrow{\sim} \Pi_{\mathcal{G}_{2/1}^{[a-1,a]}}$ and $\Pi_{2/1} \xrightarrow{\sim} \Pi_{\mathcal{G}_{2/1}^{[a,a+1]}}$ fixed above, and the fifth arrows are the isomorphisms induced by the *natural specialization outer isomorphisms* [cf. [CbTpI], Definition 2.10]—*factor* through

$$\mathrm{Aut}(\mathcal{G}_{2/1}^{[a]}) \subseteq \mathrm{Out}(\Pi_{\mathcal{G}_{2/1}^{[a]}}).$$

Now we turn to the verification of the *surjectivity* of the homomorphism ρ_2^{brch}. Let $\alpha_v \in \mathrm{Glu}(\Pi_2)$ ($\subseteq \mathrm{Out}^{\mathrm{FC}}((\Pi_v)_2)^{\mathcal{G}\text{-node}}$). Write $(\alpha_v)_1 \in \mathrm{Glu}(\Pi_1)$ for the image of $\alpha_v \in \mathrm{Glu}(\Pi_2)$ via the injection of Lemma 4.10, (i). Let $\alpha_1 \in \mathrm{Aut}^{|\mathrm{Brch}(\mathcal{G})|}(\mathcal{G})$ be such that $\rho_1^{\mathrm{brch}}(\alpha_1) = (\alpha_v)_1 \in \mathrm{Glu}(\Pi_1)$ [cf. Theorem 4.2, (iii); Definition 4.11]. Now, by applying Lemma 4.12, (v), in the case where we take the pair "$(\widetilde{v}, \widetilde{w})$" to be

$(\widetilde{v}(0), \widetilde{v}(1))$, we obtain an outomorphism $\beta_{[0,1]} \stackrel{\text{def}}{=} \beta_{\widetilde{v}(0),\widetilde{v}(1)}[\alpha_1]$ [cf. Lemma 4.12, (v)] of $\Pi_2|_{[0,1]}$ [cf. the notation of the discussion preceding Claim 4.13.A]. Let

- $\widetilde{\beta}^{\dagger}_{[0,1]} \in \mathrm{Aut}(\Pi_2|_{[0,1]})$ be an automorphism that *lifts* $\beta_{[0,1]} \in \mathrm{Out}(\Pi_2|_{[0,1]})$ and *preserves* the subgroup $\Pi_{\widetilde{\pi}(0,1)} \subseteq \Pi_{[0,1]}$ [cf. condition (4) of Lemma 4.12, (v)] and
- $\widetilde{\widetilde{\gamma}}_{\infty} \in \Pi_2$ a *lifting* of $\widetilde{\gamma}_{\infty} \in \Pi_1$.

Then since [as is easily verified] $\Pi_2|_{[1,2]}$ [cf. the notation of the discussion preceding Claim 4.13.A] is the conjugate of $\Pi_2|_{[0,1]}$ by $\widetilde{\widetilde{\gamma}}_{\infty}$, by conjugating $\widetilde{\beta}^{\dagger}_{[0,1]}$ by the inner automorphism determined by $\widetilde{\widetilde{\gamma}}_{\infty}$, we obtain an automorphism $\widetilde{\beta}^{\dagger}_{[1,2]}$ of $\Pi_2|_{[1,2]}$, whose associated outomorphism we denote by $\beta_{[1,2]}$. Now we claim that the following assertion holds:

Claim 4.13.B: There exist automorphisms $\widetilde{\beta}_{[0,1]}, \widetilde{\beta}_{[1,2]}$ of $\Pi_2|_{[0,1]}, \Pi_2|_{[1,2]}$ that *lift* $\beta_{[0,1]}$, $\beta_{[1,2]}$, respectively, such that

(i) the outomorphisms of $\Pi_{2/1}$ ($\subseteq \Pi_2|_{[0,1]}, \Pi_2|_{[1,2]}$) determined by $\widetilde{\beta}_{[0,1]}, \widetilde{\beta}_{[1,2]}$ *coincide*;
(ii) the automorphism of $\Pi_{[0,1]}$ determined by the automorphism $\widetilde{\beta}_{[0,1]}$ *preserves* the subgroups $\Pi_{\widetilde{\pi}(0,1)}, \Pi_{[0]}, \Pi_{[1]} \subseteq \Pi_{[0,1]}$;
(iii) $\widetilde{\beta}_{[0,1]} = \widetilde{\beta}^{\dagger}_{[0,1]}$, and $\widetilde{\beta}_{[1,2]}$ is the post-composite of $\widetilde{\beta}^{\dagger}_{[1,2]}$ with an inner automorphism arising from an element of $\Pi_2|_{[1]}$.

Indeed, observe that there exist automorphisms $\widetilde{\beta}_{[0,1]}, \widetilde{\beta}_{[1,2]}$ [e.g., $\widetilde{\beta}^{\dagger}_{[0,1]}, \widetilde{\beta}^{\dagger}_{[1,2]}$] of $\Pi_2|_{[0,1]}, \Pi_2|_{[1,2]}$ that *lift* $\beta_{[0,1]}, \beta_{[1,2]}$, respectively, such that

- the outomorphisms $(\widetilde{\beta}_{[0,1]})_{2/1}, (\widetilde{\beta}_{[1,2]})_{2/1}$ of $\Pi_{2/1}$ determined by $\widetilde{\beta}_{[0,1]}, \widetilde{\beta}_{[1,2]}$ are *contained* in

$$\mathrm{Aut}^{|\mathrm{Brch}(\mathcal{G}^{[0,1]}_{2/1})|}(\mathcal{G}^{[0,1]}_{2/1}), \quad \mathrm{Aut}^{|\mathrm{Brch}(\mathcal{G}^{[1,2]}_{2/1})|}(\mathcal{G}^{[1,2]}_{2/1}) \quad (\subseteq \mathrm{Out}(\Pi_{2/1})),$$

  respectively, and,
- conditions (ii), (iii) of Claim 4.13.B are satisfied

[cf. the discussion of the final portion of Lemma 4.12, (v); Lemma 4.12, (v), (1); [CmbGC], Proposition 1.5, (i)]. In particular, it follows that, relative to the *specialization outer isomorphisms* $\Pi_{\mathcal{G}^{[1]}_{2/1}} \xrightarrow{\sim} \Pi_{\mathcal{G}^{[0,1]}_{2/1}}, \Pi_{\mathcal{G}^{[1]}_{2/1}} \xrightarrow{\sim} \Pi_{\mathcal{G}^{[1,2]}_{2/1}}$ that appeared in the discussion following the proof of Claim 4.13.A, together with the natural inclusion of [CbTpI], Proposition 2.9, (ii),

$$(\widetilde{\beta}_{[0,1]})_{2/1}, \quad (\widetilde{\beta}_{[1,2]})_{2/1} \in \mathrm{Aut}^{|\mathrm{Brch}(\mathcal{G}^{[1]}_{2/1})|}(\mathcal{G}^{[1]}_{2/1}) \; (\subseteq \mathrm{Out}(\Pi_{2/1})).$$

Moreover, it follows immediately from condition (3) of Lemma 4.12, (v), applied in the case of $\beta_{[0,1]}$, together with the definition of $\beta_{[1,2]}$, that the outomorphisms of the configuration space subgroup

$$\left(\Pi_2 \supseteq \Pi_2|_{[0,1]} \supseteq\right) \quad (\Pi_{\widetilde{v}(1)})_2 \quad \left(\subseteq \Pi_2|_{[1,2]} \subseteq \Pi_2\right)$$

associated to the vertex $\widetilde{v}(1)$ determined by $\beta_{[0,1]}$, $\beta_{[1,2]}$ *coincide* with α_v. Now let us recall from the above discussion that the composite

$$\Pi_{[1]} \hookrightarrow \Pi_1 \to \mathrm{Out}(\Pi_{2/1}) \xrightarrow{\sim} \mathrm{Out}(\Pi_{\mathcal{G}_{2/1}^{[1]}})$$

factors through

$$\mathrm{Aut}(\mathcal{G}_{2/1}^{[1]}) \subseteq \mathrm{Out}(\Pi_{\mathcal{G}_{2/1}^{[1]}}).$$

Thus, it follows immediately from the displayed exact sequence of Theorem 4.2, (iii) [cf. also Remark 4.9.1], that—after *possibly replacing* $\widetilde{\beta}_{[1,2]}$ by the post-composite of $\widetilde{\beta}_{[1,2]}$ with an inner automorphism arising from a suitable element of $\Pi_2|_{[1]}$ [which does *not affect* the validity of conditions (ii), (iii) of Claim 4.13.B]—if we write

$$\delta \stackrel{\mathrm{def}}{=} (\widetilde{\beta}_{[0,1]})_{2/1} \circ (\widetilde{\beta}_{[1,2]})_{2/1}^{-1} \in \mathrm{Aut}^{|\mathrm{Brch}(\mathcal{G}_{2/1}^{[1]})|}(\mathcal{G}_{2/1}^{[1]}) \ (\subseteq \mathrm{Out}(\Pi_{2/1})),$$

then it holds that $\delta \in \mathrm{Dehn}(\mathcal{G}_{2/1}^{[1]})$.

Next, let us observe that, for $a \in \{0, 1\}$, since $\widetilde{\beta}_{[a,a+1]}$ *preserves* the $\Pi_{2/1}$-conjugacy class of cuspidal inertia subgroups associated to the *diagonal cusp* [cf. condition (3) of Lemma 4.12, (v)], it follows from a similar argument to the argument applied in the proof of [CmbCsp], Proposition 1.2, (iii), that the outomorphism $(\widetilde{\beta}_{[a,a+1]})_{\{2\}}$ of $\Pi_{\{2\}}$ induced by $\widetilde{\beta}_{[a,a+1]}$ on the quotient

$$\Pi_{\mathcal{G}_{2/1}^{[1]}} \xleftarrow{\sim} \Pi_{2/1} \hookrightarrow \Pi_2 \overset{p^{\Pi}_{\{1,2\}/\{2\}}}{\twoheadrightarrow} \Pi_{\{2\}}$$

is *compatible*, relative to the natural inclusion $\Pi_{[a,a+1]} \hookrightarrow \Pi_1 \xrightarrow{\sim} \Pi_{\{2\}}$, with the outomorphism $\alpha_1|_{\Pi_{[a,a+1]}}$ [cf. condition (4) of Lemma 4.12, (v)]. Since an element of $\mathrm{Aut}^{|\mathrm{Brch}(\mathcal{G})|}(\mathcal{G})$ is *completely determined* by its restriction to $\mathrm{Aut}(\mathcal{G}_{[a,a+1]})$ [cf. [CbTpI], Definition 4.4; [CbTpI], Remark 4.8.1], we thus conclude that, relative to the natural outer isomorphisms $\Pi_{\{2\}} \xrightarrow{\sim} \Pi_1 \xrightarrow{\sim} \Pi_{\mathcal{G}}$, it holds that

$$(\widetilde{\beta}_{[a,a+1]})_{\{2\}} = \alpha_1.$$

In particular, it follows that the element of $\mathrm{Aut}^{|\mathrm{Brch}(\mathcal{G})|}(\mathcal{G})$ induced by $\delta \in \mathrm{Aut}^{|\mathrm{Brch}(\mathcal{G}_{2/1}^{[1]})|}(\mathcal{G}_{2/1}^{[1]})$ on the quotient $\Pi_{\mathcal{G}_{2/1}^{[1]}} \xleftarrow{\sim} \Pi_{2/1} \hookrightarrow \Pi_2 \overset{p^{\Pi}_{\{1,2\}/\{2\}}}{\twoheadrightarrow} \Pi_{\{2\}} \xrightarrow{\sim} \Pi_{\mathcal{G}}$ is *trivial*. On the other hand, let us observe that one verifies easily from [CbTpI],

Theorem 4.8, (iii), (iv), that this composite $\Pi_{\mathcal{G}^{[1]}_{2/1}} \overset{\sim}{\leftarrow} \Pi_{2/1} \hookrightarrow \Pi_2 \overset{p^{\Pi}_{\{1,2\}/\{2\}}}{\twoheadrightarrow} \Pi_{\{2\}} \overset{\sim}{\rightarrow} \Pi_{\mathcal{G}}$ determines an *isomorphism*

$$\mathrm{Dehn}(\mathcal{G}^{[1]}_{2/1}) \overset{\sim}{\longrightarrow} \mathrm{Dehn}(\mathcal{G}).$$

Thus, we conclude that δ is the *identity outomorphism* of $\Pi_{2/1}$. In particular, condition (i) of Claim 4.13.B is satisfied. This completes the proof of Claim 4.13.B.

Next, let us *fix* an automorphism $\widetilde{\alpha}_1 \in \mathrm{Aut}(\Pi_1)$ that *lifts* $\alpha_1 \in \mathrm{Aut}^{|\mathrm{Brch}(\mathcal{G})|}(\mathcal{G}) \subseteq \mathrm{Out}(\Pi_{\mathcal{G}}) \overset{\sim}{\leftarrow} \mathrm{Out}(\Pi_1)$ and *preserves* the subgroup $\Pi_{\widetilde{n}(0,1)} \subseteq \Pi_1$ [hence also the subgroups $\Pi_{[0]}, \Pi_{[1]}, \Pi_{[0,1]} \subseteq \Pi_1$], and whose restriction to $\Pi_{[0,1]} \subseteq \Pi_1$ *coincides* with the automorphism of $\Pi_{[0,1]}$ determined by the automorphism $\widetilde{\beta}_{[0,1]}$ of $\Pi_2|_{[0,1]}$. [One verifies easily that such an $\widetilde{\alpha}_1$ always exists—cf. Lemma 4.12, (v), (4); Claim 4.13.B, (ii).] Write $\beta_{2/1} \in \mathrm{Out}(\Pi_{2/1})$ for the outomorphism of $\Pi_{2/1} \subseteq \Pi_2|_{[0,1]}$ determined by $\widetilde{\beta}_{[0,1]}$ [or, equivalently, $\widetilde{\beta}_{[1,2]}$—cf. Claim 4.13.B, (i)]. Now we claim that the following assertion holds:

Claim 4.13.C: Write $\rho \colon \Pi_1 \to \mathrm{Out}(\Pi_{2/1})$ for the homomorphism determined by the exact sequence $1 \to \Pi_{2/1} \to \Pi_2 \overset{p^{\Pi}_{2/1}}{\to} \Pi_1 \to 1$. Then

$$\rho(\widetilde{\alpha}_1(\widetilde{\gamma}_\infty)) = \beta_{2/1} \circ \rho(\widetilde{\gamma}_\infty) \circ \beta_{2/1}^{-1} \in \mathrm{Out}(\Pi_{2/1}).$$

Indeed, let us first observe that it follows from conditions (i) and (iii) of Claim 4.13.B, together with the definition of $\widetilde{\beta}^{\dagger}_{[1,2]}$, that there exists an element $\epsilon \in \Pi_{[1]}$ such that

$$\rho(\widetilde{\gamma}_\infty) \circ \beta_{2/1} \circ \rho(\widetilde{\gamma}_\infty^{-1}) \circ \beta_{2/1}^{-1} = \rho(\epsilon^{-1}) \qquad (*_1).$$

Next, let us observe that if we write

$$\eta \overset{\mathrm{def}}{=} \widetilde{\alpha}_1(\widetilde{\gamma}_\infty) \cdot \widetilde{\gamma}_\infty^{-1} \in \Pi_1 \qquad (*_2),$$

then it follows immediately from the *commensurable terminality* of $\Pi_{[1]}$ in Π_1 [cf. [CmbGC], Proposition 1.2, (ii)], together with our choices of $\widetilde{\alpha}_1$ and $\widetilde{\gamma}_\infty$—which imply that

$$\begin{aligned}
\widetilde{\alpha}_1(\widetilde{\gamma}_\infty) \cdot \widetilde{\gamma}_\infty^{-1} \cdot \Pi_{[1]} \cdot \widetilde{\gamma}_\infty \cdot \widetilde{\alpha}_1(\widetilde{\gamma}_\infty)^{-1} &= \widetilde{\alpha}_1(\widetilde{\gamma}_\infty) \cdot \Pi_{[0]} \cdot \widetilde{\alpha}_1(\widetilde{\gamma}_\infty)^{-1} \\
&= \widetilde{\alpha}_1(\widetilde{\gamma}_\infty) \cdot \widetilde{\alpha}_1(\Pi_{[0]}) \cdot \widetilde{\alpha}_1(\widetilde{\gamma}_\infty)^{-1} \\
&= \widetilde{\alpha}_1(\widetilde{\gamma}_\infty \cdot \Pi_{[0]} \cdot \widetilde{\gamma}_\infty^{-1}) \\
&= \widetilde{\alpha}_1(\Pi_{[1]}) \\
&= \Pi_{[1]}
\end{aligned}$$

—that $\eta \in \Pi_{[1]}$. Thus, to verify Claim 4.13.C, it suffices to verify that

$$\rho(\epsilon) = \rho(\eta).$$

To this end, let $\zeta \in \Pi_{[0]}$. Then, by our choice of $\widetilde{\gamma}_\infty$, it follows that $\widetilde{\gamma}_\infty \cdot \zeta \cdot \widetilde{\gamma}_\infty^{-1} \in \Pi_{[1]}$. In particular, since the outomorphism $\beta_{2/1}$ arises from an *automorphism* $\widetilde{\beta}_{[0,1]}$ *of* $\Pi_2|_{[0,1]}$, which is an automorphism over the restriction of $\widetilde{\alpha}_1$ to $\Pi_{[0,1]}$, it follows immediately that

$$\beta_{2/1} \circ \rho(\zeta) = \rho(\widetilde{\alpha}_1(\zeta)) \circ \beta_{2/1} \qquad (*_3).$$

$$\beta_{2/1} \circ \rho(\widetilde{\gamma}_\infty \cdot \zeta \cdot \widetilde{\gamma}_\infty^{-1}) = \rho(\widetilde{\alpha}_1(\widetilde{\gamma}_\infty \cdot \zeta \cdot \widetilde{\gamma}_\infty^{-1})) \circ \beta_{2/1} \qquad (*_4).$$

Thus, if we write

$$\Theta_\epsilon \stackrel{\text{def}}{=} \rho(\epsilon \cdot \widetilde{\gamma}_\infty \cdot \widetilde{\alpha}_1(\zeta) \cdot \widetilde{\gamma}_\infty^{-1} \cdot \epsilon^{-1}) \circ \beta_{2/1} \in \mathrm{Out}(\Pi_{2/1}),$$

$$\Theta_\eta \stackrel{\text{def}}{=} \rho(\eta \cdot \widetilde{\gamma}_\infty \cdot \widetilde{\alpha}_1(\zeta) \cdot \widetilde{\gamma}_\infty^{-1} \cdot \eta^{-1}) \circ \beta_{2/1} \in \mathrm{Out}(\Pi_{2/1}),$$

then

$$\begin{aligned}
\Theta_\epsilon &= \rho(\epsilon \cdot \widetilde{\gamma}_\infty \cdot \widetilde{\alpha}_1(\zeta)) \circ \beta_{2/1} \circ \rho(\widetilde{\gamma}_\infty^{-1}) && [\text{cf. } (*_1)]\\
&= \rho(\epsilon \cdot \widetilde{\gamma}_\infty) \circ \beta_{2/1} \circ \rho(\zeta \cdot \widetilde{\gamma}_\infty^{-1}) && [\text{cf. } (*_3)]\\
&= \beta_{2/1} \circ \rho(\widetilde{\gamma}_\infty \cdot \zeta \cdot \widetilde{\gamma}_\infty^{-1}) && [\text{cf. } (*_1)]\\
&= \rho(\widetilde{\alpha}_1(\widetilde{\gamma}_\infty \cdot \zeta \cdot \widetilde{\gamma}_\infty^{-1})) \circ \beta_{2/1} && [\text{cf. } (*_4)]\\
&= \Theta_\eta && [\text{cf. } (*_2)]
\end{aligned}$$

—which thus implies that $\rho(\eta^{-1} \cdot \epsilon)$ *commutes* with $\rho(\widetilde{\gamma}_\infty \cdot \widetilde{\alpha}_1(\zeta) \cdot \widetilde{\gamma}_\infty^{-1})$. In particular, since $\widetilde{\gamma}_\infty \cdot \widetilde{\alpha}_1(\Pi_{[0]}) \cdot \widetilde{\gamma}_\infty^{-1} = \widetilde{\gamma}_\infty \cdot \Pi_{[0]} \cdot \widetilde{\gamma}_\infty^{-1} = \Pi_{[1]}$, by allowing "$\zeta$" to *vary* among the elements of $\Pi_{[0]}$, it follows that $\rho(\eta^{-1} \cdot \epsilon)$ *centralizes* $\rho(\Pi_{[1]})$. On the other hand, it follows from [Asd], Theorem 1; [Asd], the Remark following the proof of Theorem 1, that ρ is *injective*. Thus, since $\epsilon, \eta \in \Pi_{[1]}$, we conclude that $\eta^{-1} \cdot \epsilon \in Z(\Pi_{[1]}) = \{1\}$ [cf. [CmbGC], Remark 1.1.3]. This completes the proof of Claim 4.13.C.

Now let us recall that the outomorphism $\beta_{2/1}$ of $\Pi_{2/1}$ of Claim 4.13.C arises from an *automorphism* $\widetilde{\beta}_{[0,1]}$ *of* $\Pi_2|_{[0,1]}$. Thus, it follows immediately from Claims 4.13.A, 4.13.C that the outomorphism $\beta_{2/1}$ of $\Pi_{2/1}$ is *compatible* with the automorphism $\widetilde{\alpha}_1 \in \mathrm{Aut}(\Pi_1)$ relative to the homomorphism $\Pi_1 \to \mathrm{Out}(\Pi_{2/1})$ determined by the exact sequence $1 \to \Pi_{2/1} \to \Pi_2 \xrightarrow{p^{\Pi}_{2/1}} \Pi_1 \to 1$. In particular—by considering the natural isomorphism $\Pi_2 \xrightarrow{\sim} \Pi_{2/1} \overset{\text{out}}{\rtimes} \Pi_1$ [cf. the discussion entitled "*Topological groups*" in [CbTpI], §0]—we conclude that the outomorphism $\beta_{2/1} \in \mathrm{Out}(\Pi_{2/1})$ *extends* to an outomorphism α_2 of Π_2. On the other hand, it follows immediately from the various definitions involved that $\rho_2^{\mathrm{brch}}(\alpha_2) = \alpha_v \in \mathrm{Glu}(\Pi_2)$ [cf. condition (3) of Lemma 4.12, (v)], and that $\alpha_2 \in \mathrm{Out}^{\mathrm{FC}}(\Pi_2)^{\mathrm{brch}}$ [cf. condition (2) of Lemma 4.12, (v); [CmbCsp], Proposition 1.2, (i)]. This completes the proof of Lemma 4.13 in the case where $\mathcal{G}$ is *cyclically primitive*, hence also of Lemma 4.13.

□

Theorem 4.14 (Glueability of Combinatorial Cuspidalizations) *Let* (g, r) *be a pair of nonnegative integers such that* $2g - 2 + r > 0$*;* n *a positive integer;* Σ *a set of prime numbers which is either equal to the set of all prime numbers or of cardinality one;* k *an algebraically closed field of characteristic* $\notin \Sigma$*;* $(\mathrm{Spec}\, k)^{\log}$ *the log scheme obtained by equipping* $\mathrm{Spec}\, k$ *with the log structure determined by the fs chart* $\mathbb{N} \to k$ *that maps* $1 \mapsto 0$*;* $X^{\log} = X_1^{\log}$ *a* **stable log curve** *of type* (g, r) *over* $(\mathrm{Spec}\, k)^{\log}$*. Write* $\mathcal{G}$ *for the semi-graph of anabelioids of pro-*Σ *PSC-type determined by the stable log curve* $X^{\log}$*. For each positive integer* i*, write* $X_i^{\log}$ *for the* i*-th* **log configuration space** *of the stable log curve* $X^{\log}$ *[cf. the discussion entitled "Curves" in "Notations and Conventions"];* Π_i *for the maximal pro-*Σ *quotient of the kernel of the natural surjection* $\pi_1(X_i^{\log}) \twoheadrightarrow \pi_1((\mathrm{Spec}\, k)^{\log})$*. Then the following hold:*

(i) There exists a natural **commutative diagram** *of profinite groups*

$$\begin{array}{ccc} \mathrm{Out}^{\mathrm{FC}}(\Pi_{n+1})^{\mathrm{brch}} & \xrightarrow{\rho_{n+1}^{\mathrm{brch}}} & \mathrm{Glu}(\Pi_{n+1}) \\ \downarrow & & \downarrow \\ \mathrm{Out}^{\mathrm{FC}}(\Pi_n)^{\mathrm{brch}} & \xrightarrow{\rho_n^{\mathrm{brch}}} & \mathrm{Glu}(\Pi_n) \end{array}$$

[cf. Definition 4.6, (i); Definition 4.9; Lemma 4.10, (i); Definition 4.11]— where the vertical arrows are **injective***.*

(ii) The closed subgroup $\mathrm{Dehn}(\mathcal{G}) \subseteq (\mathrm{Aut}(\mathcal{G}) \subseteq) \mathrm{Out}(\Pi_1)$ *[cf. [CbTpI], Definition 4.4] is* **contained** *in the image of the injection* $\mathrm{Out}^{\mathrm{FC}}(\Pi_n)^{\mathrm{brch}} \hookrightarrow \mathrm{Out}^{\mathrm{FC}}(\Pi_1)^{\mathrm{brch}}$ *[cf. the left-hand vertical arrows of the diagrams of (i), for varying* n*]. Thus, one may regard* $\mathrm{Dehn}(\mathcal{G})$ *as a closed subgroup of* $\mathrm{Out}^{\mathrm{FC}}(\Pi_n)^{\mathrm{brch}}$*, i.e.,* $\mathrm{Dehn}(\mathcal{G}) \subseteq \mathrm{Out}^{\mathrm{FC}}(\Pi_n)^{\mathrm{brch}}$.

(iii) The homomorphism $\rho_n^{\mathrm{brch}} : \mathrm{Out}^{\mathrm{FC}}(\Pi_n)^{\mathrm{brch}} \to \mathrm{Glu}(\Pi_n)$ *of (i) and the inclusion* $\mathrm{Dehn}(\mathcal{G}) \hookrightarrow \mathrm{Out}^{\mathrm{FC}}(\Pi_n)^{\mathrm{brch}}$ *of (ii) fit into an* **exact sequence** *of profinite groups*

$$1 \longrightarrow \mathrm{Dehn}(\mathcal{G}) \longrightarrow \mathrm{Out}^{\mathrm{FC}}(\Pi_n)^{\mathrm{brch}} \xrightarrow{\rho_n^{\mathrm{brch}}} \mathrm{Glu}(\Pi_n) \longrightarrow 1.$$

In particular, the commutative diagram of (i) is **cartesian***, and the horizontal arrows of this diagram are* **surjective***.*

Proof Assertion (i) follows immediately from Lemma 4.10, (i), together with the *injectivity portion* of [NodNon], Theorem B. Assertion (ii) follows immediately from Proposition 3.24, (ii); Theorem 4.2, (i).

Finally, we verify assertion (iii). First, we claim that the following assertion holds:

Claim 4.14.A: $\mathrm{Ker}(\rho_n^{\mathrm{brch}}) = \mathrm{Dehn}(\mathcal{G})$ [cf. assertion (ii)].

Indeed, it follows immediately from Theorem 4.2, (iii) [cf. also Remark 4.9.1], together with assertion (i), that we have a natural commutative diagram

$$\begin{array}{ccccccccc} 1 & \longrightarrow & \mathrm{Ker}(\rho_n^{\mathrm{brch}}) & \longrightarrow & \mathrm{Out}^{\mathrm{FC}}(\Pi_n)^{\mathrm{brch}} & \xrightarrow{\rho_n^{\mathrm{brch}}} & \mathrm{Glu}(\Pi_n) & & \\ & & \downarrow & & \downarrow & & \downarrow & & \\ 1 & \longrightarrow & \mathrm{Dehn}(\mathcal{G}) & \longrightarrow & \mathrm{Out}^{\mathrm{FC}}(\Pi_1)^{\mathrm{brch}} & \xrightarrow{\rho_1^{\mathrm{brch}}} & \mathrm{Glu}(\Pi_1) & \longrightarrow & 1 \end{array}$$

—where the horizontal sequences are *exact*, and the vertical arrows are *injective*. Thus, Claim 4.14.A follows immediately. In particular, to complete the verification of assertion (iii), it suffices to verify the *surjectivity* of ρ_n^{brch}. The remainder of the proof of assertion (iii) is devoted to verifying this *surjectivity*.

Next, we claim that the following assertion holds:

Claim 4.14.B: If $n = 2$, then ρ_n^{brch} is *surjective*.

We verify Claim 4.14.B by *induction on* $\#\mathrm{Node}(\mathcal{G})$. If $\#\mathrm{Node}(\mathcal{G}) = 0$, then Claim 4.14.B is immediate. If $\#\mathrm{Node}(\mathcal{G}) = 1$, then Claim 4.14.B follows from Lemma 4.13. Now suppose that $\#\mathrm{Node}(\mathcal{G}) > 1$, and that the *induction hypothesis* is in force. Let $(\alpha_v)_{v \in \mathrm{Vert}(\mathcal{G})} \in \mathrm{Glu}(\Pi_2)$. Write $((\alpha_v)_1)_{v \in \mathrm{Vert}(\mathcal{G})} \in \mathrm{Glu}(\Pi_1)$ for the element of $\mathrm{Glu}(\Pi_1)$ determined by $(\alpha_v)_{v \in \mathrm{Vert}(\mathcal{G})}$ [i.e., the image of $(\alpha_v)_{v \in \mathrm{Vert}(\mathcal{G})}$ via the right-hand vertical arrow of the diagram of assertion (i) in the case where $n = 1$]. Let $e \in \mathrm{Node}(\mathcal{G})$. Write $\mathbb{H}$ for the *unique* sub-semi-graph of *PSC-type* [cf. [CbTpI], Definition 2.2, (i)] of the underlying semi-graph of $\mathcal{G}$ whose set of vertices is $\mathcal{V}(e)$. Then one verifies easily that $S \stackrel{\mathrm{def}}{=} \mathrm{Node}(\mathcal{G}|_{\mathbb{H}}) \setminus \{e\}$ [cf. [CbTpI], Definition 2.2, (ii)] is *not of separating type* [cf. [CbTpI], Definition 2.5, (i)] as a subset of $\mathrm{Node}(\mathcal{G}|_{\mathbb{H}})$. Thus, since $(\mathcal{G}|_{\mathbb{H}})_{\succ S}$ [cf. [CbTpI], Definition 2.5, (ii)] has *precisely one* node, and $(\alpha_v)_{v \in \mathcal{V}(e)}$ may be regarded as an element of $\mathrm{Glu}((\Pi_{\mathbb{H},S})_2)$—where we use the notation $(\Pi_{\mathbb{H},S})_2$ to denote a configuration space subgroup of Π_2 associated to $(\mathbb{H}, S)$ [cf. Definition 4.3], to which the notation "Glu(−)" is applied in the evident sense—it follows from Lemma 4.13 that there exists an outomorphism $\beta_{\mathbb{H},S}$ of $(\Pi_{\mathbb{H},S})_2 \subseteq \Pi_2$ that *lifts* $(\alpha_v)_{v \in \mathcal{V}(e)} \in \mathrm{Glu}((\Pi_{\mathbb{H},S})_2)$.

Next, let us observe that it follows immediately from the various definitions involved that

$$\gamma \stackrel{\mathrm{def}}{=} (\beta_{\mathbb{H},S}, (\alpha_v)_{v \notin \mathcal{V}(e)}) \in \mathrm{Out}((\Pi_{\mathbb{H},S})_2) \times \prod_{v \notin \mathcal{V}(e)} \mathrm{Out}((\Pi_v)_2)$$

may be regarded as an element of the "Glu(Π_2)" that occurs in the case where we take the stable log curve "$X^{\log}$" to be a stable log curve over $(\mathrm{Spec}\, k)^{\log}$ obtained by *deforming* the node corresponding to e. Thus, since the number of nodes of such a stable log curve is $= \#\mathrm{Node}(\mathcal{G}) - 1 < \#\mathrm{Node}(\mathcal{G})$, by applying the *induction hypothesis*, we conclude that the above γ arises from an outomorphism $\alpha_\gamma \in \mathrm{Out}^{\mathrm{FC}}(\Pi_2)^{\mathrm{brch}}$. On the other hand, it follows immediately from the various definitions involved that the image of α_γ via ρ_2^{brch} *coincides* with $(\alpha_v)_{v \in \mathrm{Vert}(\mathcal{G})}$. This completes the proof of Claim 4.14.B.

Finally, we verify the *surjectivity* of ρ_n^{brch} [for arbitrary n] by *induction on* n. If $n \leq 2$, then the *surjectivity* of ρ_n^{brch} follows from Theorem 4.2, (iii) [cf. also Remark 4.9.1], Claim 4.14.B. Now suppose that $n \geq 3$, and that the *induction hypothesis* is in force. Let $(\alpha_v)_{v \in \mathrm{Vert}(\mathcal{G})} \in \mathrm{Glu}(\Pi_n)$. First, let us observe that it follows from the *induction hypothesis* that there exists an element $\alpha_{n-1} \in \mathrm{Out}^{\mathrm{FC}}(\Pi_{n-1})^{\mathrm{brch}}$ such that $\rho_{n-1}^{\mathrm{brch}}(\alpha_{n-1})$ *coincides with* the element of $\mathrm{Glu}(\Pi_{n-1})$ determined by $(\alpha_v)_{v \in \mathrm{Vert}(\mathcal{G})} \in \mathrm{Glu}(\Pi_n)$ [cf. assertion (i)]. Let $\widetilde{\alpha}_{n-1}$ be an automorphism of Π_{n-1} that lifts α_{n-1}. Write $\alpha_{n-1/n-2}$ for the outomorphism of $\Pi_{n-1/n-2}$ determined by $\widetilde{\alpha}_{n-1}$ and $\widetilde{\alpha}_{n-2}$ for the automorphism of Π_{n-2} determined by $\widetilde{\alpha}_{n-1}$.

Next, let us observe that one verifies easily from the various definitions involved that $\Pi_{n/n-2} \subseteq \Pi_n$ may be regarded as the "Π_2" associated to some stable log curve "$X^{\log}$" over $(\mathrm{Spec}\, k)^{\log}$. Moreover, this stable log curve may be taken to be a *geometric fiber* of the sort discussed in Definition 3.1, (iii), in the case of the projection $p_{n-1/n-2}^{\log}$, relative to a point "$x \in X_n(k)$" that maps to the interior of the *same* irreducible component of $X^{\log}$, relative to the n projections to $X^{\log}$. In particular, by fixing such a stable log curve, together with a *suitable choice* of lifting $\widetilde{\alpha}_{n-1}$ [cf. Theorem 4.7], it makes sense to speak of $\mathrm{Glu}(\Pi_{n/n-2})$. Moreover, it follows immediately from our choice of "x" that *every configuration space subgroup* that appears in the definition [cf. Definition 4.9, (ii)] of $\mathrm{Glu}(\Pi_{n/n-2})$ *either*

- occurs as the *intersection* with $\Pi_{n/n-2}$ of some configuration space subgroup that appears in the definition [cf. Definition 4.9, (iii)] of $\mathrm{Glu}(\Pi_n)$ *or*
- projects *isomorphically*, via the projection $\Pi_n \twoheadrightarrow \Pi_2$ to the factors labeled n and $n-1$, to a configuration space subgroup of Π_2, i.e., a configuration space subgroup that appears in the definition [cf. Definition 4.9, (ii)] of $\mathrm{Glu}(\Pi_2)$.

In particular, *every tripod* that appears in the definition [cf. Definition 4.9, (ii)] of $\mathrm{Glu}(\Pi_{n/n-2})$ occurs as a tripod of a configuration space subgroup that appears *either* in the definition [cf. Definition 4.9, (iii)] of $\mathrm{Glu}(\Pi_n)$ *or* in the definition [cf. Definition 4.9, (ii)] of $\mathrm{Glu}(\Pi_2)$. Moreover, it follows from Theorem 4.7; Lemma 3.2, (iv); Lemma 4.8, (i), that the various α_v's *preserve* the conjugacy classes of these configuration space subgroups and tripods—as well as each conjugacy class of cuspidal inertia subgroups of each of these tripods!—that appear in the definition [cf. Definition 4.9, (ii)] of $\mathrm{Glu}(\Pi_{n/n-2})$. Thus, we conclude from Theorem 3.18, (ii), together with Definition 4.9, (iii), in the case of $\mathrm{Glu}(\Pi_n)$, and Definition 4.9, (ii), in the case of $\mathrm{Glu}(\Pi_2)$, that $(\alpha_v)_{v \in \mathrm{Vert}(\mathcal{G})}$ determines an element $\in \mathrm{Glu}(\Pi_{n/n-2})$, hence, by Claim 4.14.B, an element

$$\alpha_{n/n-2} \in \mathrm{Out}^{\mathrm{FC}}(\Pi_{n/n-2})$$

that lifts the element $\alpha_{n-1/n-2} \in \mathrm{Out}(\Pi_{n-1/n-2})$.

Now we claim that the following assertion holds:

Claim 4.14.C: This outomorphism $\alpha_{n/n-2}$ of $\Pi_{n/n-2}$ is *compatible* with the automorphism $\widetilde{\alpha}_{n-2}$ of Π_{n-2} relative to the homomorphism $\Pi_{n-2} \to \mathrm{Out}(\Pi_{n/n-2})$ induced by the natural exact sequence of profinite groups

$$1 \longrightarrow \Pi_{n-2/n} \longrightarrow \Pi_n \stackrel{p^{\Pi}_{n/n-2}}{\longrightarrow} \Pi_{n-2} \longrightarrow 1.$$

Indeed, this follows immediately from the corresponding fact for $\alpha_{n-1/n-2}$ [which follows from the existence of $\widetilde{\alpha}_{n-1}$], together with the *injectivity* of the natural homomorphism $\mathrm{Out}^{\mathrm{FC}}(\Pi_{n/n-2}) \to \mathrm{Out}^{\mathrm{FC}}(\Pi_{n-1/n-2})$ [cf. [NodNon], Theorem B]. This completes the proof of Claim 4.14.C.

Thus, by applying Claim 4.14.C and the natural isomorphism $\Pi_n \stackrel{\sim}{\to} \Pi_{n/n-2} \stackrel{\mathrm{out}}{\rtimes} \Pi_{n-2}$ [cf. the discussion entitled "*Topological groups*" in [CbTpI], §0], we obtain an outomorphism α_n of Π_n that lifts the outomorphism α_{n-1} of Π_{n-1}. Thus, it follows immediately from Lemma 4.10, (i), that $\rho^{\mathrm{brch}}_n(\alpha_n) = (\alpha_v)_{v \in \mathrm{Vert}(\mathcal{G})}$. This completes the proof of the *surjectivity* of ρ^{brch}_n, hence also of assertion (iii). □

Remark 4.14.1 In the notation of Theorem 4.14, observe that the data of collections of smooth log curves that [by gluing at prescribed cusps] give rise to a stable log curve whose associated semi-graph of anabelioids [of pro-Σ PSC-type] is isomorphic to $\mathcal{G}$ form a *smooth, connected* moduli stack. In particular, by considering a suitable *path* in the *étale fundamental groupoid* of this moduli stack, one verifies immediately that one may reduce the verification of an *"isomorphism version"*—i.e., concerning PFC-admissible [cf. [CbTpI], Definition 1.4, (iii)] outer isomorphisms between the pro-Σ fundamental groups of the configuration spaces associated to two *a priori distinct* stable log curves "$X^{\log}$" and "$Y^{\log}$"—of Theorem 4.14 to the *"automorphism version"* given in Theorem 4.14 [cf. [CmbCsp], Remark 4.1.4]. A similar statement may be made concerning Theorem 4.7. We leave the routine details to the interested reader. In the present monograph, we restricted our attention to the "automorphism versions" of these results in order to simplify the [already somewhat complicated!] notation.

Remark 4.14.2 One may regard [CmbCsp], Corollary 3.3, as a *special case* of the *surjectivity* of ρ^{brch}_n discussed in Theorem 4.14, i.e., the case in which $X^{\log}$ is obtained by gluing a tripod to a smooth log curve along a cusp of the smooth log curve.

Corollary 4.15 (Surjectivity Result) *In the notation of Theorem 3.16, suppose that* $n \geq 3$*. If* $r = 0$*, then we suppose further that* $n \geq 4$*. Then the tripod homomorphism*

$$\mathfrak{T}_{\Pi^{\mathrm{tpd}}} : \mathrm{Out}^{\mathrm{F}}(\Pi_n) \longrightarrow \mathrm{Out}^{\mathrm{C}}(\Pi^{\mathrm{tpd}})^{\Delta+}$$

[cf. Definition 3.19] is **surjective**.

Proof Let $\alpha \in \mathrm{Out}^{\mathrm{C}}(\Pi^{\mathrm{tpd}})^{\Delta+}$. First, let us observe that—by considering a suitable stable log curve of type (g, r) over $(\mathrm{Spec}\, k)^{\log}$ and applying a suitable *specialization isomorphism* [cf. Proposition 3.24, (i); the discussion preceding [CmbCsp], Definition 2.1, as well as [CbTpI], Remark 5.6.1]—to verify Corollary 4.15, we may assume without loss of generality that $\mathcal{G}$ is *totally degenerate* [cf. [CbTpI], Definition 2.3, (iv)], i.e., that every vertex of $\mathcal{G}$ is a tripod of $X_n^{\log}$ [cf. Definition 3.1, (v)]. Then since $\alpha \in \mathrm{Out}^{\mathrm{C}}(\Pi^{\mathrm{tpd}})^{\Delta+}$, it follows immediately from [CmbCsp], Corollary 4.2, (i), (ii) [cf. also [CmbCsp], Definition 1.11, (i)], that there exists an element $\alpha_n \in \mathrm{Out}^{\mathrm{FC}}(\Pi_n^{\mathrm{tpd}})$—where we write Π_n^{tpd} for the "Π_n" that occurs in the case where we take "$X^{\log}$" to be a *tripod*—such that α arises as the image of α_n via the natural injection $\mathrm{Out}^{\mathrm{FC}}(\Pi_n^{\mathrm{tpd}}) \hookrightarrow \mathrm{Out}^{\mathrm{FC}}(\Pi^{\mathrm{tpd}})$ of [NodNon], Theorem B. Thus, it follows immediately from Theorem 4.14, (iii), that there exists an element $\beta \in \mathrm{Out}^{\mathrm{FC}}(\Pi_n)^{\mathrm{brch}}$ that *lifts*—relative to ρ_n^{brch}—the element of $\mathrm{Glu}(\Pi_n)$ [cf. Theorems 3.16, (v); 3.18, (ii)] determined by $\alpha_n \in \mathrm{Out}^{\mathrm{FC}}(\Pi_n^{\mathrm{tpd}})$. [Here, recall that we have assumed that $\mathcal{G}$ is *totally degenerate*.] Finally, it follows from Theorems 3.16, (v); 3.18, (ii), that $\mathfrak{T}_{\Pi^{\mathrm{tpd}}}(\beta) = \alpha$, i.e., that α is *contained* in the image of $\mathfrak{T}_{\Pi^{\mathrm{tpd}}}$. This completes the proof of Corollary 4.15. □

Corollary 4.16 (Absolute Anabelian Cuspidalization for Stable Log Curves over Finite Fields) *Let* p*,* l_X*,* l_Y *be prime numbers such that* $p \notin \{l_X, l_Y\}$*;* (g_X, r_X)*,* (g_Y, r_Y) *pairs of nonnegative integers such that* $2g_X - 2 + r_X$*,* $2g_Y - 2 + r_Y > 0$*;* k_X*,* k_Y **finite fields** *of characteristic* p*;* $\overline{k}_X$*,* $\overline{k}_Y$ *algebraic closures of* k_X*,* k_Y*;* $(\mathrm{Spec}\, k_X)^{\log}$*,* $(\mathrm{Spec}\, k_Y)^{\log}$ *the log schemes obtained by equipping* $\mathrm{Spec}\, k_X$*,* $\mathrm{Spec}\, k_Y$ *with the log structures determined by the fs charts* $\mathbb{N} \to k_X$*,* $\mathbb{N} \to k_Y$ *that map* $1 \mapsto 0$*;* $X^{\log}$*,* $Y^{\log}$ **stable log curves** *of type* (g_X, r_X)*,* (g_Y, r_Y) *over* $(\mathrm{Spec}\, k_X)^{\log}$*,* $(\mathrm{Spec}\, k_Y)^{\log}$*;*

$$G_{k_X}^{\log} \stackrel{\mathrm{def}}{=} \pi_1((\mathrm{Spec}\, k_X)^{\log}) \twoheadrightarrow G_{k_X} \stackrel{\mathrm{def}}{=} \mathrm{Gal}(\overline{k}_X/k_X),$$

$$G_{k_Y}^{\log} \stackrel{\mathrm{def}}{=} \pi_1((\mathrm{Spec}\, k_Y)^{\log}) \twoheadrightarrow G_{k_Y} \stackrel{\mathrm{def}}{=} \mathrm{Gal}(\overline{k}_Y/k_Y)$$

the natural surjections [well-defined up to composition with an inner automorphism]; $s_X : G_{k_X} \to G_{k_X}^{\log}$*,* $s_Y : G_{k_Y} \to G_{k_Y}^{\log}$ **sections** *of the above natural surjections* $G_{k_X}^{\log} \twoheadrightarrow G_{k_X}$*,* $G_{k_Y}^{\log} \twoheadrightarrow G_{k_Y}$*. For each positive integer n, write* $X_n^{\log}$*,* $Y_n^{\log}$ *for the n-th* **log configuration spaces** *[cf. the discussion entitled "Curves" in "Notations and Conventions"] of* $X^{\log}$*,* $Y^{\log}$*;* ${}^X\Pi_n$*,* ${}^Y\Pi_n$ *for the maximal pro-*

l_X, pro-l_Y quotients of the kernels of the natural surjections $\pi_1(X_n^{\log}) \twoheadrightarrow G_{k_X}^{\log}$, $\pi_1(Y_n^{\log}) \twoheadrightarrow G_{k_Y}^{\log}$. Then the sections s_X, s_Y determine outer actions of G_{k_X}, G_{k_Y} on ${}^X\Pi_n$, ${}^Y\Pi_n$. Thus, we obtain profinite groups

$$ {}^X\Pi_n \overset{\text{out}}{\rtimes}_{s_X} G_{k_X}, \quad {}^Y\Pi_n \overset{\text{out}}{\rtimes}_{s_Y} G_{k_Y} $$

[cf. [MzTa], Proposition 2.2, (ii); the discussion entitled "Topological groups" in [CbTpI], §0]. Let

$$ \alpha_1 : {}^X\Pi_1 \overset{\text{out}}{\rtimes}_{s_X} G_{k_X} \xrightarrow{\sim} {}^Y\Pi_1 \overset{\text{out}}{\rtimes}_{s_Y} G_{k_Y} $$

be an **isomorphism** *of profinite groups. Then $l_X = l_Y$; there exists a* **unique** *collection of* **isomorphisms** *of profinite groups*

$$ \left\{ \alpha_n : {}^X\Pi_n \overset{\text{out}}{\rtimes}_{s_X} G_{k_X} \xrightarrow{\sim} {}^Y\Pi_n \overset{\text{out}}{\rtimes}_{s_Y} G_{k_Y} \right\}_{n \geq 1} $$

—well-defined up to composition with an inner automorphism of ${}^Y\Pi_n \overset{\text{out}}{\rtimes}_{s_Y} G_{k_Y}$ by an element of the intersection ${}^Y\Xi_n \subseteq {}^Y\Pi_n$ of the fiber subgroups of ${}^Y\Pi_n$ of co-length 1 [cf. [CmbCsp], Definition 1.1, (iii)]—such that each diagram

$$ \begin{array}{ccc} {}^X\Pi_{n+1} \overset{\text{out}}{\rtimes}_{s_X} G_{k_X} & \xrightarrow{\alpha_{n+1}} & {}^Y\Pi_{n+1} \overset{\text{out}}{\rtimes}_{s_Y} G_{k_Y} \\ \downarrow & & \downarrow \\ {}^X\Pi_n \overset{\text{out}}{\rtimes}_{s_X} G_{k_X} & \xrightarrow{\alpha_n} & {}^Y\Pi_n \overset{\text{out}}{\rtimes}_{s_Y} G_{k_Y} \end{array} $$

*—where the vertical arrows are the surjections induced by the projections $X_{n+1}^{\log} \to X_n^{\log}$, $Y_{n+1}^{\log} \to Y_n^{\log}$ obtained by forgetting the factors labeled j, for some $j \in \{1, \cdots, n+1\}$—***commutes***, up to composition with a ${}^Y\Xi_n$-inner automorphism.*

Proof First, let us observe that it follows from Corollary 4.18, (ii), below that $l_X = l_Y$. Write $l \overset{\text{def}}{=} l_X = l_Y$. Moreover, it follows from Corollary 4.18, (viii), below [i.e., in the case where condition (viii-1) is satisfied] that α_1 *maps* ${}^X\Pi_1 \subseteq {}^X\Pi_1 \overset{\text{out}}{\rtimes}_{s_X} G_{k_X}$ *bijectively* onto ${}^Y\Pi_1 \subseteq {}^Y\Pi_1 \overset{\text{out}}{\rtimes}_{s_Y} G_{k_Y}$. In particular, α_1 induces isomorphisms of profinite groups

$$ \alpha_1^{\Pi} : {}^X\Pi_1 \xrightarrow{\sim} {}^Y\Pi_1, \quad \alpha_0 : G_{k_X} \xrightarrow{\sim} G_{k_Y}. $$

For $\square \in \{X, Y\}$, write $\mathcal{G}_\square$ for the semi-graph of anabelioids of pro-l PSC-type determined by $\square^{\log}$; $\Pi_{\mathcal{G}_\square}$ for the [pro-l] fundamental group of $\mathcal{G}_\square$; $G_{k_\square}^{(l)} \subseteq G_{k_\square}$

for the maximal pro-l closed subgroup of $G_{k_\square}$; $G_{k_\square}^{(\neq l)}$ for the maximal pro-prime-to-l closed subgroup of $G_{k_\square}$. Thus, we have a natural $\pi_1(\square^{\log})$-orbit, i.e., relative to composition with automorphisms induced by conjugation by elements of $\pi_1(\square^{\log})$, of isomorphisms ${}^\square\Pi_1 \xrightarrow{\sim} \Pi_{\mathcal{G}_\square}$; fix an isomorphism ${}^\square\Pi_1 \xrightarrow{\sim} \Pi_{\mathcal{G}_\square}$ that belongs to the collection of isomorphisms that constitutes this $\pi_1(\square^{\log})$-orbit of isomorphisms. Moreover, since $G_{k_\square}$ is *isomorphic* to $\widehat{\mathbb{Z}}$ as an abstract profinite group, we have a natural decomposition

$$G_{k_\square}^{(l)} \times G_{k_\square}^{(\neq l)} \xrightarrow{\sim} G_{k_\square}.$$

Thus, the isomorphism α_0 naturally decomposes into a pair of isomorphisms

$$\alpha_0^{(l)} : G_{k_X}^{(l)} \xrightarrow{\sim} G_{k_Y}^{(l)}, \quad \alpha_0^{(\neq l)} : G_{k_X}^{(\neq l)} \xrightarrow{\sim} G_{k_Y}^{(\neq l)}.$$

Next, let us observe that since ${}^\square\Pi_1$ is *topologically finitely generated* [cf. [MzTa], Proposition 2.2, (ii)] and *pro-l*, one verifies easily that [by replacing $G_{k_\square}$ by a suitable open subgroup and applying the *injectivity* portion of [NodNon], Theorem B, together with [CmbGC], Corollary 2.7, (i)] we may assume without loss of generality that the outer action of $G_{k_\square}$ on ${}^\square\Pi_1$—hence [cf. the *injectivity* portion of [NodNon], Theorem B] also on ${}^\square\Pi_n$ for each positive integer n—*factors* through the quotient $G_{k_\square} \xleftarrow{\sim} G_{k_\square}^{(l)} \times G_{k_\square}^{(\neq l)} \twoheadrightarrow G_{k_\square}^{(l)}$.

Next, let us recall the following well-known Facts:

(1) Some positive tensor power of the l-adic cyclotomic character of $G_{k_\square}$ *factors* through the outer action of $G_{k_\square}$ on ${}^\square\Pi_1$ [cf. Corollary 4.18, (vii), below].
(2) The restriction to $G_{k_\square}^{(l)} \subseteq G_{k_\square}$ of any positive tensor power of the l-adic cyclotomic character of $G_{k_\square}$ is *injective*.

Thus, it follows from Facts (1), (2), that

(3) the resulting outer action of $G_{k_\square}^{(l)}$ on ${}^\square\Pi_1$—hence also on ${}^\square\Pi_n$ for each positive integer n—is *injective*.

In particular, it follows immediately from the *slimness* of ${}^\square\Pi_n$ [cf. [MzTa], Proposition 2.2, (ii)] that the composite

$$Z_{{}^\square\Pi_n \overset{\text{out}}{\rtimes}_{s_\square} G_{k_\square}}({}^\square\Pi_n) \hookrightarrow {}^\square\Pi_n \overset{\text{out}}{\rtimes}_{s_\square} G_{k_\square} \twoheadrightarrow G_{k_\square}$$

determines an isomorphism

$$Z_{{}^\square\Pi_n \overset{\text{out}}{\rtimes}_{s_\square} G_{k_\square}}({}^\square\Pi_n) \xrightarrow{\sim} G_{k_\square}^{\neq l}.$$

Thus, if we identify $Z_{{}^{\square}\Pi_n \overset{\mathrm{out}}{\rtimes}_{s_\square} G_{k_\square}}({}^{\square}\Pi_n)$ with $G^{\neq l}_{k_\square}$ by means of this isomorphism, then we obtain a natural isomorphism

$$\left({}^{\square}\Pi_n \overset{\mathrm{out}}{\rtimes}_{s_\square} G^{(l)}_{k_\square}\right) \times G^{(\neq l)}_{k_\square} \xrightarrow{\sim} {}^{\square}\Pi_n \overset{\mathrm{out}}{\rtimes}_{s_\square} G_{k_\square}.$$

Next, let us observe that the following assertion holds:

Claim 4.16.A: There exists a positive power q of p such that $\log_p(q)$ is divisible by $\log_p(\#k_X)$, $\log_p(\#k_Y)$, and, moreover,

$$\alpha_0^{(l)}((\mathrm{Fr}_q)^{(l)}_{k_X}) = (\mathrm{Fr}_q)^{(l)}_{k_Y}$$

—where we write $(\mathrm{Fr}_q)_{k_X} \in G_{k_X}$, $(\mathrm{Fr}_q)_{k_Y} \in G_{k_Y}$ for the q-power Frobenius elements of G_{k_X}, G_{k_Y}; $(\mathrm{Fr}_q)^{(l)}_{k_X} \in G^{(l)}_{k_X}$, $(\mathrm{Fr}_q)^{(l)}_{k_Y} \in G^{(l)}_{k_Y}$ for the respective images of $(\mathrm{Fr}_q)_{k_X} \in G_{k_X}$, $(\mathrm{Fr}_q)_{k_Y} \in G_{k_Y}$ in $G^{(l)}_{k_X}$, $G^{(l)}_{k_Y}$.

Indeed, this follows immediately from Corollary 4.18, (vii), below, together with Fact (2).

Write $H_{k_X} \subseteq G_{k_X}$, $H_{k_Y} \subseteq G_{k_Y}$ for the open subgroups of G_{k_X}, G_{k_Y} topologically generated by $(\mathrm{Fr}_q)_{k_X} \in G_{k_X}$, $(\mathrm{Fr}_q)_{k_Y} \in G_{k_Y}$ [cf. Claim 4.16.A]; $U_{k_Y} \subseteq G_{k_Y}$ for the open subgroup of G_{k_Y} topologically generated by $\alpha_0((\mathrm{Fr}_q)_{k_X}) \in G_{k_Y}$; $H^{(l)}_{k_X} \subseteq G^{(l)}_{k_X}$ for the image of $H_{k_X} \subseteq G_{k_X}$ in $G^{(l)}_{k_X}$; $H^{(l)}_{k_Y}$, $U^{(l)}_{k_Y} \subseteq G^{(l)}_{k_Y}$ for the images of H_{k_Y}, $U_{k_Y} \subseteq G_{k_Y}$ in $G^{(l)}_{k_Y}$. Then it follows from Claim 4.16.A that we have an equality $H^{(l)}_{k_Y} = U^{(l)}_{k_Y}$, and, moreover, that the isomorphism $H_{k_X} \xrightarrow{\sim} U_{k_Y}$ induced by α_0 induces an isomorphism $H^{(l)}_{k_X} \xrightarrow{\sim} U^{(l)}_{k_Y} = H^{(l)}_{k_Y}$. In particular, one verifies easily that there exists an isomorphism of profinite groups $\alpha_0^H : H_{k_X} \xrightarrow{\sim} H_{k_Y}$ that

(a) *maps* $(\mathrm{Fr}_q)_{k_X} \in G_{k_X}$ to $(\mathrm{Fr}_q)_{k_Y} \in G_{k_Y}$,

which thus implies that

(b) the isomorphism $H^{(l)}_{k_X} \xrightarrow{\sim} H^{(l)}_{k_Y}$ induced by α_0^H *coincides* with the above isomorphism $H^{(l)}_{k_X} \xrightarrow{\sim} U^{(l)}_{k_Y} = H^{(l)}_{k_Y}$ induced by α_0.

Moreover, it follows immediately from (b), together with the existence of the natural isomorphisms

$$\left({}^{X}\Pi_n \overset{\mathrm{out}}{\rtimes}_{s_X} G^{(l)}_{k_X}\right) \times G^{(\neq l)}_{k_X} \xrightarrow{\sim} {}^{X}\Pi_n \overset{\mathrm{out}}{\rtimes}_{s_X} G_{k_X},$$

$$\left({}^{Y}\Pi_n \overset{\mathrm{out}}{\rtimes}_{s_Y} G^{(l)}_{k_Y}\right) \times G^{(\neq l)}_{k_Y} \xrightarrow{\sim} {}^{Y}\Pi_n \overset{\mathrm{out}}{\rtimes}_{s_Y} G_{k_Y}$$

[cf. the discussion preceding Claim 4.16.A], that there exists an isomorphism

$$\alpha_1^H : {}^{X}\Pi_1 \overset{\mathrm{out}}{\rtimes}_{s_X} H_{k_X} \xrightarrow{\sim} {}^{Y}\Pi_1 \overset{\mathrm{out}}{\rtimes}_{s_Y} H_{k_Y}$$

such that

(c) the isomorphism "α_0" of H_{k_X} with H_{k_Y} that occurs in the case where we take the "α_1" to be α_1^H *coincides* with α_0^H [i.e., roughly speaking, α_1^H lies over α_0^H], and, moreover,
(d) the isomorphism "α_1^Π" of ${}^X\Pi_1$ with ${}^Y\Pi_1$ that occurs in the case where we take the "α_1" to be α_1^H *coincides* with [the original] α_1^Π [i.e., roughly speaking, α_1^H restricts to α_1^Π on ${}^X\Pi_1$].

In particular, we conclude, again by the existence of the natural isomorphisms

$$\left({}^X\Pi_n \overset{\text{out}}{\rtimes}_{s_X} G_{k_X}^{(l)}\right) \times G_{k_X}^{(\neq l)} \overset{\sim}{\longrightarrow} {}^X\Pi_n \overset{\text{out}}{\rtimes}_{s_X} G_{k_X},$$

$$\left({}^Y\Pi_n \overset{\text{out}}{\rtimes}_{s_Y} G_{k_Y}^{(l)}\right) \times G_{k_Y}^{(\neq l)} \overset{\sim}{\longrightarrow} {}^Y\Pi_n \overset{\text{out}}{\rtimes}_{s_Y} G_{k_Y},$$

together with the *injectivity portion* of [NodNon], Theorem B, and [CmbGC], Corollary 2.7, (i), that, to verify Corollary 4.16—by replacing $G_{k_X}, G_{k_Y}, \alpha_1$ by H_{k_X}, H_{k_Y}, α_1^H—we may assume without loss of generality that $\#k_X = \#k_Y$, and that α_0 *maps* the $\#k_X$-power Frobenius element of G_{k_X} to the $\#k_Y$-power Frobenius element of G_{k_Y}. We may also assume without loss of generality—by replacing $G_{k_\Box}$, where $\Box \in \{X, Y\}$, by a suitable open subgroup of $G_{k_\Box}$ if necessary—that the following condition holds:

(e) for $\Box \in \{X, Y\}$, $G_{k_\Box}$ acts *trivially* on the underlying semi-graph of $\mathcal{G}_\Box$.

Next, let us observe that the *uniqueness* portion of Corollary 4.16 follows immediately from the *injectivity portion* of [NodNon], Theorem B, and [CmbGC], Corollary 2.7, (i). Thus, it remains to verify the *existence* of a collection of α_n's as in the statement of Corollary 4.16. To this end, for each positive integer i, $\Box \in \{X, Y\}$, and $v \in \mathrm{Vert}(\mathcal{G}_\Box)$, write $({}^\Box\Pi_v)_i \subseteq {}^\Box\Pi_i$ for the configuration space subgroup of ${}^\Box\Pi_i$ associated to $v \in \mathrm{Vert}(\mathcal{G}_\Box)$ [well-defined up to ${}^\Box\Pi_i$-conjugation—cf. Definition 4.3].

Next, let us observe that

(f) the isomorphism $\Pi_{\mathcal{G}_X} \overset{\sim}{\to} \Pi_{\mathcal{G}_Y}$ determined by α_1^Π and the fixed isomorphisms ${}^X\Pi_1 \overset{\sim}{\to} \Pi_{\mathcal{G}_X}$, ${}^Y\Pi_1 \overset{\sim}{\to} \Pi_{\mathcal{G}_Y}$ is *graphic* [cf. Corollary 4.18, (iii), (iv), below].

Write $\alpha^{\mathrm{Vert}} \colon \mathrm{Vert}(\mathcal{G}_X) \overset{\sim}{\to} \mathrm{Vert}(\mathcal{G}_Y)$ for the bijection determined by the *graphic* isomorphism $\Pi_{\mathcal{G}_X} \overset{\sim}{\to} \Pi_{\mathcal{G}_Y}$ of (f). Thus, for each $v \in \mathrm{Vert}(\mathcal{G}_X)$, the isomorphism $\Pi_{\mathcal{G}_X} \overset{\sim}{\to} \Pi_{\mathcal{G}_Y}$ of (f) determines an outer isomorphism $\beta_v \colon ({}^X\Pi_v)_1 \overset{\sim}{\to} ({}^Y\Pi_{\alpha^{\mathrm{Vert}}(v)})_1$ [cf. [CmbGC], Proposition 1.2, (ii); [CbTpI], Lemma 2.12, (i), (ii), (iii)], which is *compatible* with the respective natural outer actions of G_{k_X}, G_{k_Y} [cf. (e)]. In particular, by applying [Wkb], Theorem C, to this outer isomorphism $\beta_v \colon ({}^X\Pi_v)_1 \overset{\sim}{\to}$ $({}^Y\Pi_{\alpha^{\mathrm{Vert}}(v)})_1$, we obtain [cf. [CmbGC], Corollary 2.7, (i)] a PFC-admissible [cf.

[CbTpI], Definition 1.4, (iii)] outomorphism $\beta_{v,n}\colon ({}^X\Pi_v)_n \xrightarrow{\sim} ({}^Y\Pi_{\alpha^{\mathrm{Vert}}(v)})_n$, which is *compatible* with the respective natural outer actions of G_{k_X}, G_{k_Y} [cf. (e)]. Moreover, since the β_v's arise from a single isomorphism $\Pi_{\mathcal{G}_X} \xrightarrow{\sim} \Pi_{\mathcal{G}_Y}$, one verifies immediately from [CbTpI], Corollary 3.9, (ii), (v), and the *injectivity* discussed in [Hsh], Remark 6, (iv) [i.e., applied to the difference between the various outer isomorphisms, determined by $\beta_{v,n}$, between tripods of $({}^X\Pi_v)_n$ and tripods of $({}^Y\Pi_{\alpha^{\mathrm{Vert}}(v)})_n$], that the collection $(\beta_{v,n})_{v\in\mathrm{Vert}(\mathcal{G}_X)}$ is *contained* in the set which corresponds—in the *"isomorphism version"* of Theorem 4.14 discussed in Remark 4.14.1—to the set "Glu(Π_n)" in the statement of Theorem 4.14. In particular, it follows from the *"isomorphism version"* of Theorem 4.14, (i), (iii), discussed in Remark 4.14.1 that the outer isomorphism determined by the isomorphism $\alpha_1^{\Pi}\colon {}^X\Pi_1 \xrightarrow{\sim} {}^Y\Pi_1$ and the collection $(\beta_{v,n})_{v\in\mathrm{Vert}(\mathcal{G})}$ uniquely determine a PFC-admissible outer isomorphism $\beta_n\colon {}^X\Pi_n \xrightarrow{\sim} {}^Y\Pi_n$ which—by the *injectivity portion* of [NodNon], Theorem B—is *compatible* with the respective outer actions of G_{k_X}, G_{k_Y}. Finally, one verifies immediately that one may construct a collection of α_n's as in the statement of Corollary 4.16 from the collection of the β_n's. This completes the proof of the *existence* of α_n's, hence also of Corollary 4.16. □

Remark 4.16.1 Corollary 4.16 may be regarded as a *generalization* of [AbsCsp], Theorem 3.1; [Hsh], Theorem 0.1; [Wkb], Theorem C.

Corollary 4.17 (Commensurator of the Image of the Absolute Galois Group of a Finite Field in the Totally Degenerate Case) *Let n be a positive integer; p, l two* **distinct** *prime numbers; (g, r) a pair of nonnegative integers $\neq (0, 3)$ such that $2g - 2 + r > 0$; k a* **finite field** *of characteristic p; $\overline{k}$ an algebraic closure of k; $(\mathrm{Spec}\, k)^{\log}$ the log scheme obtained by equipping* Spec k *with the log structure determined by the fs chart $\mathbb{N} \to k$ that maps $1 \mapsto 0$; $X^{\log}$ a* **stable log curve** *of type (g, r) over $(\mathrm{Spec}\, k)^{\log}$. Write $\mathcal{G}$ for the semi-graph of anabelioids of pro-l PSC-type associated to the stable log curve $X^{\log}$; $\mathbb{G}$ for the underlying semi-graph of $\mathcal{G}$; $\Pi_{\mathcal{G}}$ for the [pro-l] fundamental group of $\mathcal{G}$;*

$$G_k^{\log} \stackrel{\mathrm{def}}{=} \pi_1((\mathrm{Spec}\, k)^{\log}) \twoheadrightarrow G_k \stackrel{\mathrm{def}}{=} \mathrm{Gal}(\overline{k}/k)$$

for the natural surjection [well-defined up to composition with an inner automorphism]. For each positive integer i, write $X_i^{\log}$ for the i-th **log configuration space** *[cf. the discussion entitled "Curves" in "Notations and Conventions"] of $X^{\log}$; Π_i for the maximal pro-l quotient of the kernel of the natural surjection $\pi_1(X_i^{\log}) \twoheadrightarrow G_k^{\log}$. Thus, we have a natural $\pi_1(X^{\log})$-orbit, i.e., relative to composition with automorphisms induced by conjugation by elements of $\pi_1(X^{\log})$, of isomorphisms $\Pi_1 \xrightarrow{\sim} \Pi_{\mathcal{G}}$ and a natural outer action*

$$\rho_{X_i^{\log}}\colon G_k^{\log} \longrightarrow \mathrm{Out}^{\mathrm{FC}}(\Pi_i)$$

[cf. the notation of [CmbCsp], Definition 1.1, (ii)]. Fix an outer isomorphism $\Pi_1 \xrightarrow{\sim} \Pi_{\mathcal{G}}$ *whose constituent isomorphisms belong to the above* $\pi_1(X^{\log})$*-orbit of isomorphisms. Let* $H \subseteq G_k^{\log}$ *be a closed subgroup of* $G_k^{\log}$ *whose image in* G_k *is* **open**. *Write* $I_H \subseteq H$ *for the kernel of the composite* $H \hookrightarrow G_k^{\log} \twoheadrightarrow G_k$. *We shall say that* H *is of* **l-Dehn type** *if the maximal pro-l quotient of* I_H *is* **nontrivial**. *Suppose that the stable log curve* $X^{\log}$ *is* **totally degenerate** *[i.e., that the complement in* X *of the nodes and cusps is a disjoint union of* **tripods**]. *Then the following hold:*

(i) The image $\rho_{X_1^{\log}}(I_H) \subseteq \mathrm{Out}(\Pi_1)$ *is* **contained** *in* $\mathrm{Dehn}(\mathcal{G}) \subseteq \mathrm{Out}(\Pi_{\mathcal{G}}) \xleftarrow{\sim} \mathrm{Out}(\Pi_1)$ *[cf. the notation of [CbTpI], Definition 4.4]. Moreover, the image* $\rho_{X_1^{\log}}(I_H)$ *is* **nontrivial** *if and only if* H *is of* **l-Dehn type**. *Write*

$$I_H^{C(\rho)} \stackrel{\mathrm{def}}{=} (\rho_{X_1^{\log}}(I_H) \otimes_{\mathbb{Z}_l} \mathbb{Q}_l) \cap \mathrm{Dehn}(\mathcal{G}) \subseteq \mathrm{Dehn}(\mathcal{G})$$

[considered in $\mathrm{Dehn}(\mathcal{G}) \otimes_{\mathbb{Z}_l} \mathbb{Q}_l$*—cf. [CbTpI], Theorem 4.8, (iv)].*

(ii) For any positive integer $m \leq n$, *the natural injection* $\mathrm{Out}^{\mathrm{FC}}(\Pi_n) \hookrightarrow \mathrm{Out}^{\mathrm{FC}}(\Pi_m)$ *of [NodNon], Theorem B, induces* **isomorphisms**

$$Z_{\mathrm{Out}^{\mathrm{FC}}(\Pi_n)}(\rho_{X_n^{\log}}(H)) \xrightarrow{\sim} Z_{\mathrm{Out}^{\mathrm{FC}}(\Pi_m)}(\rho_{X_m^{\log}}(H)),$$

$$Z^{\mathrm{loc}}_{\mathrm{Out}^{\mathrm{FC}}(\Pi_n)}(\rho_{X_n^{\log}}(H)) \xrightarrow{\sim} Z^{\mathrm{loc}}_{\mathrm{Out}^{\mathrm{FC}}(\Pi_m)}(\rho_{X_m^{\log}}(H))$$

[cf. the discussion entitled "Topological groups" in "Notations and Conventions"],

$$N_{\mathrm{Out}^{\mathrm{FC}}(\Pi_n)}(\rho_{X_n^{\log}}(H)) \xrightarrow{\sim} N_{\mathrm{Out}^{\mathrm{FC}}(\Pi_m)}(\rho_{X_m^{\log}}(H)),$$

$$C_{\mathrm{Out}^{\mathrm{FC}}(\Pi_n)}(\rho_{X_n^{\log}}(H)) \xrightarrow{\sim} C_{\mathrm{Out}^{\mathrm{FC}}(\Pi_m)}(\rho_{X_m^{\log}}(H)).$$

(iii) Relative to the natural inclusion $\mathrm{Aut}(\mathcal{G})$ $(\subseteq \mathrm{Out}(\Pi_{\mathcal{G}}) \xleftarrow{\sim} \mathrm{Out}(\Pi_1))$, *the following equality holds:*

$$C_{\mathrm{Out}^{\mathrm{FC}}(\Pi_1)}(\rho_{X_1^{\log}}(H)) = C_{\mathrm{Aut}(\mathcal{G})}(\rho_{X_1^{\log}}(H)).$$

In particular, we have natural homomorphisms of profinite groups

$$C_{\mathrm{Out}^{\mathrm{FC}}(\Pi_n)}(\rho_{X_n^{\log}}(H)) \xrightarrow{\sim} C_{\mathrm{Out}^{\mathrm{FC}}(\Pi_1)}(\rho_{X_1^{\log}}(H)) \to \mathrm{Aut}(\mathbb{G}),$$

$$C_{\mathrm{Out}^{\mathrm{FC}}(\Pi_n)}(\rho_{X_n^{\log}}(H)) \xrightarrow{\sim} C_{\mathrm{Out}^{\mathrm{FC}}(\Pi_1)}(\rho_{X_1^{\log}}(H)) \xrightarrow{\chi_{\mathcal{G}}} \mathbb{Z}_l^*$$

[cf. the notation of [CbTpI], Definition 3.8, (ii)]—where the first arrow in each line is the isomorphism of (ii). By abuse of notation [i.e., since $\rho_{X_n^{\log}}(H)$ is not necessarily contained in $\mathrm{Aut}^{|\mathrm{grph}|}(\mathcal{G})$*—cf. the notation of [CbTpI], Definition 2.6, (i); Remark 4.1.2 of the present monograph], write*

$$Z_{\mathrm{Aut}^{|\mathrm{grph}|}(\mathcal{G})}(\rho_{X_n^{\log}}(H)) \subseteq Z_{\mathrm{Out}^{\mathrm{FC}}(\Pi_n)}(\rho_{X_n^{\log}}(H)),$$

$$Z^{\mathrm{loc}}_{\mathrm{Aut}^{|\mathrm{grph}|}(\mathcal{G})}(\rho_{X_n^{\log}}(H)) \subseteq Z^{\mathrm{loc}}_{\mathrm{Out}^{\mathrm{FC}}(\Pi_n)}(\rho_{X_n^{\log}}(H)),$$

$$N_{\mathrm{Aut}^{|\mathrm{grph}|}(\mathcal{G})}(\rho_{X_n^{\log}}(H)) \subseteq N_{\mathrm{Out}^{\mathrm{FC}}(\Pi_n)}(\rho_{X_n^{\log}}(H)),$$

$$C_{\mathrm{Aut}^{|\mathrm{grph}|}(\mathcal{G})}(\rho_{X_n^{\log}}(H)) \subseteq C_{\mathrm{Out}^{\mathrm{FC}}(\Pi_n)}(\rho_{X_n^{\log}}(H))$$

for the kernels of the restrictions of the composite homomorphism of the first line of the second display [of the present (iii)] to

$$Z_{\mathrm{Out}^{\mathrm{FC}}(\Pi_n)}(\rho_{X_n^{\log}}(H)), \quad Z^{\mathrm{loc}}_{\mathrm{Out}^{\mathrm{FC}}(\Pi_n)}(\rho_{X_n^{\log}}(H)),$$

$$N_{\mathrm{Out}^{\mathrm{FC}}(\Pi_n)}(\rho_{X_n^{\log}}(H)), \quad C_{\mathrm{Out}^{\mathrm{FC}}(\Pi_n)}(\rho_{X_n^{\log}}(H)),$$

respectively.

(iv) *Suppose that H is* **not of l-Dehn type**. *Then we have equalities*

$$\begin{aligned} Z_{\mathrm{Aut}^{|\mathrm{grph}|}(\mathcal{G})}(\rho_{X_n^{\log}}(H)) &= Z^{\mathrm{loc}}_{\mathrm{Aut}^{|\mathrm{grph}|}(\mathcal{G})}(\rho_{X_n^{\log}}(H)) \\ &= N_{\mathrm{Aut}^{|\mathrm{grph}|}(\mathcal{G})}(\rho_{X_n^{\log}}(H)) \\ &= C_{\mathrm{Aut}^{|\mathrm{grph}|}(\mathcal{G})}(\rho_{X_n^{\log}}(H)) \end{aligned}$$

[cf. the notation of (iii)]. Moreover, each of the four groups appearing in these equalities is, in fact, **independent** *of n [cf. (ii)].*

(v) *Suppose that H is of* **l-Dehn type**. *Then the composite homomorphism of the first line of the second display of (iii) determines an* **injection** *of profinite groups*

$$Z^{\mathrm{loc}}_{\mathrm{Out}^{\mathrm{FC}}(\Pi_n)}(\rho_{X_n^{\log}}(H)) \hookrightarrow \mathrm{Aut}(\mathbb{G}).$$

(vi) *Write $k_{|\mathrm{grph}|}$ ($\subseteq \overline{k}$) for the [finite] subfield of $\overline{k}$ consisting of the invariants of $\overline{k}$ with respect to [the natural action on $\overline{k}$ of] the* **kernel** *of the natural action of H on $\mathbb{G}$. Then the composite homomorphism of the second line of the second display of (iii) determines* **natural exact sequences** *of profinite groups*

$$1 \longrightarrow I_H^{N(\rho)} \longrightarrow N_{\mathrm{Aut}^{|\mathrm{grph}|}(\mathcal{G})}(\rho_{X_n^{\log}}(H)) \longrightarrow \mathbb{Z}_l^*,$$

$$1 \longrightarrow I_H^{C(\rho)} \longrightarrow C_{\mathrm{Aut}^{|\mathrm{grph}|}(\mathcal{G})}(\rho_{X_n^{\log}}(H)) \longrightarrow \mathbb{Z}_l^*$$

[cf. the notation of (i), (iii)]—where $\rho_{X_n^{\log}}(I_H)$*, hence also*

$$(\rho_{X_n^{\log}}(I_H) \subseteq)\ I_H^{N(\rho)} \stackrel{\mathrm{def}}{=} N_{\mathrm{Aut}^{|\mathrm{grph}|}(\mathcal{G})}(\rho_{X_n^{\log}}(H)) \cap \mathrm{Dehn}(\mathcal{G})$$

[cf. (i), (ii), (iii)], is an **open subgroup** *of* $I_H^{C(\rho)}$*; the image of the third arrow in each line* **contains** $\#k_{|\mathrm{grph}|} \in \mathbb{Z}_l^*$ *and does* **not depend** *on the choice of n. In particular, these images are* **open***; if, moreover,* $\#k_{|\mathrm{grph}|} \in \mathbb{Z}_l^*$ **topologically generates** $\mathbb{Z}_l^*$*, then the third arrows in each line are* **surjective***.*

(vii) The closed subgroup $\rho_{X_n^{\log}}(H) \subseteq C_{\mathrm{Out}^{\mathrm{FC}}(\Pi_n)}(\rho_{X_n^{\log}}(H))$*, hence also* $N_{\mathrm{Out}^{\mathrm{FC}}(\Pi_n)}(\rho_{X_n^{\log}}(H))$ $(\subseteq C_{\mathrm{Out}^{\mathrm{FC}}(\Pi_n)}(\rho_{X_n^{\log}}(H)))$*, is* **open** *in* $C_{\mathrm{Out}^{\mathrm{FC}}(\Pi_n)}(\rho_{X_n^{\log}}(H))$*.*

(viii) Consider the following conditions [cf. Remark 4.17.1 below]:

(1) Write $\mathrm{Aut}_{(\mathrm{Spec}\,k)^{\log}}(X^{\log})$ *for the group of automorphisms of* $X^{\log}$ *over* $(\mathrm{Spec}\,k)^{\log}$*. Then the natural homomorphism*

$$\mathrm{Aut}_{(\mathrm{Spec}\,k)^{\log}}(X^{\log}) \longrightarrow \mathrm{Aut}(\mathbb{G})$$

is **surjective***.*

(2) $\#k_{|\mathrm{grph}|} \in \mathbb{Z}_l^*$ **topologically generates** $\mathbb{Z}_l^*$*.*

If condition (1) is satisfied, and H is of **l-Dehn type***, then we have an equality*

$$Z_{\mathrm{Out}^{\mathrm{FC}}(\Pi_n)}(\rho_{X_n^{\log}}(H)) = Z^{\mathrm{loc}}_{\mathrm{Out}^{\mathrm{FC}}(\Pi_n)}(\rho_{X_n^{\log}}(H)),$$

and, moreover, the composite homomorphism of the first line of the second display of (iii) determines an isomorphism

$$Z^{\mathrm{loc}}_{\mathrm{Out}^{\mathrm{FC}}(\Pi_n)}(\rho_{X_n^{\log}}(H)) \stackrel{\sim}{\longrightarrow} \mathrm{Aut}(\mathbb{G}).$$

If conditions (1) and (2) are satisfied, then the composite homomorphisms of the two lines of the second display of (iii) determine **natural exact sequences** *of profinite groups*

$$1 \longrightarrow I_H^{N(\rho)} \longrightarrow N_{\mathrm{Out}^{\mathrm{FC}}(\Pi_n)}(\rho_{X_n^{\log}}(H)) \longrightarrow \mathrm{Aut}(\mathbb{G}) \times \mathbb{Z}_l^* \longrightarrow 1,$$

$$1 \longrightarrow I_H^{C(\rho)} \longrightarrow C_{\mathrm{Out}^{\mathrm{FC}}(\Pi_n)}(\rho_{X_n^{\log}}(H)) \longrightarrow \mathrm{Aut}(\mathbb{G}) \times \mathbb{Z}_l^* \longrightarrow 1.$$

Proof Assertion (i) follows immediately from the various definitions involved, together with [CbTpI], Lemma 5.4, (ii); [CbTpI], Proposition 5.6, (ii). Assertion (ii) follows immediately from Corollary 4.16, together with the *slimness* of Π_i for each positive integer i [cf. [MzTa], Proposition 2.2, (ii)] and the *openness* of the image of H in G_k. Assertion (iii) follows immediately from [CmbGC], Corollary 2.7, (ii) [cf. also the proof of [CmbGC], Proposition 2.4, (v)], together with the *openness* of the image of H in G_k.

For $\square \in \{Z, Z^{\mathrm{loc}}, N, C\}$ and $v \in \mathrm{Vert}(\mathcal{G})$, write

$$\square \stackrel{\mathrm{def}}{=} \square_{\mathrm{Out}^{\mathrm{FC}}(\Pi_1)}(\rho_{X_1^{\log}}(H)) \subseteq \mathrm{Out}(\Pi_1) \stackrel{\sim}{\to} \mathrm{Out}(\Pi_{\mathcal{G}});$$

$$\square_{|\mathrm{grph}|} \stackrel{\mathrm{def}}{=} \square \cap \mathrm{Aut}^{|\mathrm{grph}|}(\mathcal{G}) \subseteq \mathrm{Out}(\Pi_{\mathcal{G}})$$

[cf. the notation of [CbTpI], Definition 2.6, (i); Remark 4.1.2 of the present monograph];

$$\mathrm{pr}_v : \mathrm{Aut}^{|\mathrm{grph}|}(\mathcal{G}) \longrightarrow \mathrm{Aut}^{|\mathrm{grph}|}(\mathcal{G}|_v)$$

for the homomorphism determined by *restriction* to $\mathcal{G}|_v$ [cf. [CbTpI], Definition 2.14, (ii); [CbTpI], Remark 2.5.1, (ii)];

$$\square_v \subseteq \mathrm{Aut}^{|\mathrm{grph}|}(\mathcal{G}|_v)$$

for the image of $\square_{|\mathrm{grph}|} \subseteq \mathrm{Aut}^{|\mathrm{grph}|}(\mathcal{G})$ via pr_v. Then we claim that the following assertion holds:

Claim 4.17.A: Let $v \in \mathrm{Vert}(\mathcal{G})$. Then

$$C_v \cap \mathrm{Ker}(\chi_{\mathcal{G}|_v}) = \{1\}$$

[cf. the notation of [CbTpI], Definition 3.8, (ii)].

Indeed, let us first observe that since ${}^{\square}\Pi_1$ is *topologically finitely generated* [cf. [MzTa], Proposition 2.2, (ii)] and *pro-l*, one verifies easily that the image of the outer action $\rho_{X_1^{\log}}$ admits a *pro-l* open subgroup. Thus, since the image of H in G_k is *open*, it follows immediately from Corollary 4.18, (vii), below that $C_v \subseteq \mathrm{Aut}^{|\mathrm{grph}|}(\mathcal{G}|_v)$ is *contained* in the *local centralizer* [cf. the discussion entitled "*Topological groups*" in "Notations and Conventions"] of the natural image of G_k in $\mathrm{Aut}^{|\mathrm{grph}|}(\mathcal{G}|_v)$ [cf. the fact that $\mathcal{G}|_v$ is of *type* $(0, 3)$]. Thus, Claim 4.17.A follows immediately from the *injectivity* discussed in [Hsh], Remark 6, (iv). This completes the proof of Claim 4.17.A.

Next, we claim that the following assertion holds:

Claim 4.17.B: Let $v \in \mathrm{Vert}(\mathcal{G})$. Then

$$C_{|\mathrm{grph}|} \cap \mathrm{Ker}(\mathrm{pr}_v) = C_{|\mathrm{grph}|} \cap \mathrm{Ker}(\chi_{\mathcal{G}}) = C_{|\mathrm{grph}|} \cap \mathrm{Dehn}(\mathcal{G});$$

$$Z_{|\mathrm{grph}|} \cap \mathrm{Ker}(\mathrm{pr}_v) = Z^{\mathrm{loc}}_{|\mathrm{grph}|} \cap \mathrm{Ker}(\mathrm{pr}_v) = \{1\}.$$

In particular, we obtain *natural isomorphisms*

$$Z_{|\mathrm{grph}|} \stackrel{\sim}{\longrightarrow} Z_v, \quad Z^{\mathrm{loc}}_{|\mathrm{grph}|} \stackrel{\sim}{\longrightarrow} Z^{\mathrm{loc}}_v$$

and a natural exact sequence of profinite groups

$$1 \longrightarrow C_{|\mathrm{grph}|} \cap \mathrm{Dehn}(\mathcal{G}) \longrightarrow C_{|\mathrm{grph}|} \xrightarrow{\chi_{\mathcal{G}}} \mathbb{Z}_l^*.$$

Indeed, let us first observe that the equalities of the first line of the first display of Claim 4.17.B follow immediately from Claim 4.17.A, together with [CbTpI], Corollary 3.9, (iv). Moreover, since the image of H in G_k is *open*, the equalities of the second line of the first display of Claim 4.17.B follow immediately from [CbTpI], Theorem 4.8, (iv), (v), together with the equalities of the first line of the first display of Claim 4.17.B. This completes the proof of Claim 4.17.B.

Next, we verify assertion (iv). Let us first observe that it follows from assertion (ii) that it suffices to verify assertion (iv) in the case where $n = 1$. Next, let us observe that it follows from Lemma 3.9, (ii), that $C_{|\mathrm{grph}|} \subseteq N_{\mathrm{Out}^{\mathrm{FC}}(\Pi_1)}(Z^{\mathrm{loc}})$, which thus implies that we have a natural action [by conjugation] of $C_{|\mathrm{grph}|}$ on Z^{loc}, hence also on $Z^{\mathrm{loc}}_{|\mathrm{grph}|}$, as well as a natural [*trivial*!] action of $C_{|\mathrm{grph}|}$ on $\mathrm{Aut}(\mathbb{G})$. Moreover, by considering the inclusion

$$(C_{|\mathrm{grph}|} \supseteq)\ Z^{\mathrm{loc}}_{|\mathrm{grph}|} \xrightarrow{\sim} Z^{\mathrm{loc}}_v \hookrightarrow \mathbb{Z}_l^*$$

induced by $\chi_{\mathcal{G}|_v}$ [cf. Claims 4.17.A, 4.17.B], we conclude that the homomorphisms of the two lines of the second display of assertion (iii) determine a natural [$C_{|\mathrm{grph}|}$-*equivariant*!] *injection*

$$Z^{\mathrm{loc}} \hookrightarrow \mathrm{Aut}(\mathbb{G}) \times \mathbb{Z}_l^*.$$

Thus, since $\mathbb{Z}_l^*$ is *abelian*, it follows that $C_{|\mathrm{grph}|}$ acts *trivially* on Z^{loc}, i.e., that $C_{|\mathrm{grph}|} \subseteq Z_{\mathrm{Out}^{\mathrm{FC}}(\Pi_1)}(Z^{\mathrm{loc}})$. On the other hand, since H is *not of l-Dehn type*, one verifies easily from assertion (i) that $\rho_{X_1^{\log}}(H)$ is *abelian*, hence that $\rho_{X_1^{\log}}(H) \subseteq Z \subseteq Z^{\mathrm{loc}}$. Thus, we conclude that

$$\begin{aligned} C_{|\mathrm{grph}|} &\subseteq Z_{\mathrm{Out}^{\mathrm{FC}}(\Pi_1)}(Z^{\mathrm{loc}}) \cap \mathrm{Aut}^{|\mathrm{grph}|}(\mathcal{G}) \\ &\subseteq Z_{\mathrm{Out}^{\mathrm{FC}}(\Pi_1)}(\rho_{X_1^{\log}}(H)) \cap \mathrm{Aut}^{|\mathrm{grph}|}(\mathcal{G}) \\ &= Z \cap \mathrm{Aut}^{|\mathrm{grph}|}(\mathcal{G}) = Z_{|\mathrm{grph}|}. \end{aligned}$$

This completes the proof of assertion (iv).

Next, we verify assertion (v). First, let us observe that it follows from assertion (ii) that, to verify assertion (v), it suffices to verify that $Z^{\mathrm{loc}}_{|\mathrm{grph}|} = \{1\}$, hence, by Claim 4.17.B, that $\chi_{\mathcal{G}}(Z^{\mathrm{loc}}_{|\mathrm{grph}|}) = \{1\}$. On the other hand, since H is of *l-Dehn type*, by considering the conjugation action of $Z^{\mathrm{loc}}_{|\mathrm{grph}|}$ on $\rho_{X_1^{\log}}(I_H)$ [which is *nontrivial* by assertion (i)], we conclude from [CbTpI], Theorem 4.8, (iv), (v), that $\chi_{\mathcal{G}}(Z^{\mathrm{loc}}_{|\mathrm{grph}|}) = \{1\}$, as desired. This completes the proof of assertion (v).

Next, we verify assertion (vi). First, we observe that it follows from assertions (ii), (iii) that the definition of $I_H^{N(\rho)}$ is indeed *independent* of n [as the notation suggests!]. Next, we claim that the following assertion holds:

Claim 4.17.C:

$$\rho_{X_1^{\log}}(I_H) \subseteq N_{|\mathrm{grph}|} \cap \mathrm{Dehn}(\mathcal{G}) = I_H^{N(\rho)} \subseteq C_{|\mathrm{grph}|} \cap \mathrm{Dehn}(\mathcal{G}) = I_H^{C(\rho)}.$$

Indeed, the final equality follows immediately from an elementary computation [in which we apply [CbTpI], Theorem 4.8, (iv), (v)], together with assertion (i); the remainder of Claim 4.17.C follows immediately from the various definitions involved, together with assertion (i). This completes the proof of Claim 4.17.C. Now it follows immediately from Claims 4.17.B, 4.17.C, together with assertion (ii), that the composite homomorphism of the second line of the second display of (iii) determines the two displayed exact sequences of assertion (vi), and that $\rho_{X_1^{\log}}(I_H)$, hence also $I_H^{N(\rho)}$, is an *open subgroup* of $I_H^{C(\rho)}$. Moreover, since [it is immediate that] the image, via $\rho_{X_n^{\log}}$, of the kernel of the natural action of H on $\mathbb{G}$ is *contained* in $N_{|\mathrm{grph}|}$, the image of the third arrow in each line of the displayed sequences of assertion (vi) *contains* $\#k_{|\mathrm{grph}|} \in \mathbb{Z}_l^*$. Finally, it follows from assertion (ii) that the image of the third arrow in each line of the displayed sequences of assertion (vi) does *not depend* on the choice of n. This completes the proof of assertion (vi).

Assertion (vii) follows immediately from assertions (iii) and (vi), together with the *finiteness* of $\mathrm{Aut}(\mathbb{G})$. Assertion (viii) follows immediately from assertions (v) and (vi). This completes the proof of Corollary 4.17. □

Remark 4.17.1

(i) One verifies easily that condition (1) of Corollary 4.17, (viii), holds if, for instance, $k = k_{|\mathrm{grph}|}$, and, moreover, the *lengths* [cf. [CbTpI], Definition 5.3, (ii)] of the various nodes of $X^{\log}$ [whose base-change from k to $\overline{k}$ may be thought of as the special fiber stable log curve of [CbTpI], Definition 5.3] *coincide*.

(ii) In a similar vein, one verifies easily that condition (2) of Corollary 4.17, (viii), holds if, for instance, $k_{|\mathrm{grph}|} = \mathbb{F}_p$, and, moreover, p *remains prime* in the cyclotomic extension $\mathbb{Q}(e^{2\pi i/l^2})$, where $i = \sqrt{-1}$, and we assume that l is *odd*.

Remark 4.17.2 The computation, in the case where $n = 1$, of the *centralizer* (respectively, *normalizer* and *commensurator*) in Corollary 4.17, (viii), may be thought of as a sort of **relative geometrically pro-*l*** (respectively, **[semi-]absolute geometrically pro-*l***) version of the **Grothendieck Conjecture** for **totally degenerate** stable log curves over **finite fields**. In fact, the proofs of these computations of Corollary 4.17, (viii), in the case where $n = 1$, only involve the theory of [CmbGC] and [CbTpI]. On the other hand, these computations of Corollary 4.17, (viii), can only be performed under certain *relatively restrictive conditions* [cf. Remark 4.17.1]. It is precisely for this reason that Corollary 4.17, (ii), which may be thought of as an *application of the theory of the present monograph*, is of interest in the context of these computations of Corollary 4.17, (viii).

Corollary 4.18 (Compatibility with Geometric Subgroups) *Let p, l_X, l_Y be prime numbers such that $p \notin \{l_X, l_Y\}$; (g_X, r_X), (g_Y, r_Y) pairs of nonnegative integers such that $2g_X - 2 + r_X, 2g_Y - 2 + r_Y > 0$; k_X, k_Y* **finite fields** *of characteristic p; $\overline{k}_X$, $\overline{k}_Y$ algebraic closures of k_X, k_Y; $(\mathrm{Spec}\, k_X)^{\log}$, $(\mathrm{Spec}\, k_Y)^{\log}$ the log schemes obtained by equipping $\mathrm{Spec}\, k_X$, $\mathrm{Spec}\, k_Y$ with the log structures determined by the fs charts $\mathbb{N} \to k_X$, $\mathbb{N} \to k_Y$ that map $1 \mapsto 0$; $X^{\log}$, $Y^{\log}$* **stable log curves** *of type (g_X, r_X), (g_Y, r_Y) over $(\mathrm{Spec}\, k_X)^{\log}$, $(\mathrm{Spec}\, k_Y)^{\log}$;*

$$G_{k_X}^{\log} \stackrel{\mathrm{def}}{=} \pi_1((\mathrm{Spec}\, k_X)^{\log}) \twoheadrightarrow G_{k_X} \stackrel{\mathrm{def}}{=} \mathrm{Gal}(\overline{k}_X/k_X),$$

$$G_{k_Y}^{\log} \stackrel{\mathrm{def}}{=} \pi_1((\mathrm{Spec}\, k_Y)^{\log}) \twoheadrightarrow G_{k_Y} \stackrel{\mathrm{def}}{=} \mathrm{Gal}(\overline{k}_Y/k_Y)$$

the natural surjections [well-defined up to composition with an inner automorphism]; ${}^X H \subseteq G_{k_X}^{\log}$, ${}^Y H \subseteq G_{k_Y}^{\log}$ closed subgroups of $G_{k_X}^{\log}$, $G_{k_Y}^{\log}$; ${}^X I \subseteq {}^X H$, ${}^Y I \subseteq {}^Y H$ the kernels of the composites ${}^X H \hookrightarrow G_{k_X}^{\log} \twoheadrightarrow G_{k_X}$, ${}^Y H \hookrightarrow G_{k_Y}^{\log} \twoheadrightarrow G_{k_Y}$; ${}^X\Pi$, ${}^Y\Pi$ the maximal pro-l_X, pro-l_Y quotients of the kernels of the natural surjections $\pi_1(X^{\log}) \twoheadrightarrow G_{k_X}^{\log}$, $\pi_1(Y^{\log}) \twoheadrightarrow G_{k_Y}^{\log}$; $\mathcal{G}_X$, $\mathcal{G}_Y$ the semi-graphs of anabelioids of pro-l PSC-type determined by $X^{\log}$, $Y^{\log}$; $\Pi_{\mathcal{G}_X}$, $\Pi_{\mathcal{G}_Y}$ the [pro-l] fundamental groups of $\mathcal{G}_X$, $\mathcal{G}_Y$ [so we have natural $\pi_1(X^{\log})$-, $\pi_1(Y^{\log})$-orbits—i.e., relative to composition with automorphisms induced by conjugation by elements of $\pi_1(X^{\log})$, $\pi_1(Y^{\log})$—of isomorphisms ${}^X\Pi \xrightarrow{\sim} \Pi_{\mathcal{G}_X}$, ${}^Y\Pi \xrightarrow{\sim} \Pi_{\mathcal{G}_Y}$]. Then the natural outer actions of $G_{k_X}^{\log}$, $G_{k_Y}^{\log}$ on ${}^X\Pi$, ${}^Y\Pi$ determine outer actions of ${}^X I \subseteq {}^X H$, ${}^Y I \subseteq {}^Y H$ on ${}^X\Pi$, ${}^Y\Pi$. Thus, we obtain profinite groups

$${}^X\Pi \overset{\mathrm{out}}{\rtimes} {}^X I \subseteq {}^X\Pi \overset{\mathrm{out}}{\rtimes} {}^X H, \quad {}^Y\Pi \overset{\mathrm{out}}{\rtimes} {}^Y I \subseteq {}^Y\Pi \overset{\mathrm{out}}{\rtimes} {}^Y H$$

[cf. the discussion entitled "Topological groups" in [CbTpI], §0]. Suppose that, for each $\square \in \{X, Y\}$, **one** *of the following two conditions is satisfied:*

(a) *The* **equality** ${}^\square H = G_{k_\square}^{\log}$ *holds.*
(b) *The composite* ${}^\square H \hookrightarrow G_{k_\square}^{\log} \twoheadrightarrow G_{k_\square}$ *is an* **isomorphism**.

We shall refer to a closed subgroup of ${}^X\Pi$, ${}^Y\Pi$ obtained by forming the image—by the inverse of an element of the $\pi_1(X^{\log})$-, $\pi_1(Y^{\log})$-orbits of isomorphisms ${}^X\Pi \xrightarrow{\sim} \Pi_{\mathcal{G}_X}$, ${}^Y\Pi \xrightarrow{\sim} \Pi_{\mathcal{G}_Y}$ discussed above—in ${}^X\Pi$, ${}^Y\Pi$ of a verticial (respectively, cuspidal; nodal; edge-like) subgroup of $\Pi_{\mathcal{G}_X}$, $\Pi_{\mathcal{G}_Y}$ as a **verticial** *(respectively,* **cuspidal**; **nodal**; **edge-like***)* **subgroup** *of ${}^X\Pi \overset{\mathrm{out}}{\rtimes} {}^X H$, ${}^Y\Pi \overset{\mathrm{out}}{\rtimes} {}^Y H$. We shall refer to a closed subgroup of ${}^X\Pi \overset{\mathrm{out}}{\rtimes} {}^X I$, ${}^Y\Pi \overset{\mathrm{out}}{\rtimes} {}^Y I$ obtained by forming the normalizer in*

${}^X\Pi \overset{\mathrm{out}}{\rtimes} {}^X I$, ${}^Y\Pi \overset{\mathrm{out}}{\rtimes} {}^Y I$ *[i.e., as opposed to* ${}^X\Pi \overset{\mathrm{out}}{\rtimes} {}^X H$, ${}^Y\Pi \overset{\mathrm{out}}{\rtimes} {}^Y H$*] of a verticial (respectively, cuspidal; nodal; edge-like) subgroup of* ${}^X\Pi \overset{\mathrm{out}}{\rtimes} {}^X H$, ${}^Y\Pi \overset{\mathrm{out}}{\rtimes} {}^Y H$ *as a* **verticial** *(respectively,* **cuspidal***;* **nodal***;* **edge-like***)* ***I*-decomposition subgroup** *of* ${}^X\Pi \overset{\mathrm{out}}{\rtimes} {}^X H$, ${}^Y\Pi \overset{\mathrm{out}}{\rtimes} {}^Y H$. *[In particular, for each* $\square \in \{X, Y\}$*, it follows from [CmbGC], Proposition 1.2, (ii), that if* ${}^\square H$ *satisfies condition (b)—which thus implies that* ${}^\square\Pi = {}^\square\Pi \overset{\mathrm{out}}{\rtimes} {}^\square I$*—then it holds that a closed subgroup of* ${}^\square\Pi = {}^\square\Pi \overset{\mathrm{out}}{\rtimes} {}^\square I$ *is a* **verticial** *(respectively,* **cuspidal***;* **nodal***;* **edge-like***)* **subgroup** *of* ${}^\square\Pi \overset{\mathrm{out}}{\rtimes} {}^\square H$ *if and only if it is a* **verticial** *(respectively,* **cuspidal***;* **nodal***;* **edge-like***)* ***I*-decomposition subgroup** *of* ${}^\square\Pi \overset{\mathrm{out}}{\rtimes} {}^\square H$.*] Let*

$$\alpha : {}^X\Pi \overset{\mathrm{out}}{\rtimes} {}^X H \xrightarrow{\sim} {}^Y\Pi \overset{\mathrm{out}}{\rtimes} {}^Y H$$

be an **isomorphism** *of profinite groups. Then the following hold:*

(i) It holds that ${}^X H$ *satisfies condition (a) (respectively, (b)) if and only if* ${}^Y H$ *satisfies condition (a) (respectively, (b)).*
(ii) The equality $l_X = l_Y$ *holds.*
(iii) The isomorphism α *induces a* **bijection** *between the set of* **verticial *I*-decomposition subgroups** *of* ${}^X\Pi \overset{\mathrm{out}}{\rtimes} {}^X H$ *and the set of* **verticial *I*-decomposition subgroups** *of* ${}^Y\Pi \overset{\mathrm{out}}{\rtimes} {}^Y H$.
(iv) The isomorphism α *induces a* **bijection** *between the set of* **cuspidal** *(respectively,* **nodal***;* **edge-like***)* ***I*-decomposition subgroups** *of* ${}^X\Pi \overset{\mathrm{out}}{\rtimes} {}^X H$ *and the set of* **cuspidal** *(respectively,* **nodal***;* **edge-like***)* ***I*-decomposition subgroups** *of* ${}^Y\Pi \overset{\mathrm{out}}{\rtimes} {}^Y H$.
(v) The isomorphism α **restricts** *to an isomorphism*

$$({}^X\Pi \overset{\mathrm{out}}{\rtimes} {}^X H \supseteq) \quad {}^X\Pi \overset{\mathrm{out}}{\rtimes} {}^X I \xrightarrow{\sim} {}^Y\Pi \overset{\mathrm{out}}{\rtimes} {}^Y I \quad (\subseteq {}^Y\Pi \overset{\mathrm{out}}{\rtimes} {}^Y H).$$

(vi) There exists a positive integer n_χ *such that the diagram*

$$\begin{array}{ccccccc} {}^X\Pi \overset{\mathrm{out}}{\rtimes} {}^X H & \longrightarrow & {}^X H & \longrightarrow & \mathrm{Aut}(\mathcal{G}_X) & \xrightarrow{\chi_{\mathcal{G}_X}^{\otimes n_\chi}} & \mathbb{Z}_{l_X}^* \\ \alpha \downarrow \wr & & & & & & \| \\ {}^Y\Pi \overset{\mathrm{out}}{\rtimes} {}^Y H & \longrightarrow & {}^Y H & \longrightarrow & \mathrm{Aut}(\mathcal{G}_Y) & \xrightarrow[\chi_{\mathcal{G}_Y}^{\otimes n_\chi}]{} & \mathbb{Z}_{l_Y}^* \end{array}$$

—where $\chi_{\mathcal{G}_X}$, $\chi_{\mathcal{G}_Y}$ *are as in [CbTpI], Definition 3.8, (ii), and the right-hand vertical equality is the equality that arises from the equality* $l_X = l_Y$ *of (ii)—* **commutes***.*

(vii) *The composite of the upper (respectively, lower) three horizontal arrows of the diagram of (vi)* **coincides** *with the composite of the upper (respectively, lower) three horizontal arrows of the diagram*

$$\begin{array}{ccccccc} {}^X\Pi \overset{\text{out}}{\rtimes} {}^X H & \longrightarrow & {}^X H & \longrightarrow & G_{k_X} & \xrightarrow{\chi_{k_X}^{\otimes n_X}} & \mathbb{Z}_{l_X}^* \\ \alpha \downarrow \wr & & & & & & \| \\ {}^Y\Pi \overset{\text{out}}{\rtimes} {}^Y H & \longrightarrow & {}^Y H & \longrightarrow & G_{k_Y} & \xrightarrow[\chi_{k_Y}^{\otimes n_X}]{} & \mathbb{Z}_{l_Y}^* \end{array}$$

—where the integer n_X is the positive integer of (vi); the right-hand vertical equality is the equality that arises from the equality $l_X = l_Y$ of (ii); the third upper (respectively, lower) horizontal arrow is the n_X-th power of the l_X- (respectively, l_Y-) adic cyclotomic character χ_{k_X} of G_{k_X} (respectively, χ_{k_Y} of G_{k_Y}). In particular, the diagram of the preceding display **commutes**.

(viii) *Suppose that* **one** *of the following three conditions is satisfied:*

(viii-1) *Either ${}^X H$ or ${}^Y H$ satisfies condition (b).*
(viii-2) *It holds that $0 \in \{r_X, r_Y\}$.*
(viii-3) *The isomorphism α induces a* **bijection** *between the set of* **cuspidal subgroups** *of ${}^X\Pi \overset{\text{out}}{\rtimes} {}^X H$ and the set of* **cuspidal subgroups** *of ${}^Y\Pi \overset{\text{out}}{\rtimes} {}^Y H$.*

Then the isomorphism α **restricts** *to an isomorphism*

$$({}^X\Pi \overset{\text{out}}{\rtimes} {}^X H \supseteq) \quad {}^X\Pi \xrightarrow{\sim} {}^Y\Pi \quad (\subseteq {}^Y\Pi \overset{\text{out}}{\rtimes} {}^Y H).$$

Proof First, we verify assertions (i), (ii). Let $\square \in \{X, Y\}$. Write $\widehat{\mathbb{Z}}^{(p')}$ for the pro-prime-to-p completion of the ring $\mathbb{Z}$ of rational integers. The following Facts are well-known:

(1) The profinite group $G_{k_\square}$ is *isomorphic* to $\widehat{\mathbb{Z}}$ as an abstract profinite group.
(2) The kernel of the natural surjection $G_{k_\square}^{\log} \twoheadrightarrow G_{k_\square}$ admits a *natural structure of free $\widehat{\mathbb{Z}}^{(p')}$-module of rank* 1.
(3) The natural action by conjugation of $G_{k_\square}$ on the kernel of the natural surjection $G_{k_\square}^{\log} \twoheadrightarrow G_{k_\square}$ is given by the *cyclotomic character* [cf. (2)]. In particular, for each prime number $q \neq p$, every maximal pro-q subgroup of $G_{k_\square}^{\log}$ admits a *natural structure of extension of $\mathbb{Z}_q$ by $\mathbb{Z}_q$* [cf. (1), (2)]. Moreover, the image of the action $\mathbb{Z}_q \to \mathrm{Aut}(\mathbb{Z}_q) = \mathbb{Z}_q^*$ determined by such an extension is *open*.

Moreover, let us recall [cf., e.g., [AbsTpI], Proposition 2.3, (i)] that the following holds:

(4) The *pro-*$l_\Box$ group $^\Box\Pi$ is *nontrivial*, *center-free*, and *elastic* [cf. [AbsTpI], Definition 1.1, (ii)].

Thus, we conclude that $^\Box H$ satisfies condition (b) if and only if the set of prime numbers q such that every maximal pro-q subgroup of $^\Box\Pi \overset{\text{out}}{\rtimes} {}^\Box H$ is *nonabelian* is *of cardinality* 1. Moreover, the prime number $l_\Box$ may be characterized as the *unique* prime number q such that there exists a maximal pro-q subgroup of $^\Box\Pi \overset{\text{out}}{\rtimes} {}^\Box H$ that is *not isomorphic to a closed subgroup of an extension of* $\mathbb{Z}_q$ *by* $\mathbb{Z}_q$. This completes the proofs of assertions (i), (ii). In the remainder of the proof of Corollary 4.18, we shall write

$$l \stackrel{\text{def}}{=} l_X = l_Y$$

[cf. assertion (ii)].

Next, we verify assertion (iii). For $\Box \in \{X, Y\}$ and $J \subseteq {}^\Box\Pi \overset{\text{out}}{\rtimes} {}^\Box H$ an open subgroup, write

$$J^{\text{RTF}}$$

for the *maximal pro-RTF-quotient* of the profinite group J [cf. [AbsTpI], Proposition 1.2, (iv)]; $^\Box\Pi^J \stackrel{\text{def}}{=} J \cap {}^\Box\Pi \subseteq {}^\Box\Pi$; $^\Box H^J \subseteq {}^\Box H$ for the image of the composite $J \hookrightarrow {}^\Box\Pi \overset{\text{out}}{\rtimes} {}^\Box H \twoheadrightarrow {}^\Box H$ [so we have a commutative diagram of profinite groups

$$\begin{array}{ccccccccc}
1 & \longrightarrow & ^\Box\Pi^J & \longrightarrow & J & \longrightarrow & ^\Box H^J & \longrightarrow & 1 \\
 & & \downarrow & & \downarrow & & \downarrow & & \\
1 & \longrightarrow & ^\Box\Pi & \longrightarrow & ^\Box\Pi \overset{\text{out}}{\rtimes} {}^\Box H & \longrightarrow & ^\Box H & \longrightarrow & 1
\end{array}$$

—where the horizontal sequences are *exact*, and the vertical arrows are the natural inclusions]; $G^J_{k_\Box} \subseteq G_{k_\Box}$ for the image of the composite $J \hookrightarrow {}^\Box\Pi \overset{\text{out}}{\rtimes} {}^\Box H \twoheadrightarrow {}^\Box H \hookrightarrow G^{\log}_{k_\Box} \twoheadrightarrow G_{k_\Box}$;

$$({}^\Box\Pi^J)^{\text{comb}}$$

for the "*combinatorial quotient*" of $^\Box\Pi^J$, i.e., the quotient of $^\Box\Pi^J$ by the normal closed subgroup normally topologically generated by the closed subgroups of $^\Box\Pi^J$ obtained by forming the intersections of $^\Box\Pi^J$ with the *verticial* subgroups of $^\Box\Pi \overset{\text{out}}{\rtimes} {}^\Box H$.

Now we claim that the following assertion holds:

Claim 4.18.A: For $\square \in \{X, Y\}$ and $J \subseteq {}^{\square}\Pi \overset{\text{out}}{\rtimes} {}^{\square}H$ an open subgroup, the quotient of J^{RTF} by the image of the normal closed subgroup ${}^{\square}\Pi^{J} \subseteq J$ in J^{RTF} is $G_{k_{\square}}^{J}$.

Indeed, this assertion follows immediately from Facts (1), (2), (3).

Next, we claim that the following assertion holds:

Claim 4.18.B: Let $\square \in \{X, Y\}$, $J \subseteq {}^{\square}\Pi \overset{\text{out}}{\rtimes} {}^{\square}H$ an open subgroup, Q a *torsion-free abelian* profinite group, and $J \to Q$ a homomorphism of profinite groups. Then the composite

$$ {}^{\square}\Pi^{J} \hookrightarrow J \to Q $$

factors through the natural surjection ${}^{\square}\Pi^{J} \twoheadrightarrow ({}^{\square}\Pi^{J})^{\text{comb}}$.

To this end, let us first observe that since [it is well-known that] the image, in $\mathbb{Z}_l^*$, of the l-adic cyclotomic character of $G_{k_{\square}}$ is *open*, one verifies immediately that the image by the composite ${}^{\square}\Pi^{J} \hookrightarrow J \to Q$ of any *edge-like* subgroup of ${}^{\square}\Pi \overset{\text{out}}{\rtimes} {}^{\square}H$ [i.e., any intersection of ${}^{\square}\Pi^{J}$ with any *edge-like* subgroup of ${}^{\square}\Pi \overset{\text{out}}{\rtimes} {}^{\square}H$] is *trivial* [cf., e.g., [CmbGC], Remark 1.1.3]. In a similar vein, it follows immediately from the "*Riemann hypothesis for abelian varieties over finite fields*" [cf., e.g., [Mumf], pp. 190–191] that the image by the composite ${}^{\square}\Pi^{J} \hookrightarrow J \twoheadrightarrow Q$ of any *verticial* subgroup of ${}^{\square}\Pi \overset{\text{out}}{\rtimes} {}^{\square}H$ [i.e., any intersection of ${}^{\square}\Pi^{J}$ with any *verticial* subgroup of ${}^{\square}\Pi \overset{\text{out}}{\rtimes} {}^{\square}H$] is *trivial*. This completes the proof of Claim 4.18.B.

Next, we claim that the following assertion holds:

Claim 4.18.C: For $\square \in \{X, Y\}$ and $J \subseteq {}^{\square}\Pi \overset{\text{out}}{\rtimes} {}^{\square}H$ an open subgroup, the natural exact sequence $1 \to {}^{\square}\Pi^{J} \to J \to {}^{\square}H^{J} \to 1$ fits into a *commutative* diagram of profinite groups

$$
\begin{array}{ccccccccc}
1 & \longrightarrow & {}^{\square}\Pi^{J} & \longrightarrow & J & \longrightarrow & {}^{\square}H^{J} & \longrightarrow & 1 \\
 & & \downarrow & & \downarrow & & \downarrow & & \\
 & & ({}^{\square}\Pi^{J})^{\text{comb}} & \longrightarrow & J^{\text{RTF}} & \longrightarrow & G_{k_{\square}}^{J} & \longrightarrow & 1
\end{array}
$$

—where the horizontal sequences are *exact*, and the vertical arrows are the natural surjections.

Indeed, Claim 4.18.C follows immediately, in light of Claim 4.18.A, by applying Claim 4.18.B to the various RTF-subgroups of J [cf. [AbsTpI], Definition 1.1, (i)].

Next, we claim that the following assertion holds:

Claim 4.18.D: For $\square \in \{X, Y\}$, there exists an open subgroup $J_0 \subseteq {}^{\square}\Pi \overset{\text{out}}{\rtimes} {}^{\square}H$ that satisfies the following condition: For $J \subseteq J_0$ an arbitrary open subgroup, there exists an open subgroup $J_H^{\dagger} \subseteq {}^{\square}H^{J}$ such that if we write $J^{\dagger} \overset{\text{def}}{=} J \times_{{}^{\square}H^{J}} J_H^{\dagger}$, then the corresponding left-hand lower horizontal arrow $({}^{\square}\Pi^{J^{\dagger}})^{\text{comb}} \to (J^{\dagger})^{\text{RTF}}$ of the diagram of Claim 4.18.C is *injective*.

To this end, let $J_0 \subseteq {}^{\Box}\Pi \overset{\text{out}}{\rtimes} {}^{\Box}H$ be an open subgroup such that, for every open subgroup $J \subseteq J_0$, the quotient $({}^{\Box}\Pi^J)^{\text{comb}}$ is a *center-free free pro-l* group [where we note that one verifies easily [cf. [CmbGC], Remark 1.1.3] that such a J_0 always exists]. Next, let us observe that, to verify Claim 4.18.D, we may assume without loss of generality, by replacing ${}^{\Box}H^J$ by a suitable open subgroup of ${}^{\Box}H^J$, that the outer action of J on $({}^{\Box}\Pi^J)^{\text{comb}}$ by conjugation is *trivial* [where we note that one verifies easily that such an open subgroup of ${}^{\Box}H^J$ always exists]. Since, as discussed above, $({}^{\Box}\Pi^J)^{\text{comb}}$ is *center-free*, if one writes $J^{\text{comb}} \overset{\text{def}}{=} J/\text{Ker}({}^{\Box}\Pi^J \twoheadrightarrow ({}^{\Box}\Pi^J)^{\text{comb}})$, then this *triviality* implies that the inclusions

$$({}^{\Box}\Pi^J)^{\text{comb}} \hookrightarrow J^{\text{comb}} \hookleftarrow Z_{J^{\text{comb}}}(({}^{\Box}\Pi^J)^{\text{comb}})$$

[cf. the discussion entitled "*Topological groups*" in [CbTpI], §0] determine an *isomorphism*

$$({}^{\Box}\Pi^J)^{\text{comb}} \times Z_{J^{\text{comb}}}(({}^{\Box}\Pi^J)^{\text{comb}}) \xrightarrow{\sim} J^{\text{comb}}.$$

On the other hand, since $({}^{\Box}\Pi^J)^{\text{comb}}$ is a *free pro-l* group, the natural surjection $({}^{\Box}\Pi^J)^{\text{comb}} \twoheadrightarrow (({}^{\Box}\Pi^J)^{\text{comb}})^{\text{RTF}}$ is an *isomorphism*. In particular, the composite of natural homomorphisms $({}^{\Box}\Pi^J)^{\text{comb}} \xrightarrow{\sim} (({}^{\Box}\Pi^J)^{\text{comb}})^{\text{RTF}} \hookrightarrow (J^{\text{comb}})^{\text{RTF}}$ is *injective*. Thus, since the natural surjection $J \twoheadrightarrow (J^{\text{comb}})^{\text{RTF}}$ *factors* through J^{RTF}, Claim 4.18.D follows immediately. This completes the proof of Claim 4.18.D.

Next, we claim that the following assertion holds:

Claim 4.18.E: Let $\Box \in \{X, Y\}$ and $A \subseteq {}^{\Box}\Pi \overset{\text{out}}{\rtimes} {}^{\Box}H$ a closed subgroup. Then the following two conditions are equivalent:

(E-1) The closed subgroup $A \subseteq {}^{\Box}\Pi \overset{\text{out}}{\rtimes} {}^{\Box}H$ is *contained* in a *verticial I-decomposition subgroup* of ${}^{\Box}\Pi \overset{\text{out}}{\rtimes} {}^{\Box}H$.

(E-2) For $J \subseteq {}^{\Box}\Pi \overset{\text{out}}{\rtimes} {}^{\Box}H$ an arbitrary open subgroup, the composite $A \cap J \hookrightarrow J \twoheadrightarrow J^{\text{RTF}}$ is *trivial*.

To this end, let us first observe that the implication (E-1) $\Rightarrow$ (E-2) follows immediately from Claim 4.18.C, together with Facts (1), (2), (3). On the other hand, by applying Claims 4.18.C, 4.18.D to the various open subgroups of ${}^{\Box}\Pi \overset{\text{out}}{\rtimes} {}^{\Box}H$ for each $\Box \in \{X, Y\}$, one verifies immediately from Proposition 1.5 that the implication (E-2) $\Rightarrow$ (E-1) holds. This completes the proof of Claim 4.18.E. On the other hand, since any inclusion of *verticial I-decomposition subgroups* is an *equality* [cf. [CmbGC], Proposition 1.2, (i), (ii)], assertion (iii) follows immediately from Claim 4.18.E. This completes the proof of assertion (iii).

Next, we verify assertion (iv). We begin the proof of assertion (iv) with the following claim:

Claim 4.18.F: Let $\square \in \{X, Y\}$. Suppose that $\square^{\log}$ is a *smooth log curve* over $(\mathrm{Spec}\, k_{\square})^{\log}$ [cf. the discussion entitled "*Curves*" in [CbTpI], §0]. Then the inclusions

$$^{\square}\Pi \hookrightarrow {}^{\square}\Pi \overset{\mathrm{out}}{\rtimes} {}^{\square}I \hookleftarrow Z({}^{\square}\Pi \overset{\mathrm{out}}{\rtimes} {}^{\square}I)$$

[cf. the discussion entitled "*Topological groups*" in [CbTpI], §0] determine an *isomorphism*

$$^{\square}\Pi \times Z({}^{\square}\Pi \overset{\mathrm{out}}{\rtimes} {}^{\square}I) \overset{\sim}{\longrightarrow} {}^{\square}\Pi \overset{\mathrm{out}}{\rtimes} {}^{\square}I.$$

Moreover, the composite $Z({}^{\square}\Pi \overset{\mathrm{out}}{\rtimes} {}^{\square}I) \hookrightarrow {}^{\square}\Pi \overset{\mathrm{out}}{\rtimes} {}^{\square}I \twoheadrightarrow {}^{\square}I$ is an *isomorphism*. In particular, if ${}^{\square}H$ satisfies condition (a) (respectively, (b)), then $Z({}^{\square}\Pi \overset{\mathrm{out}}{\rtimes} {}^{\square}I)$ admits a *structure of free* $\widehat{\mathbb{Z}}^{(p')}$*-module of rank* 1 (respectively, is *trivial*) [cf. Fact (2)].

Indeed, since [we have assumed that] $\square^{\log}$ is a *smooth log curve* over $(\mathrm{Spec}\, k_{\square})^{\log}$, this assertion follows immediately from the *slimness* of ${}^{\square}\Pi$ [cf. [CmbGC], Remark 1.1.3], together with the various definitions involved.

Next, let us observe that it follows from [CmbGC], Proposition 1.2, (ii), that,

(5) for each $\square \in \{X, Y\}$, if A is a VCN-subgroup of ${}^{\square}\Pi$, then the intersection of ${}^{\square}\Pi$ with the normalizer, in ${}^{\square}\Pi \overset{\mathrm{out}}{\rtimes} {}^{\square}I$, of A *coincides* with A.

Moreover, let us also observe that it follows from [NodNon], Remark 2.4.2; [NodNon], Remark 2.7.1, that,

(6) for each $\square \in \{X, Y\}$, any *inclusion* of VCN-subgroups of ${}^{\square}\Pi$ gives rise to an *inclusion* of the normalizers, in ${}^{\square}\Pi \overset{\mathrm{out}}{\rtimes} {}^{\square}I$, of the respective VCN-subgroups.

Next, we claim that the following assertion holds:

Claim 4.18.G: The isomorphism α induces a *bijection* between the set of *edge-like* I*-decomposition subgroups* of ${}^{X}\Pi \overset{\mathrm{out}}{\rtimes} {}^{X}H$ and the set of *edge-like* I*-decomposition subgroups* of ${}^{Y}\Pi \overset{\mathrm{out}}{\rtimes} {}^{Y}H$.

To this end, let us first observe that it follows immediately—in light of Facts (5), (6)—from assertion (iii) that, to verify Claim 4.18.G, we may assume without loss of generality—by replacing ${}^{\square}\Pi \overset{\mathrm{out}}{\rtimes} {}^{\square}H$ by the normalizer, in ${}^{\square}\Pi \overset{\mathrm{out}}{\rtimes} {}^{\square}H$, of a *verticial* I*-decomposition subgroup* of ${}^{\square}\Pi \overset{\mathrm{out}}{\rtimes} {}^{\square}H$ for each $\square \in \{X, Y\}$—that $X^{\log}$, $Y^{\log}$ are *smooth log curves* over $(\mathrm{Spec}\, k_X)^{\log}$, $(\mathrm{Spec}\, k_Y)^{\log}$, and that the isomorphism α *restricts* to an isomorphism of ${}^{X}\Pi \overset{\mathrm{out}}{\rtimes} {}^{X}I$ $(\subseteq {}^{X}\Pi \overset{\mathrm{out}}{\rtimes} {}^{X}H)$ with ${}^{Y}\Pi \overset{\mathrm{out}}{\rtimes} {}^{Y}I$ $(\subseteq {}^{Y}\Pi \overset{\mathrm{out}}{\rtimes} {}^{Y}H)$.

Next, let us observe that if ${}^{X}H$, hence also ${}^{Y}H$ [cf. assertion (i)], satisfies condition (b), then since [it is well-known that] the image, in $\mathbb{Z}_l^*$, of the l-adic cyclotomic character of $G_{k_{\square}}$ is *open* for each $\square \in \{X, Y\}$, Claim 4.18.G follows immediately from [CmbGC], Corollary 2.7, (i).

Thus, in the remainder of the proof of Claim 4.18.G, we may assume without loss of generality that ${}^{X}H$, hence also ${}^{Y}H$ [cf. assertion (i)], satisfies condition (a). Then, by applying a similar argument to the argument in the proof of Claim 4.18.G in the case where ${}^{X}H$ satisfies condition (b) to the isomorphism

$$({}^{X}\Pi \overset{\text{out}}{\rtimes} {}^{X}H)/Z({}^{X}\Pi \overset{\text{out}}{\rtimes} {}^{X}I) \overset{\sim}{\longrightarrow} ({}^{Y}\Pi \overset{\text{out}}{\rtimes} {}^{Y}H)/Z({}^{Y}\Pi \overset{\text{out}}{\rtimes} {}^{Y}I)$$

induced by α [cf. Claim 4.18.F], we conclude that this induced isomorphism determines a *bijection* between the set of images of *edge-like subgroups* of ${}^{X}\Pi \overset{\text{out}}{\rtimes} {}^{X}H$ in the quotient $({}^{X}\Pi \overset{\text{out}}{\rtimes} {}^{X}H)/Z({}^{X}\Pi \overset{\text{out}}{\rtimes} {}^{X}I)$ and the set of images of *edge-like subgroups* of ${}^{Y}\Pi \overset{\text{out}}{\rtimes} {}^{Y}H$ in the quotient $({}^{Y}\Pi \overset{\text{out}}{\rtimes} {}^{Y}H)/Z({}^{Y}\Pi \overset{\text{out}}{\rtimes} {}^{Y}I)$. Now let us observe that it follows immediately from Claim 4.18.F and [CmbGC], Proposition 1.2, (ii), that, for each $\Box \in \{X, Y\}$ and each *edge-like* subgroup $A \subseteq {}^{\Box}\Pi \overset{\text{out}}{\rtimes} {}^{\Box}H$, the *edge-like* I*-decomposition subgroup* of ${}^{\Box}\Pi \overset{\text{out}}{\rtimes} {}^{\Box}H$ obtained by forming the normalizer of A in ${}^{\Box}\Pi \overset{\text{out}}{\rtimes} {}^{\Box}I$ *coincides* with the inverse image by the natural surjection ${}^{\Box}\Pi \overset{\text{out}}{\rtimes} {}^{\Box}H \twoheadrightarrow ({}^{\Box}\Pi \overset{\text{out}}{\rtimes} {}^{\Box}H)/Z({}^{\Box}\Pi \overset{\text{out}}{\rtimes} {}^{\Box}I)$ of the image of A in $({}^{\Box}\Pi \overset{\text{out}}{\rtimes} {}^{\Box}H)/Z({}^{\Box}\Pi \overset{\text{out}}{\rtimes} {}^{\Box}I)$. Thus, Claim 4.18.G holds. This completes the proof of Claim 4.18.G. On the other hand, assertion (iv) follows—in light of Facts (5), (6)—from assertion (iii), Claim 4.18.G, and [CmbGC], Proposition 1.5, (i). This completes the proof of assertion (iv).

Next, we verify assertions (v), (vi), (vii). First, we observe that assertion (vii) is a formal consequence of assertion (vi), together with [AbsCsp], Proposition 1.2, (ii); [CbTpI], Corollary 3.9, (ii), (iii). Now suppose that there is *no nodal subgroup* of ${}^{X}\Pi \overset{\text{out}}{\rtimes} {}^{X}H$, hence also [cf. assertion (iv)] of ${}^{Y}\Pi \overset{\text{out}}{\rtimes} {}^{Y}H$. Then assertion (v) follows from assertion (iii). Moreover, by considering, for each $\Box \in \{X, Y\}$, the *cyclotome* obtained by applying the construction of "Λ_v" of [CbTpI], Definition 3.8, (i), to the collection of data consisting of

- the profinite group $({}^{\Box}\Pi \overset{\text{out}}{\rtimes} {}^{\Box}I)/Z({}^{\Box}\Pi \overset{\text{out}}{\rtimes} {}^{\Box}I)$ and
- the various images in $({}^{\Box}\Pi \overset{\text{out}}{\rtimes} {}^{\Box}I)/Z({}^{\Box}\Pi \overset{\text{out}}{\rtimes} {}^{\Box}I)$ of the edge-like I-decomposition subgroups of ${}^{\Box}\Pi \overset{\text{out}}{\rtimes} {}^{\Box}H$,

one verifies immediately from assertions (iv), (v), together with Claim 4.18.F, that assertion (vi) [i.e., in the case where one takes "n_χ" in the statement of assertion (vi) to be 1], hence also assertion (vii), holds.

Thus, in the remainder of the proofs of assertions (v), (vi), (vii), we may assume without loss of generality that both ${}^{X}\Pi \overset{\text{out}}{\rtimes} {}^{X}H$ and ${}^{Y}\Pi \overset{\text{out}}{\rtimes} {}^{Y}H$ have a *nodal* subgroup. Then one verifies immediately from assertions (iii), (iv) [cf. also Facts (5), (6)], together with Lemma 4.19 below [cf. [NodNon], Definition 2.4, (i); [NodNon], Remark 2.4.2], that assertion (vi), hence also assertion (vii), holds. On the other hand, for each $\square \in \{X, Y\}$, we conclude from Fact (2) in the proof of Corollary 4.16 that

(7) if we write

$$({}^{\square}\Pi \overset{\text{out}}{\rtimes} {}^{\square}I)^{(l)} \subseteq {}^{\square}\Pi \overset{\text{out}}{\rtimes} {}^{\square}I$$

for the [unique—cf. Fact (2)] maximal pro-l subgroup of ${}^{\square}\Pi \overset{\text{out}}{\rtimes} {}^{\square}I$, then the closed subgroup $({}^{\square}\Pi \overset{\text{out}}{\rtimes} {}^{\square}I)^{(l)} \subseteq ({}^{\square}\Pi \overset{\text{out}}{\rtimes} {}^{\square}I \subseteq) \, {}^{\square}\Pi \overset{\text{out}}{\rtimes} {}^{\square}H$ *coincides* with the closed subgroup of ${}^{\square}\Pi \overset{\text{out}}{\rtimes} {}^{\square}H$ obtained by forming the unique maximal pro-l subgroup of the *kernel* of the composite of the relevant [i.e., upper if $\square = X$; lower if $\square = Y$] three horizontal arrows of the diagram of assertion (vii).

Moreover, we also conclude immediately from Facts (1), (2), (3) that, for each $\square \in \{X, Y\}$,

(8) the kernel of the composite

$${}^{\square}\Pi \overset{\text{out}}{\rtimes} {}^{\square}H \twoheadrightarrow {}^{\square}\Pi \overset{\text{out}}{\rtimes} {}^{\square}H/({}^{\square}\Pi \overset{\text{out}}{\rtimes} {}^{\square}I)^{(l)} \twoheadrightarrow ({}^{\square}\Pi \overset{\text{out}}{\rtimes} {}^{\square}H/({}^{\square}\Pi \overset{\text{out}}{\rtimes} {}^{\square}I)^{(l)})^{\text{RTF}}$$

coincides with the closed subgroup ${}^{\square}\Pi \overset{\text{out}}{\rtimes} {}^{\square}I$.

In particular, it follows from assertion (vii) and Facts (7), (8) that the isomorphism α *restricts* to an isomorphism of ${}^{X}\Pi \overset{\text{out}}{\rtimes} {}^{X}I$ $(\subseteq {}^{X}\Pi \overset{\text{out}}{\rtimes} {}^{X}H)$ with ${}^{Y}\Pi \overset{\text{out}}{\rtimes} {}^{Y}I$ $(\subseteq {}^{Y}\Pi \overset{\text{out}}{\rtimes} {}^{Y}H)$, as desired. This completes the proof of assertion (v).

Finally, we verify assertion (viii). If condition (viii-1) is satisfied, then since [it follows from assertion (i) that] ${}^{\square}\Pi = {}^{\square}\Pi \overset{\text{out}}{\rtimes} {}^{\square}I$ for each $\square \in \{X, Y\}$, assertion (viii) follows from assertion (v). Thus, in the remainder of the proof of assertion (viii), we suppose that both ${}^{X}H$ and ${}^{Y}H$ satisfy condition (a).

Next, suppose that condition (viii-2) is satisfied. Then it follows from assertion (iv) that $(r_X, r_Y) = (0, 0)$. Write

$${}^{\square}I^{(l)} \subseteq {}^{\square}I$$

for the [unique—cf. Fact (2)] maximal pro-l subgroup of ${}^{\square}I$. Then one verifies easily that one may naturally regard ${}^{\square}I^{(l)}$ as a *quotient* of $({}^{\square}\Pi \overset{\text{out}}{\rtimes} {}^{\square}I)^{(l)}$ [cf. Fact (7)], and, moreover, that the closed subgroup ${}^{\square}\Pi$ of ${}^{\square}\Pi \overset{\text{out}}{\rtimes} {}^{\square}H$ *coincides* with the kernel of the natural surjection $({}^{\square}\Pi \overset{\text{out}}{\rtimes} {}^{\square}I)^{(l)} \twoheadrightarrow {}^{\square}I^{(l)}$. In particular, it follows from assertions (v), (vii) that, to verify assertion (viii) in the case where condition (viii-2) is satisfied, it suffices to verify the following assertion:

Claim 4.18.H: Let $\Box \in \{X, Y\}$ and $({}^{\Box}\Pi \overset{\text{out}}{\rtimes} {}^{\Box}I)^{(l)} \twoheadrightarrow A$ a quotient of $({}^{\Box}\Pi \overset{\text{out}}{\rtimes} {}^{\Box}I)^{(l)}$. Then it holds that this quotient $({}^{\Box}\Pi \overset{\text{out}}{\rtimes} {}^{\Box}I)^{(l)} \twoheadrightarrow A$ *coincides* with the quotient $({}^{\Box}\Pi \overset{\text{out}}{\rtimes} {}^{\Box}I)^{(l)} \twoheadrightarrow {}^{\Box}I^{(l)}$ if and only if the following three conditions are satisfied:

(H-1) The profinite group A is *isomorphic* to $\mathbb{Z}_l$ as an abstract profinite group.

(H-2) The kernel of the surjection $({}^{\Box}\Pi \overset{\text{out}}{\rtimes} {}^{\Box}I)^{(l)} \twoheadrightarrow A$ is *normal* in ${}^{\Box}\Pi \overset{\text{out}}{\rtimes} {}^{\Box}H$. Thus, the outer action of ${}^{\Box}\Pi \overset{\text{out}}{\rtimes} {}^{\Box}H$ on $({}^{\Box}\Pi \overset{\text{out}}{\rtimes} {}^{\Box}I)^{(l)}$ by conjugation *induces* an action [cf. (H-1)] of ${}^{\Box}\Pi \overset{\text{out}}{\rtimes} {}^{\Box}H$ on the quotient A. Moreover, the resulting character $\rho_A\colon {}^{\Box}\Pi \overset{\text{out}}{\rtimes} {}^{\Box}H \to \text{Aut}(A) = \mathbb{Z}_l^*$ [cf. (H-1)] is *trivial* on ${}^{\Box}\Pi \overset{\text{out}}{\rtimes} {}^{\Box}I \subseteq {}^{\Box}\Pi \overset{\text{out}}{\rtimes} {}^{\Box}H$.

(H-3) The n_X-th power of the character ρ_A of (H-2) *coincides* with the composite of the relevant [i.e., upper if $\Box = X$; lower if $\Box = Y$] three horizontal arrows of the diagram of assertion (vii).

First, let us observe that it follows from Facts (2), (3) that the quotient $({}^{\Box}\Pi \overset{\text{out}}{\rtimes} {}^{\Box}I)^{(l)} \twoheadrightarrow {}^{\Box}I^{(l)}$ satisfies the three conditions in the statement of Claim 4.18.H. Next, let us observe that it follows immediately from [CmbGC], Propositions 1.3, 2.6, that if a given quotient $({}^{\Box}\Pi \overset{\text{out}}{\rtimes} {}^{\Box}I)^{(l)} \twoheadrightarrow A$ satisfies conditions (H-1), (H-2), then the image in A of an arbitrary *nodal* [or, equivalently, *edge-like*—cf. the equality $(r_X, r_Y) = (0, 0)$ discussed above] subgroup of ${}^{\Box}\Pi \overset{\text{out}}{\rtimes} {}^{\Box}H$ is *trivial*. Thus, it follows immediately from the "*Riemann hypothesis for abelian varieties over finite fields*" [cf., e.g., [Mumf], pp. 190–191], together with Fact (3) and condition (H-3), that Claim 4.18.H holds. This completes the proof of Claim 4.18.H, hence also of assertion (viii) in the case where condition (viii-2) is satisfied.

Finally, one may verify assertion (viii) in the case where condition (viii-3) is satisfied by applying assertion (viii) in the case where condition (viii-2) is satisfied. Indeed, let us first observe that it follows immediately from Fact (1) [which implies that G_{k_X}, G_{k_Y} are *torsion-free*] that we may assume without loss of generality, by replacing ${}^{\Box}\Pi \overset{\text{out}}{\rtimes} {}^{\Box}H$ by a suitable open subgroup of ${}^{\Box}\Pi \overset{\text{out}}{\rtimes} {}^{\Box}H$ for each $\Box \in \{X, Y\}$, that $g_X, g_Y \geq 2$. Then we may assume without loss of generality, by replacing ${}^{\Box}\Pi \overset{\text{out}}{\rtimes} {}^{\Box}H$ by the quotient of ${}^{\Box}\Pi \overset{\text{out}}{\rtimes} {}^{\Box}H$ by the normal closed subgroup normally topologically generated by the *cuspidal* subgroups for each $\Box \in \{X, Y\}$, that $(r_X, r_Y) = (0, 0)$ [cf. (viii-3)]. Thus, it follows from assertion (viii) in the case where condition (viii-2) is satisfied that assertion (viii) holds. This completes the proof of assertion (viii), hence also of Corollary 4.18. □

Remark 4.18.1 In the situation of Corollary 4.18, (viii), if one *omits* the assumption that one of the conditions (viii-1), (viii-2), and (viii-3) holds, then the conclusion of Corollary 4.18, (viii), *no longer holds* in general. Indeed:

(i) First, we consider the case of a *smooth log curve* [cf. the discussion entitled "*Curves*" in [CbTpI], §0]. In the situation of Corollary 4.18, write $l \overset{\text{def}}{=} l_X$. Let $T^{\log}$ be a *tripod* over $(\operatorname{Spec} k_X)^{\log}$ [cf. the discussion entitled "*Curves*" in [CbTpI], §0] such that the natural action of $G_{k_X}^{\log}$ on the set of cusps of

$T^{\log}$ is *trivial*. Then, by taking "${}^T H$" to be $G_{k_X}^{\log}$, we obtain a profinite group ${}^T\Pi \overset{\text{out}}{\rtimes} {}^T H$. In the remainder of the discussion of the present (i),

> we construct an automorphism of ${}^T\Pi \overset{\text{out}}{\rtimes} {}^T H$ that does *not preserve* the closed subgroup ${}^T\Pi \subseteq {}^T\Pi \overset{\text{out}}{\rtimes} {}^T H$.

Let $C \subseteq {}^T\Pi \overset{\text{out}}{\rtimes} {}^T H$ be a cuspidal subgroup of ${}^T\Pi \overset{\text{out}}{\rtimes} {}^T H$. Write $Z \overset{\text{def}}{=} Z({}^T\Pi \overset{\text{out}}{\rtimes} {}^T I)$ for the center of ${}^T\Pi \overset{\text{out}}{\rtimes} {}^T I$ and $I_C \subseteq {}^T\Pi \overset{\text{out}}{\rtimes} {}^T H$ for the *cuspidal* $\dot{I}$*-decomposition subgroup* of ${}^T\Pi \overset{\text{out}}{\rtimes} {}^T H$ obtained by forming the normalizer in ${}^T\Pi \overset{\text{out}}{\rtimes} {}^T I$ of C. Then

(i-a) the natural inclusions ${}^T\Pi \hookrightarrow {}^T\Pi \overset{\text{out}}{\rtimes} {}^T I$ and $Z \hookrightarrow {}^T\Pi \overset{\text{out}}{\rtimes} {}^T I$ determine an *isomorphism* ${}^T\Pi \times Z \overset{\sim}{\to} {}^T\Pi \overset{\text{out}}{\rtimes} {}^T I$ [cf. Claim 4.18.F]. Moreover, the natural inclusions $C \hookrightarrow I_C$ and $Z \hookrightarrow I_C$ determine an *isomorphism* $C \times Z \overset{\sim}{\to} I_C$ [cf. Claim 4.18.F; [CmbGC], Proposition 1.2, (ii)].

Moreover, it is well-known that the following assertions hold:

(i-b) The *unique* maximal pro-l subgroup of Z admits a *structure of free* $\mathbb{Z}_l$*-module of rank* 1 [cf. Claim 4.18.F]. Moreover, the natural action of G_{k_X} on this unique maximal pro-l subgroup of Z $(= \{0\} \times Z \subseteq {}^T\Pi^{\text{ab}} \times Z \overset{\sim}{\to} ({}^{\square}\Pi \overset{\text{out}}{\rtimes} {}^{\square}I)^{\text{ab}})$ [cf. (i-a)] induced by the natural outer action of G_{k_X} on ${}^T\Pi \overset{\text{out}}{\rtimes} {}^T I$ is given by the l*-adic cyclotomic character* [cf. Fact (3) in the proof of Corollary 4.18].

(i-c) The pro-l group ${}^T\Pi^{\text{ab}}$ admits a *structure of free* $\mathbb{Z}_l$*-module of rank* 2. Moreover, the natural action of G_{k_X} on ${}^T\Pi^{\text{ab}}$ $(= {}^T\Pi^{\text{ab}} \times \{0\} \subseteq {}^T\Pi^{\text{ab}} \times Z \overset{\sim}{\to} ({}^{\square}\Pi \overset{\text{out}}{\rtimes} {}^{\square}I)^{\text{ab}})$ [cf. (i-a)] induced by the natural outer action of G_{k_X} on ${}^T\Pi \overset{\text{out}}{\rtimes} {}^T I$ is given by the l*-adic cyclotomic character*.

Thus, since C admits a *structure of free* $\mathbb{Z}_l$*-module of rank* 1 [cf. [CmbGC], Remark 1.1.3], there exists a *nontrivial* homomorphism $\phi\colon {}^T\Pi$ $(\twoheadrightarrow {}^T\Pi^{\text{ab}})$ $\to Z$ whose kernel is *topologically normally generated* by C. Now write α_I for the automorphism of the profinite group ${}^T\Pi \times Z$ $(\overset{\sim}{\to} {}^T\Pi \overset{\text{out}}{\rtimes} {}^T I)$ [cf. (i-a)] given by mapping ${}^T\Pi \times Z \ni (\sigma, z) \mapsto (\sigma, z \cdot \phi(\sigma)) \in {}^T\Pi \times Z$. Next, let us observe that the composite $H_C \overset{\text{def}}{=} N_{{}^T\Pi \overset{\text{out}}{\rtimes} {}^T H}(C) \hookrightarrow {}^T\Pi \overset{\text{out}}{\rtimes} {}^T H \twoheadrightarrow G_{k_X}$ is *surjective*, with *kernel* equal to I_C. Thus, it follows from Fact (1) in the proof of Corollary 4.18 that this composite $H_C \twoheadrightarrow G_{k_X}$ admits a *section*, which determines an isomorphism

$$({}^T\Pi \overset{\text{out}}{\rtimes} {}^T I) \rtimes G_{k_X} \overset{\sim}{\longrightarrow} {}^T\Pi \overset{\text{out}}{\rtimes} {}^T H.$$

Let us *fix* such a section. Next, observe that it follows from (i-b), (i-c) that the above automorphism α_I is *compatible* with the action of G_{k_X} on ${}^T\Pi \overset{\mathrm{out}}{\rtimes} {}^T I$ determined by the *fixed section* of $H_C \twoheadrightarrow G_{k_X}$. Thus, we conclude that the above automorphism α_I of ${}^T\Pi \overset{\mathrm{out}}{\rtimes} {}^T I$ *extends* to an automorphism α of ${}^T\Pi \overset{\mathrm{out}}{\rtimes} {}^T H$ that preserves and induces the *identity automorphism* on the image of the fixed section of $H_C \twoheadrightarrow G_{k_X}$. Now let us observe that it is immediate that

> α_I, hence also α, does *not preserve* ${}^T\Pi \subseteq {}^T\Pi \overset{\mathrm{out}}{\rtimes} {}^T I$,

as desired. Let us also observe that since $C \subseteq {}^T\Pi$ is contained in the *kernel* of ϕ, it follows from (i-a) that α_I preserves and induces the *identity automorphism* on the cuspidal I-decomposition subgroup $I_C \subseteq {}^T\Pi \overset{\mathrm{out}}{\rtimes} {}^T I$. In particular, we conclude immediately that

(i-d) the automorphism α of ${}^T\Pi \overset{\mathrm{out}}{\rtimes} {}^T H$ preserves and induces the *identity automorphism* on H_C.

(ii) Next, we consider the case of a *singular* stable log curve [i.e., a stable log curve that is *not smooth*]. In the situation of (i), let $W^{\log}$ be a stable log curve over $(\mathrm{Spec}\, k_X)^{\log}$ such that

- $W^{\log}$ has *precisely two* irreducible components each of which is a *tripod*,
- $W^{\log}$ has a *single node*, and, moreover,
- the natural action of $G_{k_X}^{\log}$ on the dual semi-graph of $W^{\log}$ is *trivial*.

[Thus, $W^{\log}$ is *of type* $(0, 4)$.] Then, by taking "${}^W H$" to be $G_{k_X}^{\log}$, we obtain a profinite group ${}^W\Pi \overset{\mathrm{out}}{\rtimes} {}^W H$ [cf. the situation and notational conventions of Corollary 4.18]. In the remainder of the discussion of the present (ii),

> we construct an automorphism of ${}^W\Pi \overset{\mathrm{out}}{\rtimes} {}^W H$ that does *not preserve* the closed subgroup ${}^W\Pi \subseteq {}^W\Pi \overset{\mathrm{out}}{\rtimes} {}^W H$.

Write v_1, v_2 for the distinct two irreducible components of $W^{\log}$. Let $V_1, V_2 \subseteq {}^W\Pi \subseteq {}^W\Pi \overset{\mathrm{out}}{\rtimes} {}^W H$ be verticial subgroups of ${}^W\Pi \overset{\mathrm{out}}{\rtimes} {}^W H$ associated to v_1, v_2 such that $N \overset{\mathrm{def}}{=} V_1 \cap V_2 \neq \{1\}$, which thus [cf. [NodNon], Lemma 1.9, (i)] implies that N is a *nodal subgroup* of ${}^W\Pi \overset{\mathrm{out}}{\rtimes} {}^W H$. For each $i \in \{1, 2\}$, write

$$H_{V_i} \overset{\mathrm{def}}{=} N_{{}^W\Pi \overset{\mathrm{out}}{\rtimes} {}^W H}(V_i), \quad H_N \overset{\mathrm{def}}{=} N_{{}^W\Pi \overset{\mathrm{out}}{\rtimes} {}^W H}(N).$$

Then one verifies immediately [cf. [CmbGC], Proposition 1.2, (ii)] that

(ii-a) there exists a *commutative diagram* of profinite groups

$$\begin{array}{ccccccc} N & \subseteq & V_i & \subseteq & H_{V_i} & \supseteq & H_N \\ \wr\downarrow & & \wr\downarrow & & \wr\downarrow & & \wr\downarrow \\ C & \subseteq & {}^T\Pi & \subseteq & {}^T\Pi \overset{\text{out}}{\rtimes} {}^T H & \supseteq & H_C \end{array}$$

—where the horizontal arrows are the *natural inclusions*, and the vertical arrows are *isomorphisms*.

Moreover, it follows immediately from a similar argument to the argument applied in the proof of [CmbCsp], Proposition 1.5, (iii) [i.e., in essence, from the evident analogue for semi-graphs of anabelioids of the "*van Kampen Theorem*" in elementary algebraic topology], that

(ii-b) the natural inclusions

$$H_{V_1} \hookrightarrow {}^W\Pi \overset{\text{out}}{\rtimes} {}^W H \hookleftarrow H_{V_2}$$

determine an *isomorphism*

$$\varinjlim(H_{V_1} \hookleftarrow H_N \hookrightarrow H_{V_2}) \xrightarrow{\sim} {}^W\Pi \overset{\text{out}}{\rtimes} {}^W H$$

—where the inductive limit is taken in the category of profinite groups—which restricts to an *isomorphism* of closed subgroups

$$\varinjlim(V_1 \hookleftarrow N \hookrightarrow V_2) \xrightarrow{\sim} {}^W\Pi$$

—where the inductive limit is taken in the category of profinite groups.

On the other hand, it follows from (i-d) and (ii-a) that, for each $i \in \{1, 2\}$, α determines an automorphism β_i of H_{V_i} that

- does *not preserve* $V_i \subseteq H_{V_i}$ but
- preserves and induces the *identity automorphism* on the closed subgroup $H_N \subseteq H_{V_i}$.

Thus, by (ii-b), β_1 and β_2 determine an automorphism γ of ${}^W\Pi \overset{\text{out}}{\rtimes} {}^W H$ that does *not preserve* the closed subgroup ${}^W\Pi \subseteq {}^W\Pi \overset{\text{out}}{\rtimes} {}^W H$, as desired.

Lemma 4.19 (An Explicit Description of a Power of the Cyclotomic Character) *Let J be a profinite group, $\rho_J \colon J \to \mathrm{Aut}(\mathcal{G}_0)$ a continuous homomorphism, and $I \subseteq J$ a* **normal** *closed subgroup of J such that either*

(a) the composite $I \hookrightarrow J \stackrel{\rho_J}{\to} \mathrm{Aut}(\mathcal{G}_0)$ is **of SNN-type** *[cf. [NodNon], Definition 2.4, (iii)], or*
(b) $I = \{1\}$.

Write

$$\Pi_I \stackrel{\mathrm{def}}{=} \Pi_{\mathcal{G}_0} \stackrel{\mathrm{out}}{\rtimes} I \subseteq \Pi_J \stackrel{\mathrm{def}}{=} \Pi_{\mathcal{G}_0} \stackrel{\mathrm{out}}{\rtimes} J$$

[cf. the discussion entitled "Topological groups" in [CbTpI], §0]. Thus, we have a commutative diagram of profinite groups

$$\begin{array}{ccccccccc} 1 & \longrightarrow & \Pi_{\mathcal{G}_0} & \longrightarrow & \Pi_I & \longrightarrow & I & \longrightarrow & 1 \\ & & \| & & \downarrow & & \downarrow & & \\ 1 & \longrightarrow & \Pi_{\mathcal{G}_0} & \longrightarrow & \Pi_J & \longrightarrow & J & \longrightarrow & 1 \end{array}$$

—where the horizontal sequences are **exact**, *and the vertical arrows are the natural inclusions. Write $\widetilde{\mathcal{G}}_0 \to \mathcal{G}_0$ for the universal covering of $\mathcal{G}_0$ corresponding to $\Pi_{\mathcal{G}_0}$. Let e_0 be a node of $\widetilde{\mathcal{G}}_0$. Write $\Pi_{e_0} \subseteq \Pi_{\mathcal{G}_0}$ for the nodal subgroup associated to e_0. Write*

$$\Pi_{e_0, J} \subseteq \Pi_J$$

for the [necessarily open] subgroup consisting of the elements $\sigma \in \Pi_J$ such that the natural action of σ on the underlying semi-graph $\mathbb{G}_0$ of $\mathcal{G}_0$ stabilizes the two branches of the node $e_0(\mathcal{G}_0)$ of $\mathcal{G}_0$ determined by e_0. Then the following hold:

(i) Let N be a positive integer and γ an element of Π_J. Then there exists a collection of data as follows

- *a normal open subgroup $H \subseteq \Pi_J$ of Π_J,*
- *a positive integer m,*
- *verticial subgroups $\Pi_{v_0}, \Pi_{v_1}, \ldots, \Pi_{v_{m-1}} \subseteq \Pi_{\mathcal{G}_0}$ of $\Pi_{\mathcal{G}_0}$ associated to vertices $v_0, v_1, \ldots, v_{m-1}$ of $\widetilde{\mathcal{G}}_0$, respectively, and*
- *nodal subgroups $\Pi_{e_1}, \ldots, \Pi_{e_m} \subseteq \Pi_{\mathcal{G}_0}$ of $\Pi_{\mathcal{G}_0}$ associated to nodes $e_1, \ldots, e_m$ of $\widetilde{\mathcal{G}}_0$, respectively,*

such that if we write

$$D_{e_j} \stackrel{\mathrm{def}}{=} N_{\Pi_I}(\Pi_{e_j})$$

for each $j \in \{0, 1, \ldots, m\}$ [cf. [NodNon], Definition 2.2, (iii)], then

(1) the inclusions $\Pi_{e_0} \subseteq \Pi_{v_0}$, $\Pi_{e_m} \subseteq \Pi_{v_{m-1}}$ [which imply that e_0, e_m **abut** *to v_0, v_{m-1}, respectively—cf. [NodNon], Lemma 1.7] hold,*

(2) if $m \geq 2$, then, for every $j \in \{1, \ldots, m-1\}$, the inclusion $\Pi_{e_j} \subseteq \Pi_{v_{j-1}} \cap \Pi_{v_j}$ [which implies that e_j **abuts** *to v_{j-1} and v_j—cf. [NodNon], Lemma 1.7] holds,*

(3) the quotient

$$D_{e_0} \twoheadrightarrow D_{e_0} \otimes_{\widehat{\mathbb{Z}}^{\Sigma_0}} (\widehat{\mathbb{Z}}^{\Sigma_0}/N\widehat{\mathbb{Z}}^{\Sigma_0})$$

$$\cong \begin{cases} (\widehat{\mathbb{Z}}^{\Sigma_0}/N\widehat{\mathbb{Z}}^{\Sigma_0}) \times (\widehat{\mathbb{Z}}^{\Sigma_0}/N\widehat{\mathbb{Z}}^{\Sigma_0}) & \text{if (a) is satisfied} \\ \widehat{\mathbb{Z}}^{\Sigma_0}/N\widehat{\mathbb{Z}}^{\Sigma_0} & \text{if (b) is satisfied} \end{cases}$$

[cf. [CmbGC], Remark 1.1.3; [NodNon], Lemma 2.5, (i); [NodNon], Remark 2.7.1] of D_{e_0} **factors** *through the quotient of D_{e_0} determined by the composite $D_{e_0} \hookrightarrow \Pi_J \twoheadrightarrow \Pi_J/H$, and, moreover,*

(4) the image of $D_{e_m} \subseteq \Pi_J$ in Π_J/H **coincides** *with the image of $\gamma \cdot D_{e_0} \cdot \gamma^{-1} \subseteq \Pi_J$ in Π_J/H.*

For each $j \in \{0, 1, \ldots, m-1\}$, write

$$D_{v_j} \stackrel{\text{def}}{=} N_{\Pi_I}(\Pi_{v_j}) \supseteq I_{v_j} \stackrel{\text{def}}{=} Z_{\Pi_I}(\Pi_{v_j}) = Z(D_{v_j})$$

[cf. [NodNon], Definition 2.2, (i); [NodNon], Lemma 2.5, (i); [NodNon], Remark 2.7.1; [CmbGC], Remark 1.1.3]; $b_{j,j}$, $b_{j+1,j}$ for the respective branches of the nodes e_j, e_{j+1} that abut to the vertex v_j determined by the inclusions $\Pi_{e_j} \subseteq \Pi_{v_j}$, $\Pi_{e_{j+1}} \subseteq \Pi_{v_j}$ [cf. (1), (2)]. Thus, for $j \in \{0, 1, \ldots, m\}$ and $s \in \{0, 1, \ldots, m-1\}$ such that $s \in \{j-1, j\}$, it follows from [NodNon], Remark 2.7.1, that we have natural inclusions

$$\begin{array}{ccccc} I_{v_s} & \subseteq & D_{e_j} & \subseteq & D_{v_s} \\ & & \cup & & \cup \\ & & \Pi_{e_j} & \subseteq & \Pi_{v_s}, \end{array}$$

which determine a **commutative diagram** *of profinite groups*

$$\begin{array}{ccc} D_{e_j}/I_{v_s} & \longrightarrow & D_{v_s}/I_{v_s} \\ \uparrow \wr & & \uparrow \wr \\ \Pi_{e_j} & \longrightarrow & \Pi_{v_s} \end{array}$$

—where the horizontal arrows are the natural inclusions, and the vertical arrows are **isomorphisms**.

(ii) *In the situation of (i), by applying the construction of "Λ_v" of [CbTpI], Definition 3.8, (i), to the collection of data consisting of*

- *the profinite group* D_{v_s}/I_{v_s} *and*
- *the various images in* D_{v_s}/I_{v_s}, *by the right-hand vertical isomorphism* $\Pi_{v_s} \xrightarrow{\sim} D_{v_s}/I_{v_s}$ *of the final display of (i), of the edge-like subgroups of* $\Pi_{\mathcal{G}_0}$ *contained in* Π_{v_s},

one may construct a **cyclotome**

$$\Lambda(D_{v_s}/I_{v_s}).$$

Moreover, by applying the construction of "$\mathfrak{syn}_b$" of [CbTpI], Corollary 3.9, (v), to the collection of data consisting of

- *the profinite groups* D_{e_j}/I_{v_s}, D_{v_s}/I_{v_s},
- *the various images in* D_{v_s}/I_{v_s}, *by the right-hand vertical isomorphism* $\Pi_{v_s} \xrightarrow{\sim} D_{v_s}/I_{v_s}$ *of the final display of (i), of the edge-like subgroups of* $\Pi_{\mathcal{G}_0}$ *contained in* Π_{v_s}, *and*
- *the upper horizontal arrow* $D_{e_j}/I_{v_s} \hookrightarrow D_{v_s}/I_{v_s}$ *of the final display of (i) [i.e., that corresponds the branch* $b_{j,s}$*],*

one may construct an **isomorphism**

$$\mathfrak{syn}_{b_{j,s}} : D_{e_j}/I_{v_s} \xrightarrow{\sim} \Lambda(D_{v_s}/I_{v_s}).$$

Write

$$M_{v_s} \stackrel{\text{def}}{=} \begin{cases} I_{v_s} \otimes_{\widehat{\mathbb{Z}}^{\Sigma_0}} \Lambda(D_{v_s}/I_{v_s}) & \textit{if (a) is satisfied} \\ \Lambda(D_{v_s}/I_{v_s}) & \textit{if (b) is satisfied} \end{cases}$$

$$M_{e_j} \stackrel{\text{def}}{=} \begin{cases} \det(D_{e_j}) & \textit{if (a) is satisfied} \\ D_{e_j}/I_{v_s} & \textit{if (b) is satisfied} \end{cases}$$

*—where the "*det*" is taken with respect to the structure of free* $\widehat{\mathbb{Z}}^{\Sigma_0}$*-module of finite rank of the profinite group* D_{e_j}*; we observe that if condition (a) is satisfied, then the exact sequence of* **free** $\widehat{\mathbb{Z}}^{\Sigma_0}$**-modules of finite rank**

$$1 \longrightarrow I_{v_s} \longrightarrow D_{e_j} \longrightarrow D_{e_j}/I_{v_s} \longrightarrow 1$$

yields a natural **identification**

$$M_{e_j} = I_{v_s} \otimes_{\widehat{\mathbb{Z}}^{\Sigma_0}} (D_{e_j}/I_{v_s})$$

of $\widehat{\mathbb{Z}}^{\Sigma_0}$-modules [cf. [CmbGC], Remark 1.1.3; [NodNon], Lemma 2.5, (i); [NodNon], Remark 2.7.1]; we observe that if condition (b) is satisfied, then since $I_{v_s} = \{1\}$, we have a natural isomorphism $D_{e_j} \xrightarrow{\sim} D_{e_j}/I_{v_s} = M_{e_j}$. If condition (a) is satisfied, then let us write

$$^{M}\mathfrak{syn}_{b_{j,s}} : M_{e_j} = I_{v_s} \otimes_{\widehat{\mathbb{Z}}^{\Sigma_0}} (D_{e_j}/I_{v_s}) \xrightarrow{\sim} I_{v_s} \otimes_{\widehat{\mathbb{Z}}^{\Sigma_0}} \Lambda(D_{v_s}/I_{v_s}) = M_{v_s}$$

for the isomorphism determined by the above isomorphism $\mathfrak{syn}_{b_{j,s}}$. If condition (b) is satisfied, then let us write

$$^{M}\mathfrak{syn}_{b_{j,s}} \stackrel{\text{def}}{=} \mathfrak{syn}_{b_{j,s}} : M_{e_j} = D_{e_j}/I_{v_s} \xrightarrow{\sim} \Lambda(D_{v_s}/I_{v_s}) = M_{v_s}.$$

(iii) In the situation of (ii), write $n_0 \stackrel{\text{def}}{=} 2$ (respectively, 1) if condition (a) (respectively, (b)) is satisfied. Write

$$\Phi_N(\gamma) \in \mathrm{Aut}(M_{e_0} \otimes_{\widehat{\mathbb{Z}}^{\Sigma_0}} (\widehat{\mathbb{Z}}^{\Sigma_0}/N\widehat{\mathbb{Z}}^{\Sigma_0})) = (\widehat{\mathbb{Z}}^{\Sigma_0}/N\widehat{\mathbb{Z}}^{\Sigma_0})^*$$

for the automorphism of the free $\widehat{\mathbb{Z}}^{\Sigma_0}$-module [of rank one] $M_{e_0} \otimes_{\widehat{\mathbb{Z}}^{\Sigma_0}} (\widehat{\mathbb{Z}}^{\Sigma_0}/N\widehat{\mathbb{Z}}^{\Sigma_0})$ obtained by forming the composite of the isomorphism

$$M_{e_0} \otimes_{\widehat{\mathbb{Z}}^{\Sigma_0}} (\widehat{\mathbb{Z}}^{\Sigma_0}/N\widehat{\mathbb{Z}}^{\Sigma_0}) \xrightarrow{\sim} M_{e_m} \otimes_{\widehat{\mathbb{Z}}^{\Sigma_0}} (\widehat{\mathbb{Z}}^{\Sigma_0}/N\widehat{\mathbb{Z}}^{\Sigma_0})$$

determined by conjugating by $\gamma \in \Pi_J$ [cf. conditions (3), (4) in (i)] with the isomorphism

$$M_{e_m} \otimes_{\widehat{\mathbb{Z}}^{\Sigma_0}} (\widehat{\mathbb{Z}}^{\Sigma_0}/N\widehat{\mathbb{Z}}^{\Sigma_0}) \xrightarrow{\sim} M_{e_0} \otimes_{\widehat{\mathbb{Z}}^{\Sigma_0}} (\widehat{\mathbb{Z}}^{\Sigma_0}/N\widehat{\mathbb{Z}}^{\Sigma_0})$$

determined by the inverse of the composite

$$M_{e_0} \xrightarrow[\sim]{^{M}\mathfrak{syn}_{b_{0,0}}} M_{v_0} \xrightarrow[\sim]{^{M}\mathfrak{syn}_{b_{1,0}}^{-1}} M_{e_1} \xrightarrow[\sim]{^{M}\mathfrak{syn}_{b_{1,1}}} M_{v_1} \xrightarrow[\sim]{^{M}\mathfrak{syn}_{b_{2,1}}^{-1}}$$

$$\cdots \xrightarrow[\sim]{^{M}\mathfrak{syn}_{b_{m-1,m-1}}} M_{v_{m-1}} \xrightarrow[\sim]{^{M}\mathfrak{syn}_{b_{m,m-1}}^{-1}} M_{e_m}.$$

Suppose that $\gamma \in \Pi_{e_0,J}$. Then the image of γ by the composite

$$\Pi_J \longrightarrow J \xrightarrow{\rho_J} \mathrm{Aut}(\mathcal{G}_0) \xrightarrow{\chi_{\mathcal{G}_0}^{\otimes 2n_0}} (\widehat{\mathbb{Z}}^{\Sigma_0})^* \longrightarrow (\widehat{\mathbb{Z}}^{\Sigma_0}/N\widehat{\mathbb{Z}}^{\Sigma_0})^*$$

[cf. [CbTpI], Definition 3.8, (ii)] **coincides** *with $\Phi_N(\gamma)^2 \in (\widehat{\mathbb{Z}}^{\Sigma_0}/N\widehat{\mathbb{Z}}^{\Sigma_0})^*$.*

(iv) *Let* $\rho\colon \Pi_J \to (\widehat{\mathbb{Z}}^{\Sigma_0})^*$ *be a* **character** *[i.e., a continuous homomorphism] and* n_ρ *a positive integer divisible by* $2[\Pi_J : \Pi_{e_0,J}]$. *Suppose that, for each positive integer* N' *and each* $\gamma' \in \Pi_{e_0,J}$, *the image of* $\rho(\gamma')^2 \in (\widehat{\mathbb{Z}}^{\Sigma_0})^*$ *in* $(\widehat{\mathbb{Z}}^{\Sigma_0}/N'\widehat{\mathbb{Z}}^{\Sigma_0})^*$ **coincides** *with* $\Phi_{N'}(\gamma')^2 \in (\widehat{\mathbb{Z}}^{\Sigma_0}/N'\widehat{\mathbb{Z}}^{\Sigma_0})^*$ *[cf. (iii)]. Then the* n_ρ*-th power of the character* ρ **coincides** *with the* n_ρ*-th power of the character obtained by forming the composite*

$$\Pi_J \longrightarrow J \xrightarrow{\rho_J} \mathrm{Aut}(\mathcal{G}_0) \xrightarrow{\chi_{\mathcal{G}_0}^{\otimes n_0}} (\widehat{\mathbb{Z}}^{\Sigma_0})^*$$

[cf. (iii)].

Proof Assertions (i), (ii) follow immediately from the various definitions involved. Next, we verify assertion (iii). Let us first observe that it follows immediately from the various definitions involved that there exist $\delta \in \Pi_{\mathcal{G}_0} \subseteq \Pi_{e_0,J}$ and $\epsilon \in N_{\Pi_J}(\Pi_{e_0}) \cap \Pi_{e_0,J}$ $(\subseteq N_{\Pi_J}(D_{e_0}) \cap \Pi_{e_0,J})$ such that $\gamma = \delta \cdot \epsilon$. Now one verifies immediately from [CbTpI], Corollary 3.9, (ii), (v); [CbTpI], Corollary 5.9, (ii), that the action of ϵ on M_{e_0} by conjugation is given by *multiplication by* $\chi_{\mathcal{G}_0}(\epsilon)^{n_0}$. Moreover, let us observe that one verifies easily that the collection of data of assertion (i) [i.e., associated to γ] satisfies conditions (1), (2), (3), (4) in assertion (i) in the case where we take "γ" to be δ. Also, let us observe that the image of δ by the composite

$$\Pi_J \longrightarrow J \xrightarrow{\rho_J} \mathrm{Aut}(\mathcal{G}_0) \xrightarrow{\chi_{\mathcal{G}_0}^{\otimes n_0}} (\widehat{\mathbb{Z}}^{\Sigma_0})^* \longrightarrow (\widehat{\mathbb{Z}}^{\Sigma_0}/N\widehat{\mathbb{Z}}^{\Sigma_0})^*$$

is *trivial*. Thus, assertion (iii) follows immediately from [CbTpI], Corollary 3.9, (ii), (v), (vi). This completes the proof of assertion (iii). Assertion (iv) is a formal consequence of assertion (iii). This completes the proof of Lemma 4.19. □

References

[Asd] M. Asada, The faithfulness of the monodromy representations associated with certain families of algebraic curves. J. Pure Appl. Algebra **159**, 123–147 (2001)

[HT] H. Hamidi-Tehrani, Groups generated by positive multi-twists and the fake lantern problem. Algebr. Geom. Topol. **2**, 1155–1178 (2002)

[Hsh] Y. Hoshi, Absolute anabelian cuspidalizations of configuration spaces of proper hyperbolic curves over finite fields. Publ. Res. Inst. Math. Sci. 45, 661–744 (2009)

[NodNon] Y. Hoshi, S. Mochizuki, On the combinatorial anabelian geometry of nodally nondegenerate outer representations. Hiroshima Math. J. **41**, 275–342 (2011)

[CbTpI] Y. Hoshi, S. Mochizuki, Topics surrounding the combinatorial anabelian geometry of hyperbolic curves I: Inertia groups and profinite Dehn twists, in *Galois-Teichmüller Theory and Arithmetic Geometry*, Adv. Stud. Pure Math., vol. 63 (Math. Soc. Japan, Tokyo, 2012), pp. 659–811

[Ishi] A. Ishida, The structure of subgroup of mapping class groups generated by two Dehn twists. Proc. Japan Acad. Ser. A Math. Sci., **72**, 240–241 (1996)

[Lch] P. Lochak, Results and conjectures in profinite Teichmüller theory, in *Galois-Teichmüller Theory and Arithmetic Geometry*, Adv. Stud. Pure Math., vol. 63 (Math. Soc. Japan, Tokyo, 2012), pp. 263–335

[LocAn] S. Mochizuki, The local pro-p anabelian geometry of curves. Invent. Math. **138**, 319–423 (1999)

[ExtFam] S. Mochizuki, Extending families of curves over log regular schemes. J. Reine Angew. Math. **511**, 43–71 (1999)

[SemiAn] S. Mochizuki, Semi-graphs of anabelioids. Publ. Res. Inst. Math. Sci. **42**, 221–322 (2006)

[CmbGC] S. Mochizuki, A combinatorial version of the Grothendieck conjecture. Tohoku Math. J. **59**, 455–479 (2007)

[AbsCsp] S. Mochizuki, Absolute anabelian cuspidalizations of proper hyperbolic curves. J. Math. Kyoto Univ. **47**, 451–539 (2007)

[AbsTpI] S. Mochizuki, Topics in absolute anabelian geometry I: generalities. J. Math. Sci. Univ. Tokyo **19**, 139–242 (2012)

[CmbCsp] S. Mochizuki, On the combinatorial cuspidalization of hyperbolic curves. Osaka J. Math. **47**, 651–715 (2010)

[MzTa] S. Mochizuki, A. Tamagawa, The algebraic and anabelian geometry of configuration spaces. Hokkaido Math. J. **37**, 75–131 (2008)

Y. Hoshi, S. Mochizuki, *Topics Surrounding the Combinatorial Anabelian Geometry of Hyperbolic Curves II*, Lecture Notes in Mathematics 2299,
https://doi.org/10.1007/978-981-19-1096-8

[Mumf] D. Mumford, *Abelian varieties*, with appendices by C. P. Ramanujam and Yuri Manin, corrected reprint of the second (1974) edition. Tata Institute of Fundamental Research Studies in Mathematics, vol. 5. Published for the Tata Institute of Fundamental Research, Bombay (Hindustan Book Agency, New Delhi, 2008)

[RZ] L. Ribes, P. Zalesskii, *Profinite groups*, Second edition, Ergebnisse der Mathematik und ihrer Grenzgebiete, vol. 3. Folge. A Series of Modern Surveys in Mathematics, vol. 40 (Springer-Verlag, Berlin, 2010)

[Wkb] Y. Wakabayashi, On the cuspidalization problem for hyperbolic curves over finite fields. Kyoto J. Math. **56**, 125–164 (2016)

[SGA1] *Revêtements étales et groupe fondamental* (SGA 1), Séminaire de Géométrie Algébrique du Bois Marie 1960–1961, directed by A. Grothendieck, with two papers by M. Raynaud, Documents Mathématiques (Paris), 3, (Société Mathématique de France, Paris, 2003)

LECTURE NOTES IN MATHEMATICS Springer

Editors in Chief: J.-M. Morel, B. Teissier;

Editorial Policy

1. Lecture Notes aim to report new developments in all areas of mathematics and their applications – quickly, informally and at a high level. Mathematical texts analysing new developments in modelling and numerical simulation are welcome.

   Manuscripts should be reasonably self-contained and rounded off. Thus they may, and often will, present not only results of the author but also related work by other people. They may be based on specialised lecture courses. Furthermore, the manuscripts should provide sufficient motivation, examples and applications. This clearly distinguishes Lecture Notes from journal articles or technical reports which normally are very concise. Articles intended for a journal but too long to be accepted by most journals, usually do not have this "lecture notes" character. For similar reasons it is unusual for doctoral theses to be accepted for the Lecture Notes series, though habilitation theses may be appropriate.

2. Besides monographs, multi-author manuscripts resulting from SUMMER SCHOOLS or similar INTENSIVE COURSES are welcome, provided their objective was held to present an active mathematical topic to an audience at the beginning or intermediate graduate level (a list of participants should be provided).

   The resulting manuscript should not be just a collection of course notes, but should require advance planning and coordination among the main lecturers. The subject matter should dictate the structure of the book. This structure should be motivated and explained in a scientific introduction, and the notation, references, index and formulation of results should be, if possible, unified by the editors. Each contribution should have an abstract and an introduction referring to the other contributions. In other words, more preparatory work must go into a multi-authored volume than simply assembling a disparate collection of papers, communicated at the event.

3. Manuscripts should be submitted either online at www.editorialmanager.com/lnm to Springer's mathematics editorial in Heidelberg, or electronically to one of the series editors. Authors should be aware that incomplete or insufficiently close-to-final manuscripts almost always result in longer refereeing times and nevertheless unclear referees' recommendations, making further refereeing of a final draft necessary. The strict minimum amount of material that will be considered should include a detailed outline describing the planned contents of each chapter, a bibliography and several sample chapters. Parallel submission of a manuscript to another publisher while under consideration for LNM is not acceptable and can lead to rejection.

4. In general, **monographs** will be sent out to at least 2 external referees for evaluation.

   A final decision to publish can be made only on the basis of the complete manuscript, however a refereeing process leading to a preliminary decision can be based on a pre-final or incomplete manuscript.

   Volume Editors of **multi-author works** are expected to arrange for the refereeing, to the usual scientific standards, of the individual contributions. If the resulting reports can be forwarded to the LNM Editorial Board, this is very helpful. If no reports are forwarded or if other questions remain unclear in respect of homogeneity etc, the series editors may wish to consult external referees for an overall evaluation of the volume.

5. Manuscripts should in general be submitted in English. Final manuscripts should contain at least 100 pages of mathematical text and should always include

   - a table of contents;
   - an informative introduction, with adequate motivation and perhaps some historical remarks: it should be accessible to a reader not intimately familiar with the topic treated;
   - a subject index: as a rule this is genuinely helpful for the reader.
   - For evaluation purposes, manuscripts should be submitted as pdf files.

6. Careful preparation of the manuscripts will help keep production time short besides ensuring satisfactory appearance of the finished book in print and online. After acceptance of the manuscript authors will be asked to prepare the final LaTeX source files (see LaTeX templates online: https://www.springer.com/gb/authors-editors/book-authors-editors/manuscriptpreparation/5636) plus the corresponding pdf- or zipped ps-file. The LaTeX source files are essential for producing the full-text online version of the book, see http://link.springer.com/bookseries/304 for the existing online volumes of LNM). The technical production of a Lecture Notes volume takes approximately 12 weeks. Additional instructions, if necessary, are available on request from lnm@springer.com.

7. Authors receive a total of 30 free copies of their volume and free access to their book on SpringerLink, but no royalties. They are entitled to a discount of 33.3 % on the price of Springer books purchased for their personal use, if ordering directly from Springer.

8. Commitment to publish is made by a *Publishing Agreement*; contributing authors of multiauthor books are requested to sign a *Consent to Publish form*. Springer-Verlag registers the copyright for each volume. Authors are free to reuse material contained in their LNM volumes in later publications: a brief written (or e-mail) request for formal permission is sufficient.

Addresses:

Professor Jean-Michel Morel, CMLA, École Normale Supérieure de Cachan, France
E-mail: moreljeanmichel@gmail.com

Professor Bernard Teissier, Equipe Géométrie et Dynamique,
Institut de Mathématiques de Jussieu – Paris Rive Gauche, Paris, France
E-mail: bernard.teissier@imj-prg.fr

Springer: Ute McCrory, Mathematics, Heidelberg, Germany,
E-mail: lnm@springer.com

Printed in the United States
by Baker & Taylor Publisher Services